Wine Flavour
Chemistry

Wine Flavour Chemistry

Second edition

Jokie Bakker
Ronald J. Clarke

WILEY-BLACKWELL

A John Wiley & Sons, Ltd., Publication

Blackwell Publishing was acquired by John Wiley & Sons in February 2007. Blackwell's publishing program has been merged with Wiley's global Scientific, Technical and Medical business to form Wiley-Blackwell.

Registered Office
John Wiley & Sons, Ltd, The Atrium, Southern Gate, Chichester, West Sussex, PO19 8SQ, UK

Editorial Offices
9600 Garsington Road, Oxford, OX4 2DQ, UK
The Atrium, Southern Gate, Chichester, West Sussex, PO19 8SQ, UK
2121 State Avenue, Ames, Iowa 50014-8300, USA

For details of our global editorial offices, for customer services and for information about how to apply for permission to reuse the copyright material in this book please see our website at www.wiley.com/wiley-blackwell.

Library of Congress Cataloging-in-Publication Data

Bakker, Jokie.
  Wine flavour chemistry / Jokie Bakker, Ronald J. Clarke.
    p. cm.
  Clarke's name appears first on the earlier edition.
  Includes bibliographical references and index.
  ISBN 978-1-4443-3042-7 (hardcover : alk. paper) 1. Wine–Flavor and odor.
2. Wine–Chemistry. I. Clarke, R. J. (Ronald James) II. Title.
  TP548.5.F55C53 2004
  663'.2–dc23
                                                        2011014370

A catalogue record for this book is available from the British Library.

This book is published in the following electronic formats: ePDF [9781444345995]; Wiley Online Library [9781444346022]; ePub [9781444346008]; Mobi [9781444346015]

Set in 9/12.5pt Interstate Light by SPi Publisher Services, Pondicherry, India
Printed and bound in Malaysia by Vivar Printing Sdn Bhd

1 2012

Contents

Preface to the Second Edition

Wine is primarily consumed for pleasure, and despite some attributed health benefits, it does not form an essential part of our diet. Therefore the sensory properties of wines are considered very important and the appreciation of its flavour arguably gives the wine drinker most pleasure. The wine is bought for its appeal in the bottle, for the knowledge the wine drinker has about the sensory properties and the anticipated enjoyment of consuming the wine. After evaluating the colour of the wine in the glass, most wine consumers will smell the wine, and judge its qualities. The release of flavours from wine continues when drinking the wine, and gives further flavour sensations in addition to the perception of many other taste and mouthfeel compounds that should be present in balanced amounts in the wine. There is a very wide range of well made wines available, so if a wine does not deliver the flavour the wine drinker desires and appreciates, a different choice can be made for the next occasion. Since wine flavour plays such a crucial role in wine, this makes a book devoted to *Wine Flavour Chemistry* particularly relevant.

The technological advances in viticulture, wine-making and the resulting improved wine flavours have been based upon the scientific exploration of vines, grapes and wines, their constituents, their chemistry and all aspects of the wine-making process by scientists in many disciplines in research establishments world-wide. The understanding of flavour chemistry and its perception is determined by numerous scientific disciplines, ranging from chemistry and laws of physics to human physiology. Since the first edition of *Wine Flavour Chemistry* more research has become available on wine flavours, with new compounds still being identified. In addition, scientists place much emphasis on determining the potential sensory contributions of flavour compounds in wine, making flavour information as relevant as possible. Advantageous in all research dealing with flavours is the increased scientific understanding of the perception of volatiles and the award of the Nobel Prize for Physiology and Medicine in 2004 to Richard Buck & Linda Axel for their pioneering research on the genetics of the perception of odour (ref. in Chapter 4) has given great impetuous to this research field.

Although aspects of wine-making may always remain an art, such as the numerous choices to be made in the wine production process in order to optimize the wine flavour, science has definitely got a very sound and well deserved foothold in the wine making industry. Many highly trained and skilled wine makers work in the wine industry, helping to adapt advances in wine science in order to successfully influence wine making and ultimately wine flavour. All wine drinkers benefit from the well made wines with a wide range of wine flavours available nowadays. This updated book, *Wine Flavour Chemistry* may attract many different readers interested in wine, ranging from wine consumers, students, academics and people working in the wine industry. Information has been gathered from scientific research, review papers and books to update this comprehensive overview of the subject.

I wish to express my grateful thanks to all the scientists who have kindly shared their research information, which was of invaluable help in preparing this updated book.

Jokie Bakker MSc (Wageningen), PhD (Bristol)

Preface to the First Edition

This volume on wine flavour chemistry has been in gestation for many years; an original draft was started some ten years ago. A number of events led to our renewed interest in getting this book published. First, in the UK, wine has become a drink enjoyed by many consumers at numerous occasions, whereas previously, wine tended to be a drink shrouded by mystique and enjoyed mainly by more knowledgeable people. In contrast, in countries where wine production has been established for a long time, a long-standing culture of wine consumption, mainly with meals, has been established. Since there are now many people in the UK as well as in other areas of the world interested in consuming wines there is also an increased quest for knowledge about wine, making a book focused on the flavour and its chemistry particularly pertinent. Second, during the last two decades, there has been an enormous development in knowledge about viticulture and the technology of wine making worldwide. This has resulted in a much-improved wine quality. Third, many 'new' wine regions have been established. These have not been inhibited by cultural pre-conceptions about wine production, and have experimented in many different ways, pushing the boundaries of both viticulture and wine making. Fourth, financial investments in vineyards and wineries, hand in hand with generally vastly improved wine production skills, have given an array of wine flavours from grapes, which wine-makers in the past could barely have believed possible. The cultivation and production of single variety/cultivar wines have given consumers an insight into the many flavours possible in wines. Of course, the technological advances in viticulture and wine making have been based upon the scientific exploration of grapes and wines, their constituents and their chemistry, by scientists in many disciplines worldwide. Advances in analytical laboratory instruments have proven to be a great help. For example, mass spectrometers have, over the last ten years, become much more sensitive, much smaller, much cheaper and easier to use. This has resulted in an explosion of new data regarding the volatile compounds of wine. Numerous other advances in analytical techniques have aided the quest for knowledge about wine flavour, colour and taste. Interestingly, alongside these techno-logical advances has been the development of sensory science. There is now an array of scientifically based sensory analytical techniques allowing scientists to determine our perception of wine, including the measurements of differences between wines and descriptions of sensory properties.

Two recently published scientific books on wines summarize much of the wine information available. In 1994, Jackson in Canada published his excellent and wide-ranging book, *Wine Science*, with a second edition in 2000. This was followed in 2000 by the equally excellent *Handbook of Enology* from Ribéreau-Gayon and his colleagues in France. This new book on the chemistry of wine flavours draws together aspects of wine making pertinent to wine flavour, and tries to link chemistry, flavour composition and sensory properties. Our volume draws much information from these antecedent books for which due acknowledgement is readily made. We make similar grateful acknowledgement to several highly perceptive and entertaining, but not overtly scientific, books on wines by British wine writers and journalists. Of course, information from many scientific papers is also used to give a comprehensive overview of the subject. This book, *Wine Flavour Chemistry*, differs from all the foregoing in that it is uniquely devoted to the subject of the *flavour* of the wine. It is based on the chemistry of the compounds, both volatile and non-volatile, together with the application of the techniques of modern sensory analysis.

The quantities of volatile compounds in head-space air, that is, the air above a glass of wine, are determined by *partition coefficients*, which express the ratio between the amount of those compounds in the liquid and in the air above, both amounts often only in parts per million or much less. The quantities present have to be above *threshold levels of detection and recognition* in order for them to be perceived effectively by the olfactory organs of the nose. There are also threshold levels for non-volatile compounds present in much larger quantities in wines (several grams per litre), and detected only by the taste buds of the tongue, which detect the basic tastes of acidity, sweetness, bitterness and saltiness. Signals from these two highly sensitive organs (the olfactory epithelium and taste buds) are transported through nerve fibres to the brain. Certain other elements of sensory analysis, such as colour, appearance and tactile sensations in the mouth, including astringency, contribute to the overall flavour assessment. Perhaps it is the naming of all the different flavour sensations in all the different wines that is of the greatest intellectual and aesthetic interest, together with their association by scientists with particular chemical substances or groups of substances.

This volume, therefore, attempts to bring together in a readable and accessible form the most recent research from this rapidly developing field. It is aimed to be of interest to consumers with an inquisitive mind about wine, and to all those involved with the production of and trade in wines with an interest in the chemical and technical aspects of wine flavour.

A considerable amount of threshold flavour/odour and other flavour compound information has been recently compiled by Flament in his comprehensive book, *Coffee Flavor Chemistry* (2002). Many wine odour compounds, now known to number around 400, are also to be found among the much larger number of volatile compounds (about 800) in green and roasted coffee brews. In wines, alkyl esters from the fermentation of the must and unchanged terpenes from the grapes are particularly characteristic, though they may undergo change during subsequent ageing; in coffee, many distinctive compounds develop from the roasting process.

Reference to coffee flavour, indeed that of other beverages, should not be surprising since one of the authors (RJC), whilst an enthusiastic wine bibber, spent some forty years in the scientific study and industrial manufacture of coffee, and is the author or co-editor of several books on that subject. The other author of this book (JB) brings a wealth of knowledge and experience directly from her research into wine, especially within fortified wines, with numerous scientific research papers published on this subject.

We wish to express our grateful thanks to all those who have helped us with this book, in particular Professor Clifford of the University of Surrey, UK, and Professor André Charrier, ENSA M, France.

Ronald J. Clarke MA(Oxon), PhD(Hon), CEng, FIChemE, FIFST
Jokie Bakker MSc(Wageningen), PhD(Bristol)

Chapter 1

Introduction

1.1 Scope of the book

The primary meaning of the word 'wine' is the product of the aqueous fermentation by yeasts of the sugars in the juice of grapes. The fermented juices of many other fruits are sometimes also called wines, though they do not enjoy the same popularity or prestige as the grape wines. Fermented liquids from materials containing starch or cereals are usually called 'beers'. The term wine is incorrectly used, for example, in rice wines, since the sugars in rice are stored as starch. Since fermenting yeasts can only convert sugars into alcohol, raw materials containing starch need to be processed so that, first, sugars are generated, for example, by hydrolytic cleavage of the starch. Uniquely, grapes contain tartaric acid, which has preservative qualities, and which, in addition to the presence of fermentable sugars, gives wines both a relatively high acid and alcohol content.

This book is solely concerned with wine from grapes. The focus is on the chemistry and flavour of table wines, which are normally consumed with meals. These wines have an alcohol content of 9-15% v/v (percent by volume), typically 11.5-14% v/v for red wine. Many red wines from hot wine regions exceed this percentage since the grapes are picked more mature with higher sugar levels. Wines consumed before a meal (aperitifs) are usually 'dry' (low sugar content) and often fortified to raise the alcohol content to about 20% v/v. Wines consumed after a meal tend to be sweet, for example, made by fortification with alcohol before the yeasts have converted all the sugar into alcohol, giving fortified wines such as Port as made in the Douro region in northern Portugal. Such fortified sweet wine styles are also made in other regions, and will be referred to as Port style. Fortification of dry wine followed by a special maturation process gives Sherry, as made in the Jerez region in Spain. Some other wine regions also use variants of this production to make Sherry style wines. Sherry can either be kept dry to be served before a meal or sweetened to be served after

Wine Flavour Chemistry, Second Edition. Jokie Bakker and Ronald J. Clarke.

a meal. The wine-making process of the classic wines Port, Sherry and the less popular Madeira will be described separately.

There is a very wide range of types of table wines, from sweet to dry and from still to sparkling, including its most famous example, Champagne. Table wines can be red, rosé or white, the colours depending on the choice of grapes and the wine-making processes used. The wines can be sweet or dry, although red wines tend to be always dry, while white wines are produced from dry to very sweet, with a range of different sweetness levels in between. A most remarkable sweet white wine is made from grapes infected by *Noble Rot* that is caused by a mould (*Botrytis cinerea*). The term 'wine' will be restricted to the main species of the vine plant, *Vitis vinifera*, which covers about 98% of the total wine production from grapes.

1.2   Historical background

There is much historical information on wine, for example, Johnson's (1989) excellent writing and McGovern (2003). It is generally considered that vines originated from the Caucasus area of Russia, between the Baltic and the Caspian Seas. After the Stone Age, some 6000 years ago, settled agricultural practices developed in the 'Fertile Crescent' of Mesopotamia and Egypt. Wild vines, botanically known as *Vitis vinifera sylvestris*, became domesticated and strictly speaking became the so-called *Vitis vinifera L. sativa*.

The *Vitis* genus contains many similar species, with other names such as *Vitis labrusca* (see the next paragraph). From those very early origins in Mesopotamia and Egypt, vines and wine-making methods were exported to the Greek- and Latin-speaking world of the Mediterranean. After the decline of these civilizations, wine production in Europe was not established until late medieval times. Wines were shipped in barrels and even countries with little or no wine production could enjoy drinking wine. Fortified wines such as Sherries (the Sack of Shakespeare's Falstaff) developed as a result of the Arabian invention of distillation, which gave the required skills to prepare fortifying spirit that could be added to wines. In the late nineteenth century wine production was fully established in France. France, Italy and Spain are still the three largest wine producers.

Speculation remains, however, about the flavour of these early wines compared with the wines we know today. Grape juice is fairly easily fermented by ubiquitous yeasts, and the fermented product can be reasonably stable due to its relatively high alcohol and acid content. However, there are longer-term storage problems and it is likely that wine spoilage was a frequent problem. This was due to a lack of understanding of the actual processes involved in vinification and their effective application. For example, it was not until the days of Louis Pasteur in the 1860s that the role of yeasts in wine fermentation, in addition to the role of some lactic acid bacteria in wine spoilage, was uncovered. The scientific achievements of Pasteur regarding the discovery of the microbiological processes involved in wine-making laid the foundations for the modern wine industry.

As late as the eighteenth century, wines were mostly sweet, although even from these more recent times we have little information regarding the sensory properties of the wines. Roman wines are thought to have been more like syrups. Wines were stored and transported in amphorae, which were long earthenware vessels fitted with stoppers, often made of waxy materials. The Romans used glass decanters to bring wine to the table but glass was too fragile for storing wine. An especial boost came with the invention of glass bottles in the early seventeenth century that were sufficiently strong to allow the transport and storage of wine. Once corks started to be generally used to seal wine bottles, it was a relatively small step to mature the wine in bottles, which had to lie at a fairly even temperature to prevent leaking. Wines have long been imported to the UK, where there was an appreciative market for so-called fine wines. For example, there was a marked interest in red wines from Bordeaux in the early twentieth century, to accompany the lengthy Edwardian dinners.

The native vine plant is confined to certain latitudes of the world and its domesticated version similarly requires favourable growing conditions. In particular, vines thrive in a climate with the right combination of sun and rain, although varieties/cultivars have been adapted to suit various climatic conditions. The type of soil is important, with adequate drainage being a prerequisite for successful vine cultivation. The areas of growth include North and South America, outside the tropics and excluding the very temperate zones. The commercial production of wines in many regions outside Europe did not really develop until the late nineteenth century. White settlers in Australia and New Zealand were interested in wine-making but only after World War II did the wine industry really develop. The spread of vine and wine is probably also closely linked to social and cultural aspects of communities.

Viticulture in Europe and elsewhere, like the production of other domesticated plants cultivated for food and drink, has been closely associated with the activities of plant breeders. Hence, over the centuries many varieties/ cultivars of the species *Vitis vinifera* have been selected, e.g. *Vitis vinifera* Pinot Noir, and are responsible for the various wines that are available in the market place. An important part of the history of wine is the disease caused by *Phylloxera*, a root louse pest accidentally imported from America that struck nearly all vines in Europe in the 1870s, devastating many vineyards by killing the vines and thereby ruining the wine industry. It was not until the discovery that grafting local European vines onto American imported root-stocks conferred resistance to the disease that the wine industry in Europe started to recover. Ironically, *Phylloxera* eventually attacked vines in California around 1980, damaging many vineyards.

1.3 Wine flavour

The smell and taste of a wine are directly associated with the chemistry of the entire wine-making process. The word flavour usually indicates the combination of smell (or odour) and taste. However, when assessing the sensory properties of wine, the word 'tasting' is used to indicate that the flavour of the wine is being judged. The flavour of wine originates from (1) the grapes, (2) the treatment of the must (grape juice) and its fermentation and (3) the maturation process of the wine. The chemistry of the flavour compounds derived from these three sources will be discussed in some detail for both non-volatile (Chapter 3) and volatile (Chapter 4) compounds.

Wine writers in numerous books and articles, many in the English language, have dealt with the subject of wine flavour. Some texts are aimed at the marketing aspects of wine and emphasize the opinions of expert wine tasters. Other texts are more critical, such as Barr (1988). Of course, there are also numerous texts in French and German, dealing with all aspects of wine. The number of technical texts which directly relate the flavour of the wine to its chemistry is much more limited, though there are some chapters in books on food and beverage flavour in general (see Bibliography). Many scientific papers describe only individual aspects of wine flavour and its chemistry. None of these texts are complete; they omit to raise many questions and fail to answer many others. An exception is the comprehensive scientific book of Jackson (1994, revised for the second and third edition in 2000 and 2008 respectively), which discusses in detail the three interrelated topics of wine science: grapevine growth, wine production and wine sensory analysis. Ribéreau et al. (2006) have published a similar work in two volumes (in English).

The term 'wine tasting' is often used and suggests ignorance of the essential nature of wine flavour, which is a combination of (a) the five taste sensations (sweet, salt, sour, bitter, umami) from non-volatile substances perceived on the tongue and (b) the aroma (or smell) sensation from volatile substances perceived by the olfactory organs behind the nose. Volatile substances reach the olfactory organs by two routes, sometimes referred to as the nasal and retronasal routes. Nasal means that volatile compounds will reach the olfactory organ through the nostrils of the nose during the period of 'nosing' the wine from the glass. Nosing is the traditional sniffing of the air space above the glass of wine, before any sample is placed in the mouth. Once in the mouth, the wine is warmed up, moved around in the mouth and there is the option of noisily sucking air through the mouth. All these actions help the volatile compounds to escape from the wine and to travel retronasally via the back of the mouth to the olfactory organ. Volatile compounds detected during nosing are often described separately, and may or may not be similar or identical to those detected on the palate. Wine tasting will be discussed in Chapter 5.

There is no consensus in the use of terms like bouquet, aroma, etc. and different wine writers may use them with different meanings. The term 'aroma' is most commonly used to describe the smell of the wine derived from the grapes, while the term 'bouquet' tends to refer to the smell of the wine formed as part of the development during maturation.

The quality and quantity of colour as well as the clarity of the wine are assessed entirely by eye, usually before the tasting. Next, our sense of smell and taste are used to assess the flavour of the wine. The depth of intensity and the multicomponent detection of flavour notes in wines (usually described in terms of flavour notes from other fruit/vegetable/mineral/animal sources) that are used to describe wine attributes by many expert wine tasters is surprising to the non-expert wine-drinker, and at times stretches credulity. In addition to flavour recognition and description, there are also the perceptions of mouthfeel, temperature, bubbles, etc., which all are registered and assessed by our senses. Over and above the enjoyment of the wine flavour, wine is also drunk for its stimulant properties, derived from up to 15% v/v ethyl alcohol, formed by the fermentation of sugars in the must by fermenting yeasts.

The flavour of wine is determined by the grape variety (or varieties), in combination with the growing conditions, such as climate, agronomic factors during growth and harvest, and these are reflected in the composition and organic chemistry of the must. Perhaps equally important is the process of vinification used; in particular, must treatment, temperature, yeast strain, use of fermentation aids, filtration and other processes used, together with any maturation (ageing) process. The relative importance of these factors is a moot point, but they are all determined by chemical causes. Interestingly, for example, French wines are essentially characterized by the region in which they are produced, as referred to in the Appellation *d'Origine Contrôlée* (AC), usually without mention of the grape varieties used (although some French wines nowadays list the grape(s) used on the label). Although the grape varieties are defined in the AC, the proportions used may vary from year to year. As from 1 May 2009 the term 'Appellation d'Origine Contrôlée AOC or AC' has, with resistance from some producers, been progressively replaced with the new European standard, Appellation d'Origine Protégée (AOP). In contrast wine makers in many other countries, especially in the 'New World', make a feature of characterizing their wines by the grape variety used. In short, currently the French emphasize the *'terroir'*, while in many relatively new wine-making countries the emphasis is on grape variety. A particular grape variety (e.g. Chardonnay, Cabernet Sauvignon) can, evidently, produce a rather different wine flavour as a result of the method of vinification and maturation, even though, usually, the characteristics of the wine flavour for the grape *variety* remain recognizable. A given variety grown in a certain region is also claimed to give a different wine than when grown in another region, even though the process of vinification is essentially the same. This subject is further explored in Chapter 5.

Therefore the flavour of the wine derived from the grape has to be considered in terms of its complex chemical composition, which is detailed in subsequent chapters. Current wine-making practice is outlined in Section 1.5 to give a better understanding of the background in which chemical changes related to flavour occur. Formation pathways of flavour compounds during vinification are discussed in Chapter 7. Some physiological aspects of wines related to wine chemistry will be described briefly.

1.4 Wine colour

Wines are primarily distinguished by their colour and fall into three groups (1) white wines, which include most sparkling wines, (2) red wines, including most fortified Port style wines and (3) rosé wines, essentially an intermediate between white and red wines. A wine's colour is determined by the choice of grape and the vinification process. White grapes, which usually have pale yellow skins, give white wines, while black grapes, which have blue, red or even black skins, depending on the amount of colouring matter in the skins, give mostly red wines. Red grapes can give a range of wine colours, from deep red to rosé, depending on the wine-making process, and by careful handling they can even yield white wines, for example the 'blanc de noirs' in Champagne production. Since most of the colour is in the grapes' skin, the choice of the vinification technique for red grapes allows a lesser or greater extraction of colour into the wine.

Within each colour, however, there will be differences between wines, which are easily perceived in the wine glass. The clarity may also differ, due to very small amounts of very finely suspended insoluble matter (not desired), although nowadays most wines are clarified before reaching the consumer. The changes in colour that occur during ageing, whether the wine is stored in-bottle or in-cask, are determined by chemical composition; the colour of red wines depends on the content and composition of anthocyanins. Colour and the chemistry of the changes in colour during maturation will be discussed briefly in Chapter 3.

1.5 Vinification

Traditionally good wines were made in regions where the conditions were frequently just right to give healthy, ripe grapes, with somewhat cooler weather during vinification. Although undoubtedly much knowledge was collected over many years, there was limited control over the process and wine-making was, to an extent, considered to be an art. With the advance of our scientific knowledge of many aspects of wine-making and the improvements of technology used in wine-making, in particular the use of refrigeration at various stages, there is now much control over the process. Nonetheless, the wine maker still faces many choices that determine the properties of the wine, and so the art still remains in making the best possible wine that is typical for the grapes and the region. The use of modern technology has also enabled good quality wines to be made in many more regions, including ones once thought to be too warm.

A basic understanding of the wine-making process ('vinification') is necessary in any study of wine chemistry and wine flavour. Grapes are the key ingredients, and they should be healthy, mature and in good condition. The choice of grape variety will influence the wine flavour and colour and to an extent depends on the region. Grape varieties and some of the main growing regions are discussed in Chapter 2. In essence the grapes are picked, crushed to form a 'must' (grape juice) and fermented by yeasts to convert the sugars present into alcohol. There are three stages in wine-making, all of which can influence the flavour and colour of the wine:

(1) Pre-fermentation, during which various pre-fermentation treatments of the grapes or must can be given (such as sulfur dioxide addition, sugar or acid adjustments, nitrogen contents and possible addition, clarification of must, contact time with the skins or 'maceration' and cooling).
(2) Fermentation of the must, during which several factors have to be managed (such as choice of fermenting yeast, fermentation temperature, maceration time and pressing conditions).
(3) Post-fermentation, during which several different treatments are available. Some are probably essential (such as racking to remove the spent yeast or 'lees'), while others are optional, depending on the desired characteristics of the wine (such as filtration, cold stabilization). Wines can be made to drink when young, or after maturation (or ageing) in different types of vessels (such as old vats, new oak barrels or bottles).

Several general rules of modern wine-making have emerged. The production of both red and white quality wines requires attention to the following, as emphasized in many wine-making publications:

(1) Grapes should be picked at optimum ripeness, in sound and healthy condition, at as low a temperature as possible (in very warm conditions, they should be cooled) and transported to the winery with minimal delay for immediate processing. The must for white wines should be cooled both before and during fermentation.
(2) Strict adherence to cleanliness of all wine-making equipment; to prevent the growth of spoilage organisms on the grapes, in the must or in the wine at all stages of fermentation and maturation. A particular risk are *Acetobacter* bacteria, which convert alcohol into acetic acid, hence spoiling the wine into vinegar.
(3) Non-oxidizing atmospheres should be used, especially in the early stages of vinification of white wines, for example by blanketing the must or wine with inert gas and/or by addition(s) of sulfur dioxide.
(4) The temperature of the fermentation should be controlled. Heat produced during fermentation (an exothermic process) in stainless steel tanks should be removed by efficient cooling and refrigeration. This is usually done by cooling the outside of the tank or sometimes by pumping the must through an external heat exchange unit. Fermentation in barrels may lose heat through the relatively larger surface area; however, barrels are difficult to cool, though ice is sometimes used.

The success of wines produced in regions such as Australia, California and South Africa has been attributed to careful attention in controlling the vinification process, especially to the factors listed above. Of course, great care is also given to planning the vineyard (site selection, considering soil, climate and choice of the most appropriate grape) and vineyard management (pruning, fertilization, preventing disease and picking at grape maturity). Modern scientific methods of wine-making analysis and control have also been adopted. Nevertheless, traditional wine makers in Europe have not been slow to adopt modern practices and the overall quality of wine now available to consumers has improved significantly.

The equipment used for fermentation and details of that used for both pre- and post-fermentation stages have been well described in detail (Jackson, 2008; Ribéreau-Gayon et al., 2006), while Robinson (1995) has given a good account for the general reader. The actual equipment used differs between wineries, depending on local conditions and the style of wine that is made.

Spoilage of wine is what all wine-makers want to avoid. This risk is present at all stages of wine-making, and can be both microbiological and chemical in nature. Under the wrong conditions, *Acetobacter* bacteria can change wine into vinegar in a very short time. Yeasts also pose a risk, as reviewed by Loureiro & Malfeito-Ferreira (2003), pointing out that understanding the ecosystems of wineries is crucial in prevention of spoilage. Excess air is also a known enemy, possibly causing chemical spoilage by oxidation of the wine, and enhancing the risk of microbial spoilage.

1.5.1 Vinification process

The basic wine-making process described here, highlighting those parts that greatly influence the resulting end product, covers many components that are common to both red and white wines. Recent textbooks by Jackson (2008) and by Ribéreau-Gayon et al. (2006) offer the reader a more in-depth treatment. The formation of specific flavour compounds during vinification (Chapter 7, devoted to reaction pathways) and volatile compounds (Chapter 4) are discussed elsewhere. A recent review on red wine-making reviews steps of wine-making affecting colour (Sacchi et al., 2005), and interestingly cold soaks and the addition of sulfur dioxide tended to have an effect on the wine only over a short time, after some maturation the differences became minimal. Other parameters, such as yeast selection and carbonic maceration showed that the grape varieties affected the results.

Flow charts for both the production of red and white wines (Figs. 1.1 and 1.2 respectively) give a helpful overview of the various wine-making steps. Figures 1.1 and 1.2 reflect the headings below, as appropriate. Specific information for the production of red (Section 1.5.2) and white (Section 1.5.3) wines is included below. These brief descriptions focus on the differences from the general vinification process and, therefore, not all captions will be used in Sections 1.5.2 and 1.5.3.

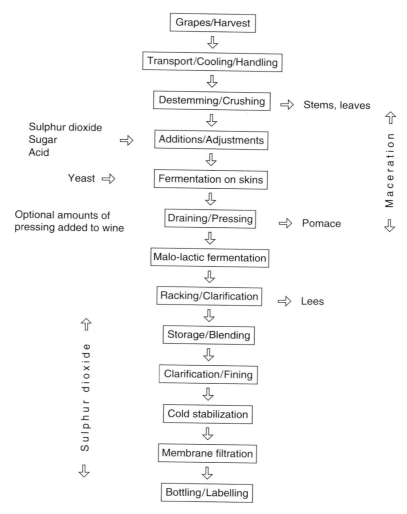

Figure 1.1 Typical processing sequence in vinification of red wines. Some steps are optional (see text). Oak contact can occur during fermentation and/or storage. Prolonged storage in-barrel or tank will mature wines.

The production of specialized wines is described separately, briefly emphasizing the part of the process that makes these wines different. As stressed above, some wine-making steps are essential, but the wine maker can make numerous decisions that will determine the overall style of the wine.

Pre-fermentation

Grapes/harvest

The first important step is the grape harvest. The grapes are picked at commercial maturity, which is usually determined by both their acid and the sugar content. The maturity of the grape will also affect the aroma and

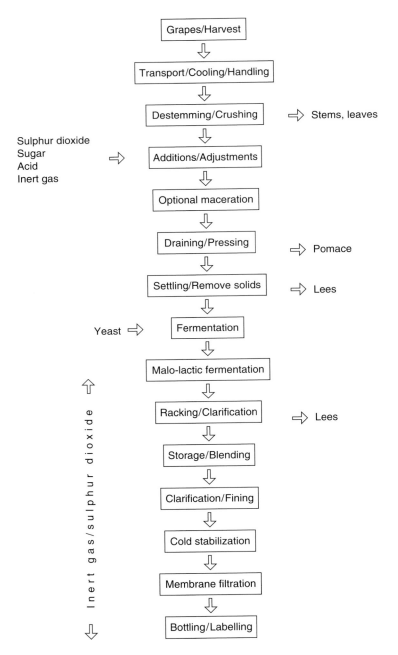

Figure 1.2 Typical processing sequence in vinification of white wines. Some steps are optional (see text). Oak contact can occur during fermentation and/ or storage. Prolonged storage in-barrel or tank will mature wines.

phenolic composition typical for the variety, an aspect of increasing importance since they affect the varietal character and quality of the wine. As grapes mature, their acid content decreases and their sugar content increases. In hotter climates, the mature grapes tend to be high in sugar and low, sometimes even too low, in acid. In cooler climates the reverse happens: the grapes may struggle to reach full maturity and remain low, sometimes even too low, in sugar content and high, sometimes too high, in acid content.

The grapes can be picked by hand or by mechanical harvester. Hand picking allows access in even the most inaccessible vineyards (steep slopes, very soft soils), does not restrict the pruning style of the vines and some selection in the vineyard can be made (rejection of immature fruit and overripe, mouldy fruit). However, hand picking has a high labour cost. Machine picking is fast, costs much less in labour and can be done at night, which is attractive in hot regions to keep the fruit as cool as possible. Fast picking may help to ensure that the grapes can all be picked at optimum maturity, especially important in hot climates where a short delay in picking may give overripe and shrivelled grapes. However, mechanical picking has specific requirements on vineyard lay-out, access and vine pruning and training, etc. No selective picking of individual grape bunches is possible and it may not be desirable to pick grape varieties with very thin skins.

Transport/handling/cooling

Grapes should be handled with care to avoid damage that would encourage the growth of undesirable micro-organisms and lead to oxidative browning. Hence, hand-picked grapes are generally placed in shallow containers to prevent them being pressed under their own weight. The fruit must be kept cool and transported to the winery for processing with minimal delay. When using a mechanical harvester, it is important to avoid damaging the fruit.

De-stemming/crushing

Next, the grapes are usually de-stemmed and crushed. De-stemming involves the removal of stems, grape stalks and leaves, thereby minimizing the extraction of phenols and other compounds considered undesirable in the wines. Some of these phenols may impart bitterness and astringency to wine (Chapter 3), whilst some of the other compounds can give off-flavours or a haze in the wine.

The grapes are crushed, usually immediately after de-stemming, as gently as possible since excess pressure may damage the seeds, which will lead to the extraction of an excess of compounds (phenols) that, due to their bitter and astringent properties, would tend to give a harsh wine. With crushing the grapes are broken open, thus releasing the juice from the grapes. This facilitates the onset of fermentation, since the yeast has easy access to the sugar-containing juice. Most modern wine-making equipment carries out both stages, and it is generally thought that de-stemming before crushing gives the least extraction of undesirable compounds in the wine. In the past the stems were, notably, not removed, and their presence in the fermentation of red wine facilitated the pressing procedure. Modern equipment for pressing the wine does not require the presence of stems. Some traditional wine makers may still choose to leave the stems in during the fermentation stage, although any vine leaves should always be removed.

Additions/adjustments

Additions of sulfur dioxide, sugar, acid, nitrogen and enzymes can be made but also acid adjustments can be made. Dependent on the soundness of the grapes, sulfur dioxide may be added at crushing. Sulfur dioxide has several functions; it protects the must against both browning reactions and, to an extent, unwanted micro-organisms. It suppresses the growth of wild yeasts and delays the onset of the fermentation. It is probably most effective in its use to protect against undesirable spoilage bacteria, especially if the pH of the must is low. However, it can also inhibit the malo-lactic fermentation, so its use should be avoided if an early malo-lactic fermentation is desired. It also damages the grape skin, and therefore enhances extraction of compounds from the skins.

Just prior to fermentation several adjustments may be made to the must, depending on local conditions and regulations. Grapes from cool regions containing insufficient sugar to ensure an adequate final alcohol content in the finished wine may receive an addition of dry sugar (sucrose, in crystal form). This process is called *chaptalization*, and was first introduced in France by M. Chaptal in 1801. Usually no more sugar than the equivalent of 2% v/v alcohol may be added. In addition, if the must contains excess acid, which tends to be a risk in cool climates, it may need to be de-acidified, often by the use of 'Acidex®', a form of calcium carbonate. In the same way, must prepared from grapes from hot regions containing insufficient acid may need to be acidified, usually by an addition of tartaric acid, otherwise the wines may taste flat and may have a high pH, which is undesirable for both chemical and microbiological stability.

When it is suspected that insufficient nitrogen is available to complete the yeast fermentation, an addition of diammonium phosphate can be made to the must to ensure an adequate concentration of nitrogen for yeast growth and maintenance. In particular highly clarified juice for white wine fermentation or must made from *Botrytis* infected grapes may not have a sufficient nitrogen concentration. There are two reasons for this addition. Firstly to avoid the fermentation getting stuck before the process is finished and secondly to avoid the production of hydrogen sulfide by yeasts, which is produced when the yeasts have prematurely depleted nitrogen in the fermenting must, as reviewed by Bell & Henschke (2005) and Ugliano et al. (2007). However, Jackson (2008) considers the addition of diammonium phosphate unnecessary and suggests that it does not prevent the formation of hydrogen sulfide. Possibly other yeast nutrients may also be inadequate in the fermenting must, not just nitrogen, for more detailed discussion see Ugliano et al. (2007). They suggest that cleaner, fruitier style wines may be obtained if wine-makers ensure that there is between 250 and 300 mg L^{-1} nitrogen available for yeast fermentation. During the last decade much more information regarding the interaction of yeast and nutrients in the must and the effect on the formation of flavour has become apparent and is further discussed in Chapter 7.

Additions of enzymes, including pectolytic enzymes to enhance colour extraction can also be made, however, the review by Sacchi et al. (2005) concluded that pectinases do not seem to increase anthocyanins in wines but only increase the extraction of other phenolic compounds in wines. Presumably, anthocyanins were rapidly lost by the naturally occurring maturation reactions these compounds participate in, as suggested by the increased polymeric pigment formation observed in pectinase treated wines.

Additions of enzymes can also be made to stimulate the hydrolyses of glycosides from terpenes, since only the non glycosilated terpenes contribute to the volatile varietal aroma characteristics of the must, or more importantly the wine. The review by Maicas & Mateo (2005) gives an in depth overview of the current enzymes available, which gives wine makers potential tools to enhance the varietal aroma. The authors also warned against some potential adverse effects, such as the production of off-flavour as a result of vinyl phenols, resulting from the presence of undesirable cinnamate decarboxylase activity. Overall, with further studies on both the definition of desirable varietal wine aroma and precisely defined action of added enzymes, the use of these glucolytic enzymes give scope in the future to manipulate the wine aroma.

Enzymes can also be added as processing aids, such as enhancing the ease of pressing or aiding juice clarification.

Maceration

This is the process of steeping or soaking the skins and seeds of the grapes in the grape juice that was released from the grapes during crushing. It forms an important part of red wine-making, since during maceration the phenols, including the coloured ones (anthocyanins) and the flavour compounds from the grape skins, the seeds and the residual grape stems are dissolved into the juice, or into the fermenting must. The efficiency of the maceration depends on the temperature, the length of time and the amount of agitation the macerating mash receives. A recent review of Sacchi et al. (2005) lists the various techniques commonly used and their effects on the extraction of colour of red wines.

The maceration of a red wine typically starts before the onset of fermentation, and continues during part or the entire fermentation, and in some instances even after fermentation. The extraction of compounds will change in the presence of an increasing concentration of ethanol. How much maceration the wine should get is decided by the wine maker. Generally, longer macerations will lead to darker, more phenolic wines, requiring longer maturation before the wine is ready to drink. Red wines made to be drunk young tend to have short maceration times and are run off the skins before the fermentation has finished. Most of the anthocyanins are extracted during the first five days, whilst prolonged extraction increases the phenolic compounds in wines. Astringency of red wines is mostly attributable to phenols, hence longer extraction of red grapes tends to give more astringent wines. Many of the flavour compounds are in the skins of the grapes; hence maceration also increases the extraction of flavour compounds into the wine, although little information is available on red wines.

Temperature of maceration can also be varied, and before the fermentation starts some wine-makers elect to do cool (about 15°C) to cold (4°C) macerations; temperature effects on colour and flavour have been reported (see Jackson, 2008), although increased risks of spoilage due to the adaptation the indigenous flora has also been reported.

As well as the temperature and duration of the maceration, during fermentation the agitation is also usually controlled. Without agitation, solids (grape skins, seeds, etc.) would float to the top with the stream of carbon dioxide produced during the fermentation and form a thick surface layer, known as the 'cap'. Obviously there would be no extraction from such a cap, and the risk of spoilage organisms growing in the cap (which is warm once the fermentation gets on the way from the heat released from the fermenting yeasts) is extremely high. Therefore it is imperative to keep the cap wet by some way of 'working' the must. Probably the most commonly used method is known as 'pumping over', in which juice is drawn from underneath the cap and quite literally pumped over it. Often such a process can be done semi-automatically every few hours, the time required depending on the size of the fermentation vessel, temperature, etc.

Of the several other methods used to keep the cap wet, some are quite simple, although labour intensive, while others rely on more sophisticated fermentation equipment. If the fermentation is performed in a shallow open vat, the cap can physically be punched down with the aid of simple punching sticks. If the fermentation takes place in a tank, a physical restraint (a metal grill or wooden beams) can be fitted just under the expected level of the must when the tank is full, to keep the cap submerged. There are also fermentation vats designed specially to keep the cap wet, such as autovinification vats (using pressure from the carbon dioxide released during fermentation to pump over the must) and roto-fermenters (using rotation of the tank, usually the tank being fitted in brackets on its side). A combination of colour extraction, must cooling and aeration can also be achieved by a technique of rack and return, a widely used method also known as *délestage*. After the start of the fermentation and the formation of the cap, the juice is drained into a holding tank, sprayed into a second tank and then pumped back into the fermentation tank. At the drainage stage seeds can be removed, desirable if grapes are immature to avoid excess extraction of extractable phenols.

Another maceration technique discussed was the use of freezing to enhance extraction (Sacchi et al., 2005). They reported that freezing tended to enhance the extraction of both anthocyanins and phenols, presumably because freezing damages both the skin of the berries, allowing easier extraction of the coloured anthocyanins and the seeds of the fruit, thus aiding the extraction of phenols. The authors also suggested that the use of dry ice has the added advantage of protecting the berries from oxygen. However, although such studies add greatly to our understanding of the processes in wine-making, currently it would seem that freezing is too expensive to have commercial applications.

Fermentation

Fermentation process

The next major step is the fermentation, in which the fermentable sugars (glucose and fructose) present in the grape juice (including any added sugar) are converted by yeasts into ethanol (ethyl alcohol) and carbon dioxide, with the generation of heat, the excess of which needs to be removed. The fermentation temperature is an important variable. The fermentation also produces many of the aromatic characteristics of the finished wine. The fermentation is usually carried out in large, closed stainless steel tanks (capacity 2000 L up to 2000 hL), which are temperature controlled so as to lower the fermentation temperature as appropriate. Open fermentation vats are also used, although not for modern white wines. The fermentation time increases with decreasing temperature. The effect of fermentation temperature on the formation of volatile flavour compounds, in particular esters, is discussed in Chapter 7.

Yeasts are unicellular micro-organisms that are classified taxonomically as 'fungi'. Yeasts have several commercial applications, and they are used also for beer brewing, baking and biomass production. Yeasts used in wine-making generally belong to the *Saccharomyces* genus, the most important species of which, *cerevisiae*, has some unique characteristics – perhaps one of the most useful ones being its tolerance to ethanol (up to 15% v/v), a very toxic compound for most other micro-organisms. Naturally occurring fermenting yeasts will start the fermentation at 18°C, though, increasingly, cultured yeasts are used. Some wine makers rely on fermenting yeasts that naturally occur in the winery, to start the fermentation and do not add any yeasts, allowing the natural yeasts to ferment the must, in order to achieve wines with a character believed to be more typical for the region. Generally, a strain of *S. cerevisiae* is added to the must, ensuring that the fermentation will start without much delay. The fermentation in wine-making may include several other types of yeast, some indigenous ones, although most yeasts will only grow at very low levels of ethanol (Chapter 7), and they tend to be more sensitive to sulfur dioxide. The fermenting yeast, inoculated or from the winery, will in most cases take over from all other organisms that may have become established during the early part of the fermentation and ferment the must to 'dryness' (low residual sugar).

Once dried fermenting yeasts became commercially available, most wine-makers deciding to inoculate the must did so mostly to avoid problems such as slow or even stuck fermentations and exert greater control over the fermentation, whilst minimizing the potential role of indigenous yeasts. An inoculum of active fermenting yeasts generally guarantees a rapid fermentation, also reducing the risk of spoilage. However wine-making knowledge has grown substantially over the last decades regarding the role of the yeasts in wine fermentation and its effect on wine aroma and quality. There are recent reviews discussing and summarizing the role of yeasts in wine fermentation, wine aroma and wine quality giving details of current knowledge (Fleet, 2008; Swiegers *et al.*, 2008; Ugliano & Henschke, 2009; Bisson & Karpel, 2010). The current view is that the fermentation is more complex than previously assumed, with indigenous wild yeasts contributing significantly to the fermentation and the resulting wine quality, whereas it was assumed that these organisms died soon after the fermenting *S. cerevisiae* took over.

Further studies on flavour modulation by yeasts and the use of a mixture of yeasts to inoculate fermentation, may give in future yeasts even better tailored to fermentation and flavour production than currently available. The selection of yeast strains with suitable characteristics for wine-making, the use of mixed strains allowing a succession of fermentations to take place, and making these commercially available will in the long run give the wine-maker many more opportunities to select the yeast according to the desired wine style and quality, and create the wine most suited to the current consumer preferences. This further future exploitation of yeasts requires detailed information regarding the attributes valued in wines, so accurate selection criteria for the fermenting yeasts can be developed. Aspects regarding the effect of yeasts on wine aroma are discussed in more detail in Chapter 7.

Draining/pressing

The juice is taken from the grape skins when sufficient maceration has occurred (the fermentation may or may not have finished; to be decided by the wine maker) either by running off the juice without exerting any pressure on the grape skins (free run juice) or by pressing further juice (pressed juice) out of the remaining grape mash, usually referred to as pomace. The pressed juice tends to contain more tannins than the free run juice. The wine maker can choose either to keep pressed juice separate from the free run juice or to mix some or all of the pressed juice with the free run juice. For lighter wines, made to drink early, the pressed juice is kept separate, and often used for distillation. White wines are pressed *before* the fermentation (Section 1.5.3), while red wines are pressed *after* (or possibly during) the fermentation stage (Section 1.5.2).

There are various designs of presses available, and there is some variation in the properties of the pressed juice depending on the press, most important variables seem to be an excess of suspended solids in the juice, requiring extra fining to remove, and increasing amounts of anthocyanins and phenols in higher press run fractions. It is up to the wine-maker how the pressed wine fractions should be used.

Malo-lactic fermentation

A type of second fermentation, usually referred to as the malo-lactic fermentation, often occurs in wine, in which lactic acid bacteria convert the harsh-tasting malic acid into the softer tasting lactic acid, producing a small amount of carbon dioxide gas and raising the pH value of the wine. This fermentation can take place either approximately concurrently with the yeast fermentation or in the young wine. It seems still a matter of debate what the optimum time for this process is. Malo-lactic fermentation tends to be encouraged in wines with an excess acidity and almost without exception is carried out for red wines. However, this acid conversion is not desirable in wines already high in pH or low in acidity. Wines can be inoculated with selected micro-organisms, or the wine maker may rely on the indigenous flora of lactic acid bacteria for the malo-lactic fermentation. Inoculation with malo-lactic organisms does not automatically mean this fermentation will occur, since this process does not appear to be controlled easily. Wines kept at 20°C or above are likely to undergo the malo-lactic fermentation, while high levels of sulfur dioxide and high acidity tend to inhibit the process. Malo-lactic fermentation also changes the flavour in the wine due to the volatile compounds produced, which may or may not be desirable, depending on the style of wine the wine maker is aiming to produce. A comprehensive review by Bartowsky & Henschke (2004) describes current knowledge on malo-lactic fermentation, and in particular its contribution to the buttery flavour in wine. Bottle ageing of red wines is also reported to affect the aroma compounds formed during malo-lactic fermentation (Boido et al., 2009).

Post-fermentation

Racking/clarification

Young wines are stored and matured in small (225 L) or large wooden barrels, or in wooden, stainless steel or concrete vats. Whatever the mode of storage, a few weeks after the fermentation has ceased the wine will, usually, clarify and the yeasts be deposited at the bottom of the container. The wine is then usually taken off the lees (i.e. the precipitate of mostly yeast cells) and placed in a clean container. Thus, racking involves drawing off the wine from the barrel or tank to just above the level of the sediment or lees and transferring the wine to clean barrels or vats. This process may need to be repeated, depending on how long the wine is left to mature. Wines should be racked as required, more frequently when the wines are very young, and possibly annually when the wines are a few years old.

Oxygen

Aeration of the wine due to racking is thought to be crucial to the development of the flavour and colour of red wine but is generally detrimental to that of white wine. The higher phenol content in red wines, in particular the flavonoids, use oxygen for complex chemical reactions, leading to a softening of the wine. In contrast, white wines have much lower concentrations of phenols, mostly based on tartrate esters of hydroxycinnamic acids, prone to oxidation and forming hydrogen peroxide in the process, which can oxidize other compounds in wine, such as ethanol being oxidized into acetaldehyde. These oxidation processes were first reported by Wildenradt & Singleton (1974). Hence care should be taken when pumping white wine around at various stages post fermentation to protect the wine from the uptake of oxygen, often achieved by using a blanket of inert gas and adding some sulfur dioxide if required. In contrast, red wine can be deliberately aerated, often by pumping over the wine or storage in wooden barrels which allow the slow ingress of oxygen, to enhance the reactions involving phenolic compounds, which lead to softening the wine and colour changes and has impact on the formation of the typical red wine aroma. Another technique used to allow a controlled amount of oxygen uptake is referred to as micro-oxygenation. This is used when the extraction of oak flavours into the wine is to be avoided and hence, inert vessels for wine storage are preferred. A specially designed cooperage can be used to allow controlled oxygen uptake of red wine, or silicone tube diffusers can be used to supply oxygen below the rate of consumption. Various instruments are also available to supply oxygen at selected rates via microporous diffusers.

An excess of oxygen in wines can lead to oxidative browning, a process avoided by ensuring minimal contact with oxygen, in particular in the case of white wines, and/or the use of sulfur dioxide as an antioxidant to control browning. Enzymic browning occurs primarily in must, whereas non-enzymic browning generally occurs in wines. A recent review (Li et al., 2008) describes the pathways involved in browning, and discusses the central role of iron and copper in initiating non-enzymic browning, suggesting that control of these metals may help to avoid browning.

Oxygen in must, during fermentation and during storage impacts on red wine aroma and quality, in addition to the above mentioned effect on phenolic compounds (Kilmartin, 2009; Toit et al., 2006). In red wine maturation it is thought that oxygen can lower undesirable vegetative aromas and enhance some of the varietal fruity aromas. However, more studies are needed to unravel the interaction between the chemical changes occurring in the group of phenolic compounds, and the accompanying changes in aroma volatiles.

Bulk storage

Wines can be stored and matured in large stainless steel tanks, although the flavours that develop on maturation are different from those formed during maturation in oak vats or barrels. Both red and white wines can be 'aged' before bottling, so-called maturation and the changes that occur are discussed in detail in Chapters 3 and 4. The chemical changes that occur affect the colour, the non-volatile and volatile compounds and contribute to the changes of flavour in the wine. This use of maturation is practiced for red wines, and to a lesser extent for white wines, to accomplish changes in flavour and colour. Oxygen uptake needs to be avoided for white wines, and controlled for red ones (see the earlier paragraph). Storage temperature is typically 18°C or below, and will affect the rate of chemical ageing changes.

Barrel ageing

The main reason to use oak is for the wine to extract aroma compounds from oak into the wine. There are various ways of achieving this. Normally maturation may take place in large wooden barrels, but if a typical wood flavour is required the process is more effective in small barrels (225 L), although this is much more expensive due to the extra cost of labour and barrels. Such barrels are usually made of oak (mainly French or American, each imparting a distinct and different oak character to the wine) and may be used when either new or old. Various pretreatments of the wood (charring) have different effects on the flavour of the matured wine. Maturation in such vessels takes place over periods ranging from three months to three years, and the maturing wines should preferably be kept at about 15°C, a temperature prevalent in underground cellars.

The addition of oak chips to wines stored in stainless steel vats as been investigated as a cheaper alternative to the used of small wooden barrels (Campbell *et al.*, 2006). However, despite the numerous experiments done on potential variables, such as time of addition, size of shavings and using planks, the resulting wines tend not to have the same sensory properties as those matured in traditional small oak barrels. More research on the relationship between the composition of wines and the sensory properties will lead to a better understanding on how the process can be influenced. Not all wine regions allow the use of oak shavings, etc.

Blending

Of course, in addition to all these possible treatments, another very important aspect of wine-making is blending. Blending can be done simply to sweeten a wine, by adding some specially prepared sweetening wine, or to colour a wine, by adding some specially prepared colouring wine. However, in many wine-making industries, highly skilled blending experts prepare blends to ensure a consistency of wine character. Blending is also done to improve the wine flavour. The character of wines such as Champagne, fortified Port and Sherry depend heavily on the preparation of blends, often drawing on wines from both different years and different locations. European table wines are blended to a much lesser extent, thereby usually showing annual differences from the same producer due to differences in growing conditions (climate) as well as differences in any particular year arising from where the grapes were grown. Some wines can even be made from one particular year and one particular vineyard, so blending is not part of the process.

In contrast, for example, Australian table wines tend to rely heavily on the preparation of blends, often using wines from regions quite far apart, giving the wine maker a range of wines with which to make the desired blend. It is important to prepare the final blend some time before the pre-bottling treatments so as to allow time for compounds that are insoluble in the blend to precipitate.

Clarification/fining

Most wines clarify themselves during storage, being racked off their lees during the maturation before bottling. However, before bottling, several treatments are used to ensure the wine's stability, so that it remains clear, without the development of unsightly hazes or deposits in the bottle. Wines may be fined, to precipitate specifically compounds that may affect the long-term stability of the wine, or they may be filtered or both. Fining involves using a coagulant, which is carefully stirred through the wine and then slowly descends to the bottom of the vat, removing suspended solids or specific compounds from the wine and forming a deposit. The wine may be racked off the thus-formed deposit and filtered or centrifuged to remove all particulate material.

Many fining agents have been used by wine-makers for a long time, and can be considered as 'traditional' wine-making aids (protein based compounds, such as egg white, gelatine, casein, or clay-based such as bentonite, or others such as activated carbon). Whatever the fining agent used, all fining agent added is removed from the wine; it is not an additive which remains in the wine, but a wine-making aid, only temporarily present in wine during its fining stage.

Care should be taken not to expose the wine to air (oxygen), which will affect the flavour and the colour of the wine. By carefully choosing the type of fining, the flavour of the wine may also be modified, for example, by using polyvinylpolypyrrolidone some of the phenolic compounds that impart excess astringency to red wines can be removed.

Clarification/cold stabilization

Without further treatment, some of the tartrate salts may not remain soluble during storage, leading to the occurrence of glass-like crystals of tartrates in the wine that are sometimes found on the cork or at the bottom of the bottle. These crystals may be considered unsightly, but are completely harmless and do not affect the quality of the wine in any way. To prevent any deposit of the salts of tartaric acid, and other compounds that may not remain soluble in the wine during short- to medium-term storage in-bottle, the wine can also be cold stabilized. Usually this involves chilling the wine to near its freezing point, keeping the wine at this low temperature up to two weeks, during which time tartrates and other compounds will generally form a deposit which are then removed by filtration or centrifugation.

Clarification/(membrane) filtration/centrifugation/pasteurization

There are numerous filtration methods used, based on thicker layers of fibrous filters generally designed to remove the larger particles, whereas membrane type filters can be employed to remove large molecules and colloids. Centrifugation is another way of removing particulate matter from a wine, and can even used instead of racking, particularly when wines are intended to be drunk young. Care has to be taken to avoid oxidation, often achieved by blanketing the wine with inert gas.

Just prior to bottling, all yeasts and other micro-organisms have to be removed to ensure the microbial stability of the wine. This may be done by sterile filtration, using a narrow pore membrane to ensure the removal of yeasts and bacteria. Bottling needs to be conducted under sterile conditions, using sterile bottles to avoid contaminating the wine with micro-organisms. In addition, the bottles may be injected with nitrogen to minimize exposure to oxygen.

Instead of sterile filtration, the wines can also be pasteurized by one of numerous heat treatments to reduce the risk of spoilage, for example, by heating for a short time (a few seconds) at a relatively high temperature (95°C), or for longer (about one minute) at a lower temperature (85°C), etc. These more brutal heat treatments are unlikely to be used for expensive quality wines.

Bottling/labelling

Often, a small addition of sulfur dioxide is made just before bottling, to protect the wine from oxygen, to prevent browning and to reduce the risk of contaminating micro-organisms. The wine bottle is sealed with either a cork or a screw cap. After the initial uptake of oxygen from the cork, wine sealed with a cork usually has an 'airtight' seal especially if the bottle is stored lying down at a cool (about 16°C) and even temperature. The use of cork as a closure has been reported to have a small effect on the oxidation of wines, and has been recently reviewed (Karbowiak *et al.*, 2010). However, faulty corks can leak (which gives a risk of wine spoilage) or give unpleasant off-flavours (see also Chapter 5). Therefore artificial 'corks' are also used; however, they are often difficult to extract.

An alternative closure is a screw cap, although it does not traditionally suit the image of the bottle of wine or its drinkers. However, a screw cap protects the wine well, since it retains the added sulfur dioxide and allows minimal uptake of oxygen, thus preventing oxidation in-bottle. Wines can be stored upright and maintain their character. One problem associated with the use of screw caps is the development of reduced sulfur odours attributed to the low redox potential developing in the wine. This problem does not occur in wines closed with high quality corks. Debate regarding formation of these off-odours and prevention of their formation is still a matter of debate (see Jackson, 2008). There is no reported risk of screw cap failure and wines developing off-flavors as a result. There is a gradual shift towards its use becoming more acceptable and not just for entry level wines but also for some of the more expensive wines produced in the New World.

Bottle labels vary from country to country (some of the information may be regulated locally) and come in various styles. The bottle label will give the year in which the grapes were harvested and the wine made, but not usually when the wine was bottled. The absence of a year on the label for a table wine usually indicates that the wine is a blend from different years. This is generally only done for cheaper table wines and helps to produce a consistent product. However, many Champagnes, most Ports and nearly all Sherries are blends of different years, which in no way reflect any inferiority in quality. Wines from the same production year from the Southern Hemisphere are half a year older than those from the Northern Hemisphere. The wine label usually gives information regarding its origin and its producer. Many New World wines tend to give information about grape variety, and sometimes even some technical information about grape growing, fermentation and maturation of the wine. Some wine-makers even give tasting notes on the back label.

Bottle ageing

A further type of maturation that may be used is bottle ageing. Many white wines and most red wines are sold ready to drink and do not require any further maturation in the bottle. However, some wines are expected to age in the bottle, resulting in changes in flavour. Some delay between bottling and the sale of the wine is inevitable, resulting in some bottle ageing. However, deliberate bottle ageing is practiced by wine merchants and consumers to affect the quality of the wine. The chemical changes in bottle maturation differ from those during in-cask maturation, primarily because there is no oxygen take up or extraction from wood in-bottle maturation. The chemical changes known to be involved are discussed in Chapters 3 and 4.

1.5.2 Red wines

Pre-fermentation

Crushing/de-stemming

White and red wine-making differ significantly at the stage immediately after crushing; for red wines the skins and seeds are not removed, and maceration (i.e. contact between the skins and the juice of the grapes after crushing) on the skins is given both before and during fermentation and occasionally even after fermentation. Red wine production, including both maturation in large stainless tanks and/or maturation in small casks is outlined in Figure 1.1. The stems can be removed, or left in the must.

Maceration

The amount of anthocyanins and other phenolic compounds extracted into red wine directly affects the colour of the wine, even after some years of maturation. It has to be borne in mind that the extraction will depend on the composition of these compounds in the grapes. The environmental factors and vineyard practices impacting on the flavonoid composition in grapes and the resulting wine has been reviewed (Downey et al., 2006), and the grape composition determines the wine quality more than the choice of extraction parameters. Numerous factors other than grape variety are thought to influence the biosynthesis of flavonoids, hence viticultural aspects should to be considered as part of the wine-making process.

   The subsequent potential of the wine to age depends on the phenol content of the red wine. Mouthfeel of red wines, one of the key differences in sensory properties between red and white wine, is thought to be determined by both the amount and composition of phenolic compounds in wine. It is therefore not surprising that there are numerous different techniques in use to aid the extraction of anthocyanins and to a lesser extent other phenolic compounds, during red wine-making. Such techniques can be typical for a wine-making region, for example carbonic maceration is typical for Beaujolais Nouveau, giving a light and fruity wine intended to be consumed young. Maceration also may be a choice made by the wine maker to an extent independent of the wine region.

A recent review by Sacchi *et al.* (2005) clarified the methods currently in use for red wine-making and summarized the effects on the extraction methods on colour due to the anthocyanins and other phenols in wines. They reported no long term effects on phenol contents as a result of levels of sulfur dioxide during maceration or a so-called *cold soak* (keeping the must cool for several days before the fermentation is started) and neither did the use of pectolytic enzymes lead to increased colour in the wine. Thermovinification, whereby the must is heated to 60–70°C, followed by cooling and fermentation, tended to extract more colour, without a significant increase in other phenolic compounds. So-called 'punch downs' or 'pump overs' are frequently used in red wine-making, since during fermentation the carbon dioxide formed brings the solids to the top of the fermentation vessel, forming the cap, thus reducing the effectiveness of extraction, risking excessive heat being trapped and spoilage due to *Acetobacter* contamination. Hence the cap is often pushed down by physical punching or by taking must from the bottom of the fermentation vessel and spraying this over the top to wet the cap.

The effectiveness of extraction of these methods varies, possibly depending on the grape variety used. Interestingly, the reviewers found that studies on carbonic maceration gave conflicting results and suggested the success of this technique may well depend on the varieties used. Longer maceration times tended to increase the phenolic content in the wine, since the anthocyanin extraction tends to peak after 4–5 days. Thus both the type and the extent of the maceration can be varied, according to the desired style of wine. It would be interesting to see studies on the sensory properties of red wines made with different maceration methods and fermentation temperatures, in particular relating the phenol composition to perceived mouthfeel and astringency.

Fermentation

Yeast fermentation

The wines may be fermented to low residual sugar content on the skins, thus allowing maceration during the entire fermentation period. The grapes may spend only 2–3 days on the skins and either before or shortly after fermentation has started the juice can be run off the skins, the pomace pressed and the fermentation allowed to continue in the absence of the skins, leading to light red wines. In the presence of skins, and possibly the stems, it is preferred to ferment at 24–28°C (no higher than 30°C) for up to two weeks. In this way, the classic and traditional red wines are produced, with dark colours, a high tannin content and flavours, which generally need some maturation to develop fully. The higher fermentation temperatures typical for red wines tend not to produce the fruity fermentation esters. Red wines made for early drinking, more typical for some red wines from, for example, Australia tend to be fermented at a lower temperature (17–24°C).

The review by Sacchi *et al.* (2005) reported that generally higher fermentation temperatures increased phenolic extraction, which the reviewers suggested tend to lead to higher polymeric pigment contents in the mature wines.

Also a new group of anthocyanins, called vitisins and currently referred to as belonging to the group of pyranoanthocyanins, was first isolated, identified and reported by Bakker & Timberlake (1997). This is a very stable group of anthocyanins, which plays an important role in the colour of wines (Rentzsch et al., 2007). Recent reseach from Morata et al. (2006) shows that the fermentation temperature, the levels of sulfur dioxide and the pH during fermentation determine to an extent the formation of these colour stable anthocyanins in wines. This is further discussed in Chapter 3.

Malo-lactic fermentation

Malo-lactic fermentation is commonly practiced in red wines, and can be allowed to take place by naturally occurring lactic acid bacteria or by a specially prepared inoculum. It is desirable in red wines, which benefit from a conversion of the sharp tasting malic acid into lactic acid.

Post-fermentation

Maturation

One good reason for red wines to be aged is to allow changes in the phenolic composition to occur due to chemical reactions and precipitation, which will affect the colour of the wine as well as the sensory properties related to astringency and mouthfeel. These colour changes, which are a consequence of complex chemical reactions between the various anthocyanins and mostly non-coloured phenolic compounds occur as long as the wine is stored, although the rate will depend on various other factors, such as storage temperature and available oxygen. In addition the fermentation aroma is gradually lost or modified and the aroma often referred to as bottle aged bouquet is formed.

Bulk storage

Red wine should be kept with low sulfur dioxide concentrations and at a relatively low storage temperature. The extent of ageing required depends on the initial maceration given and also on the grape variety, other wine characteristics, local custom, etc. Red wines are often sold with some ageing, and many are not intended to be aged further. Wines generally need a fair amount of acidity to withstand, or improve with, ageing. Only some red wines (and even fewer white wines) will improve on prolonged ageing.

Some aeration of red wines, often achieved when the wine is racked off its lees and pumped over into a clean vat, is thought to benefit their development. Generally, the darker wines with higher tannin contents are thought to improve more with maturation than the lighter, less tannic wines. The amount of maturation needed to optimize red wine quality is still a subject of research. The wine maker can influence the wine properties depending on the amount of aeration given to a wine. Excess oxygen can cause oxidative browning (Li et al., 2008), as discussed above.

Barrel ageing

Red wines can be made with some *in-cask* ageing, usually in small oak barrels, often in addition to maturation in large stainless steel vats. Traditional ageing usually means storing in oak barrels or vats from three months up to about three years, possibly even using some new oak for a short duration, followed by bottle storage. Only some 10% of wines are actually aged in casks (Robinson, 1995). For example in commercial practice Spanish Rioja red wines are traditionally aged in oak barrels, usually made of American oak, to provide characteristic flavours (see Chapter 4). The choice of oak (American or French, new or old) and duration of barrel ageing, in addition to the amount of aeration during barrel ageing, will determine the sensory properties of the wine.

Bottle ageing

Wine merchants and consumers also practice maturation by deliberate bottle ageing to affect the quality of some wines. Wines given a prolonged period of bottle ageing before consumption are of great interest to wine experts, as indicated by the charts and tables produced on the condition of wines from a range of vintages, indicating wines that need further bottle ageing, wines ready to drink and wines past their optimum drinking quality. A current, common classification system used to indicate the 'drinkability' of such wines is 'not ready', 'just ready', 'at peak' and 'fading'. These designations include information on the different qualities to be expected of the wines in given vintage years.

Tasting notes are often made using a numerical scoring system, the most commonly used systems are from 0–20 or from 0–100. All wine experts have their own defined version of making notes and scoring the wines, published in annual wine tasting books and increasingly on the web. The scores experts allocate to wines may well add to the value of the wines, depending on the status of the expert of course! One example of information on the web is the www.wineanorak.com site, which has tasting notes prepared by Jamie Goode on wines from many prestigious wine producers around the world, typically grouped by region. Wines scored on this site typically average between 85–95 points, defined as good to very good but the author stresses that only good quality wines do get recorded. Tasting comments of such wines usually include an indication regarding readiness for drinking, such as 'drinking well now', 'at its peak' or 'do not keep much longer'. An example is the list of tasting notes on 60 years of classic Clarets, dating back to 1947, with all wines assessed scoring 92 points or more. Unfortunately, no prices are given! However, top wines from Bordeaux do demand very high prices.

There are actual distinct chemical changes occurring in red wines during ageing in-bottle, discussed in Chapter 4.

1.5.3   White wines

There are a number of differences for the white wine-making process. Firstly wines tend to be fermented off the skins. Secondly, air (oxygen) must be excluded as much as possible during white wine-making to prevent the development of a brown colour, due to oxidative browning, as a result of enzymic and chemical reactions of the phenolic compounds (Li *et al.*, 2008). Oxidation also needs to be avoided to prevent losing the often delicate fruity white wine aroma. Typically, this is done by adding a small amount of sulfur dioxide to the must, combined with careful wine-making practice thus avoiding oxygen uptake.

Surprisingly, one technique often practiced to protect white wine from oxidation is making it less sensitive to oxygen, achieved by oxidation of the must. The deliberate excess oxidation of the must, so-called hyperoxidation, promotes enzymic oxidation (browning) of phenols in white musts, compounds which precipitate during the fermentation. The resulting wine is less bitter and less sensitive to accidental oxidation. Not all grape varieties are suited to the use of hyperoxidation, since there can be a negative effect on the aroma development. Also, the chemical process of hyperoxidation varies with phenolic precursor concentration present in the grape variety and is discussed in detail by Ribéreau Gayon *et al.* (2006).

Additions of sulfur dioxide can be used to protect the wine, in addition to the use of an inert gas to prevent the uptake of oxygen. Thirdly, temperature control during the entire process is necessary, especially in the warmer wine-making regions, which lack natural cooling conditions during harvest.

Pre-fermentation

Harvesting/grapes

Figure 1.2 outlines the production steps involved for a standard white wine prepared for drinking young and made without subsequent maturation in oak casks. To produce fruity white wines it is especially important to pick the grapes cool (early morning, or even at night). The grapes are, preferably, kept cool and crushed with minimal delay, often with an addition of sulfur dioxide. Some white grapes are picked and immediately cooled in the vineyard. White wines are usually produced from white grapes, but can also be made from red grapes (Section 1.5.1), when great care is needed not to damage the skins since this would lead to a red coloration of the juice (and wine).

Crushing/de-stemming/pressing

The special feature of white wine production is the removal of grape skins, seeds and stems, usually immediately after crushing and draining of the grapes. Part of the juice runs out of the crushed grapes (free run juice) without added pressure (draining) and is followed by immediate pressing. Sometimes white grapes are not crushed, but immediately pressed, to minimize extraction of compounds from the skins, seeds or stalks. The fermentation is carried out on the must or grape juice, without the skins, etc.

Maceration

Often white wines are made with minimal maceration, and the crushed grapes are pressed immediately. As harder pressing generally causes greater extraction from the skins, care is often taken to press the grapes very gently to make the wines fresh and fruity and is intended to be drunk young. The free run juice generally has the highest quality, and the addition of more or less pressed juice to the free run juice affects the chemical composition of the wine and its resulting colour and flavour.

However, with some grape varieties it may be desired to extract the maximum amount of aromatic compounds from the skins, the so-called primary aromas (mainly terpenes). Here, limited maceration before the fermentation is given to assist the extraction. It is important, however, that no excess in tannin is extracted, since this would make the wine more susceptible to browning and increase its bitterness and astringency. Generally, maintaining a low maceration temperature combined with a minimal maceration time gives fresh and fruity wines, while increasing temperatures and maturation time gives a darker wine with a less fruity character.

Settling/clarification

To obtain white wines with a fresh fruity character, the grape juice (or must) is usually clarified before fermentation by 'cold settling'. The must is kept cold (10–15°C), possibly with an addition of sulfur dioxide to prevent the onset of fermentation and the suspended solids that subsequently fall to the bottom of the tank are removed by racking and/or filtering and/or centrifugation. Usually, not all solids are removed, as that tends to hinder the subsequent fermentation. Of course, any combination of pre-fermentation treatments already described in the vinification procedure (Section 1.5.1) can be given.

Fermentation

Yeast fermentation

In modern white wine-making, the fermentation is carried out in large stainless steel vessels at between 15 and 20°C, although even lower temperatures are used. Some of the Old World white wines are fermented at slightly higher temperatures, between 20 and 25°C. Fermentation takes at least 2–3 weeks or even a month or more, depending on the temperature. Precise control of the temperature is important since it governs the production of esters, the compounds responsible for the fruity character typical for modern young white wines, with lower temperatures giving a greater concentration of esters (see also Chapters 4 and 7).

Certain fine white wines are still made by fermentation in the barrel, rather than in a large tank. They are often subjected to many of the treatments described, followed by ageing in oak barrels (e.g. aged Chardonnays, as in Meursault).

The malo-lactic fermentation can also be induced at some stage of the processing. However, changes in aroma accompany this conversion, which may not be desired in fresh white wines. The sharper acidity of white wines is usually valued as adding to the freshness of the taste, another reason to avoid malo-lactic fermentation.

Post-fermentation

Racking

Often, young wines are stored briefly in large wooden barrels or vats or in stainless steel vats. They need to be racked off the lees but aeration should be avoided since this is generally detrimental to the flavour and colour of white wine. Careful monitoring of the concentration of sulfur dioxide and making a further addition when required, helps to protect the fresh and fruity character of the wine.

Clarification

Most white wines are to be bottled and drunk young, without any ageing, in order to enjoy their fresh and fruity character and therefore need clarification, stabilization and filtering before bottling (Section 1.5.1).

Bottle ageing

The fruity and fresh character of such white wines disappears during ageing and only a few white wines will improve on ageing (Chapter 4 and Chapter 7). After bottling, some fine white wines may continue to develop during bottle ageing, although white wines generally cannot be aged in-bottle as long as some reds.

1.5.4 Specialized wines

Rosé wines

Rosé wines are made from red grapes, which are crushed and de-stemmed as usual, but the maceration period is very short, often less than 24 hours. The wine-making resembles the process used for white wines. In general the maceration is stopped before fermentation has started and the desired amount of anthocyanins (giving the red-pink colour to the must and wine) has been extracted from the skins. If the juice is run off after the start of the alcoholic fermentation, the resulting wine will have a higher phenolic content.

Next the mash is pressed and the resulting juice is processed further in the same way as white wine, using juice settling and a cool fermentation. The resulting wine is much lower in tannin than red wine and generally treated more like white wine. Ageing in oak is usually avoided, the wine is stored cool, protected from oxidation and bottled for early drinking. Rosé wine can also be made using carbonic maceration (see the section on *Wines by carbonic maceration* later in this section), ensuring free run juice does not extract the grapes, thus preventing the extraction of excess colour but allowing the development of the fruity aroma typical for this fermentation method.

An array of red grape varieties is used for rosé production, not just the lighter red varieties, including blends of red and white grapes. Blending red and white wine to produce rosé table wine is not allowed in EU countries, however, pink Champagne can be made by adding red wine during the blending (or assemblage) stage of champagne production.

Wines made from organically farmed grapes

There is an expanding consumer demand for organic produce, including wine. Although there are good quality 'organic wines' available, the label organic is by no means signifying any information regarding the quality of the wine, it just refers to the production methods used in viticulture and to a lesser extent in vinification. Indeed there is a range of qualities of wines on the market, made from organically produced grapes and non-organically produced ones.

Organic wine is wine made from organically grown grapes, which means the grapes are grown without artificial fertilizers or chemicals to control weeds, pests and diseases. The correct wine label will state 'wine made from organically grown grapes', rather than 'organic wine'. The wines are made in much the same way as non-organic wines. In order for wine to be called 'wine made from organically grown grapes' the vineyard management has to comply with EU standards, including the required conversion time during which the grapes have been produced according to the organic standard. There are organic inspection bodies to verify the organic production and only certified organic producers can legally sell wines from organically produced grapes.

Organic viticultural methods focus on keeping the soil in good condition and reducing the risk of pests and diseases. The production of healthy grapes will be stimulated by selecting grapes suitable for the climatic conditions and providing growing conditions which will encourage strong and healthy vines and stimulating natural predators to keep pests in check. No doubt all viticulturists adhere to this general principle.

The production of good quality grapes requires correct vineyard fertility, which in organic grape production is done without the use of artificial fertilizers. In order to achieve correct fertility and adequate nutrients in the soil both green manure and animal manure can be used, depending on local soil requirements. Growers are selective in the use of their source of manure to prevent accidental introduction of pests and diseases, as well as the introduction of high levels of heavy metals. Mulches have been used traditionally to control weeds and to preserve water. Generally organic manures improve the soil structure, for example influencing water retention, enrich the nutrient content of the soil and enhances biological activity, such as influencing soil bacteria aiding nitrogen availability by fixing nitrogen.

The control of pests and diseases is done without the use of manmade chemicals. Generally dryer climates may have less trouble with mildew and kinds of rot typical for grapes than wetter ones. Many techniques used for such problems are based on environmental modifications, such as ensuring light and air access towards grape bunches by having a more open canopy and the removal of leaves around the bunches. Weed control may prevent the carriers of the infections. Specific pests are controlled using biological control, either by introducing specific predators or enhancing the environment thus stimulating the occurrence of natural predators. Control of numbers of predators can also be controlled with pheromones, interfering with the reproduction of predators. In depth knowledge of the micro environment is required to ensure the right measures are taken. Knowledge regarding the weather forecast may allow preventative measures to be taken, without treating the grapes unnecessarily. Organically produced grapes can be treated with sulfur and copper to prevent moulds, but its use is regulated and restricted.

Wine-making is done much in the same way as for other wines. There are no current EU standards for organic wine-making, however, the standard set by the Soil Association in Britain requires wines to be made with lower levels of sulfur dioxide (total and free sulfur dioxide) than current EU standards for non-organic wines (organic red wine 90 mg L^{-1} total and 25 mg L^{-1} free sulfur dioxide, organic white wine 100 mg L^{-1} total and 30 mg L^{-1} free sulfur dioxide). No other processing aids are regulated, so these wines are not necessarily suitable for vegetarians. The label specifies suitability for vegetarians or vegans only if no animal derived fining aid has been used.

There are also cost implications. For example it may be easy to spray against an infection, however, if the loss of the crop without treatment would have been small, the costs associated with treatment may not have been justifiable. In addition, there may also be health aspects to be considered, for example the production of toxins by excessive mould growth if grapes were left untreated. Generally farming methods for producing of organically grown grapes is more labour intensive and results in lower yields, hence justifying the increase in the premium on the price of a bottle.

Wines with added resin

In ancient times a major problem was to keep wine drinkable, by stopping it from going vinegary and protecting it from oxidation. Since the use of sulfur dioxide to protect wine was not generally known, other methods were also used. The use of pine resin or pitch is thought to date back to the ancient Greeks and Romans, and traces of resin have been found in old Greek amphorae. Additions were made to wine to extent the shelf life of the wine, although it may also have been added to mask off-flavours. The use of these additions has been maintained until today, although now its use is for flavouring purposes only, giving wines a fresh turpentine-like flavour.

Most Retsinas are made from two grape cultivars, Savatianó and Roditis. Savatianó is a white grape, well adapted to the dry conditions and relatively easy to grow. Grown in large quantities, especially in Central Greece, it produces a rather bland and low acid wine, traditionally flavoured to produce Retsina. Roditis is a red grape, although the many clones available vary in skin colour from pink to red, grown widely in the northern Peleponese. A selection of the rosé wines are flavoured to produce Retsina.

There was a sudden surge in the popularity of Retsina in the 1960s, when Retsina suddenly became the national beverage, and the production was expanded rapidly to meet demand from increases in tourism and the local market. Currently its popularity seems to be mainly in Greece.

Wines with low alcohol content

In order to produce wines with low alcohol content, typically wines are made using standard wine-making techniques, followed by the removal of part of the formed ethanol. Older methods still widely used are vacuum distillation and reverse osmosis; techniques which remove the fruity volatile compounds and tend to accentuate the more unpleasant odours (see Jackson, 2008, for more information on these methods). Current methods gaining ground are based on membrane filtration and spinning cones, and processing at lower temperatures and are reviewed by Pickering (2000) and also discussed by Jackson (2008). It is claimed that these methods are superior and these production methods may lead to improved sensory properties of low alcohol wines. For example, the spinning cone column method is fast and the disruption of the wine flavour is minimal, however, the equipment is expensive and the method has not been legalized in all wine producing countries, also leading to potential export hurdles. Adding water to the wine would be the easiest method of reducing the alcohol content but this is usually not allowed.

The retention of aroma and the recovery of aroma compounds can prevent flavour inbalance, although no doubt ethanol acts as a solvent for typical wine flavour volatiles, and its significant reduction from typically between 11.5–13.5% v/v will affect the release and perception of wine flavour (see Chapter 4, Section 4.1.2). Alcohol at higher concentrations may give an apparent sweetness, which low alcohol wines lack, possibly making the low alcohol wines more tart than the natural level alcohol wine. There are no current sensory research data confirming perception differences between low and natural alcohol wines, with comparable flavour volatile compositions.

Further research in dealcoholization methods showed a viability study using a polypropylene membrane contractor (Diban et al., 2008), this investigation showed large and unacceptable flavour losses for total dealcolization of model and real wines but a reduction of 2% alcohol gave acceptable aroma losses that did not damage the final perceived quality of wine. Another approach is the use of engineered yeast strains, which in future may also give fermenting yeasts producing an acceptable flavour but with a reduced alcohol production (Ehsani et al., 2009).

Sweet wines

Several methods are in use to produce sweet wines, many of which have evolved into specialized products with very distinctive flavours. The fermentation can be stopped if sweet wines are required, for example by the addition of alcohol to about 18% v/v, as is done in the production of Port wine. Sweet wines can also be made by sweetening the dry wine with an especially made sweetening juice, such as the use of Süssreserve, an unfermented sterilized grape juice, in the production of some German wines. Other methods use sun-dried grapes, or late-harvested grapes infected with *Botrytis* (see below).

Sweet wines from *Botrytis cinerea* infected grapes

The more interesting of specialized wines include the sweet wines derived from grapes infected by a mould. Many wines made from mouldy grapes result in unpalatable wines, but from Sémillon given the correct conditions for mould infection, highly sought after wines can be made, with appealing sensory properties. Grapes can be infected by several micro-organisms, usually resulting in so-called bunch rot grapes that are unfit for human consumption. In several areas where climatic conditions are right, a particular mould (*Botrytis cinerea*) attacks the grapes and causes what is often referred to as *Noble Rot*. This mould can also be the cause of undesirable grape rot.

Botrytis cinerea mould generally causes desirable changes in grapes grown in river valleys, where autumn mists develop at night and linger during the morning. The mists should clear and give way to sunny afternoons. During the dry conditions the grapes will shrivel and the juice will become more concentrated, reaching 30–40% sugar w/v. These mould-infected grapes can be vinified and produce wines with desirable flavours.

Late-ripening grape varieties with relatively thick skins are less susceptible to bunch rot but are suited to developing Noble Rot. Riesling and Sémillon Chapter 2) are the main varieties used, although Chenin Blanc (Loire, France), Furmint (Hungary) and, occasionally, Gewürztraminer (Alsace, France) are all susceptible to Noble Rot and can give excellent wines. Especially, the Sauternes district in France is famous for its wines made from the Sémillon grape that has been attacked by *Botrytis cinerea*. New World wines made from *Botrytis* infected grapes are now also appearing. The chemical changes taking place in the grape are discussed further in Chapter 7.

Sparkling wine in Champagne

Champagne is the classic sparkling wine, and the name Champagne is legally protected so as to be applied only to wines from the Champagne region in Northern France. Champagne is made from one white grape variety, Chardonnay and two black ones, Pinot Noir and Pinot Meunier. Because black grapes are used for the Champagne production of white wine it is important to avoid the extraction of coloured pigments from the skins and, for all grapes, to minimize the extraction of grape solids. Hence in traditional sparkling wine production the grapes are pressed whole, without either crushing them or removing their stems.

Dry white wines are made separately from the different varieties from the numerous vineyards. Malo-lactic fermentation of these young high acidity wines is usually encouraged. Over the winter all the young wines are tasted and assessed and blends are prepared (a blend is called a cuvée). The cuvée is thought to determine to a considerable extent the character of the Champagne; hence much skill is needed to achieve the appropriate composition of wines in the cuvée. After the wines have clarified, a calculated amount of sugar and a specially selected yeast (*liqueur de tirage*) are added to the cuvée before bottling the wine in extra strong bottles, closed with a special closure. The second fermentation consumes the added sugar and raises the alcohol content from about 11% to 12.5% v/v. The released carbon dioxide needs to stay trapped to give a sparkling wine.

The dead yeast cells of this second fermentation remain in the bottle during maturation, a process that contributes significantly to the typical flavour of Champagne. Bottles are matured on their sides, in cool, dark conditions, for a legal minimum of 15 months but for great Champagnes the maturation may take 3–5 years. Finally the dead yeast cells must be removed to clarify the Champagne. The process starts with riddling (*rémuage*). The bottles are stored in inclined racks and, over time, the inclination is increased until the bottles are upside down. The yeast debris accumulates on the closure in the neck of the bottle, which is frozen and the slightly slushy yeast sediment is removed from the bottle (*disgorgement*). The wine has to be topped up and is usually sweetened with some sugar syrup (*liqueur d'expédition*).

Sparkling wine by other methods

The wine-making process developed in Champagne is used in many other sparkling wine-making regions, and is often referred to as the 'classical method' or 'bottle fermented' or, as in Spain, *método por cava*. The process is quite expensive and other methods have also been developed that differ mainly in the second fermentation, including the removal of the yeasts and adding of the sugar, without losing the carbon dioxide.

In-tank carbonation

Eugène Charmat developed this method in Bordeaux at the beginning of the twentieth century. It is cheaper, faster and less labour intensive than the traditional method. Secondary fermentation is carried out in a pressure tank and the wine is usually matured for a short period. The wine is filtered, removing the lees pumped off under pressure. This method is best used for wines made from grapes with a strong varietal flavour, which can be preserved, such as Muscat (as used in Asti).

Transfer process

This process was developed in the 1940s to reduce costs by avoiding the manual riddling process. Secondary fermentation is done in the bottle but then the contents are emptied in a pressurized tank, filtered to remove the yeast cells, sweetened and re-bottled under pressure with an inert gas. The quality of the wine should not suffer much, assuming the wines are kept cool and handled hygienically.

Bicycle pump method

This is the most elementary method, involving carbonation by injecting carbon dioxide into a tank of wine under pressure. It is only used for the cheapest sparkling wines; there is no flavour development due to maturation on the yeast cells and the bubble release of carbon dioxide is usually rapid.

Wines by carbonic maceration

A variant wine-making method for wines to be drunk young is the so-called 'carbonic maceration', which is designed to extract the red colour from the skins, but less tannin than in normal red wine-making procedures. This is the common red wine-making technique for the Gamay grape in the Beaujolais area. The bright red wines with a distinctive fruity aroma are suitable for early drinking (e.g. Beaujolais Nouveau). They are made by carefully placing the whole bunches in a tank, which is then sealed. Oxygen is excluded, usually by using carbon dioxide. Fermentation by grape enzymes occurs in the intact berries, though some grapes will be crushed by their own weight.

The metabolism in the grapes is anaerobic. Under these conditions the grape berry can produce a small amount of alcohol, with a slightly higher temperature enhancing this mechanism. After about a week the process stops. The grape bunches, containing 1.5–2% v/v alcohol are then pressed and the juice is fermented by yeasts, usually at 15–20°C. The pressed juice can be kept separate or added to the free run juice, and gives a more aromatic and darker wine than the free run juice. Carbonic maceration is most suited to red grapes, and typical aroma descriptors are fruity, with notes of cherry, plums and stone fruits.

Wines by thermovinification

Thermovinification requires the intact or crushed grapes, usually for red wine-making, to be heated to between 50–87°C. The cell structure, in particular of the grape skins, is disrupted, facilitating the extraction of skin components. The holding time of the heated pulp (maceration time) decreases with increasing temperature, ranging from 2 min at 87°C to several hours at lower temperatures. The grapes are then pressed, the must is usually inoculated with a fermenting yeast, and fermented. The wine maker can decide to leave the skin and seeds in during fermentation but usually just the juice is fermented. The fermentation is usually rapid, thus requiring the fermentation temperature to be controlled.

Thermovinification efficiently extracts the red skin pigments (anthocyanins), as reviewed by Sacchi et al. (2005) which is advantageous for lighter red grapes grown in cooler wine regions, or for the production of deeply red wines, for example some Port wines. Another advantage is that grapes attacked by mould can be processed. When grapes have been attacked by a mould (Botrytis cinerea) they may contain laccase, a very stable enzyme that causes browning. Laccase is difficult to control during wine-making but thermovinification above 60°C inactivates its enzymic action. Also naturally present polyphenol oxidase in grapes is denatured, potentially reducing the risk of browning. After holding the must, the pulp is pressed, sometimes without pre-cooling, but sometimes the hot pulp is cooled by the incoming cold grapes.

The choice of the vinification temperature, the maceration time, fermentation on or off the skins and the seeds and the selection of the fermenting yeast all affect the composition of the wine, and the wine maker can influence to an extent the sensory properties of the wine (see also Chapter 7).

Wines matured *Sur Lie*

Sur Lie is the French expression for 'on the lees'. Dry white Muscadet wines (France) made from Melon du Bourgogne grapes were traditionally given maturation on the lees, gaining a hint of yeasty flavour. Keeping wine in contact with the yeast lees for an extended period after the fermentation has finished will influence the wine aroma. Autolysis of the yeast cells, whereby the cells burst and release their contents into the wine, is thought to influence the development of aroma compounds. Chardonnay wines aged on their lees develop a less buttery and more toasty aroma. This method of maturation has become more widespread, and in red wines it has been reported to reduce astringency. Most typically, this type of maturation is used for sparkling wines, to give the toasty flavour after the second yeast fermentation in-bottle, and in Fino Sherry maturation in the solera system.

Lees bind to many undesirable compounds, so the removal of lees has many practical advantages, such as removing toxic compounds, undesirable volatiles produced by spoilage yeasts and reducing the risk of the formation of undesirable volatile sulfur compounds, as reviewed by Perez-Serradilla & Luque de Castro (2008). The lees also forms a reservoir of micro-organisms and amino acids, and it is well documented that maturation on lees can lead to higher concentrations of biogenic amines, which can have adverse effects. Ageing in the presence of lees affects the phenolic composition, and in general wines are less astringent, slightly less intensely coloured with a more yellow hue. The authors also discussed effects of released polysaccharides and lipids but there is not yet sufficient evidence to determine these effects on the sensory properties of wine. The chemistry involved in the positive effects on wine quality reported in the development of flavours of wines aged on lees does not yet seem to be well documented.

1.5.5   Fortified wines

The general processes of fortified Port wine, Sherry and Madeira are outlined here, while the underlying chemistry that affects the flavour and colour of wine, Sherry and Madeira will be discussed in Chapter 6, specifically focusing on where these processes differ from those of table wines.

Numerous countries make Port-style wines. Australia and South Africa make sweet red wines of good quality, although different grapes are used. Portuguese production of typical Port wine is outlined below. Some countries make Sherry-style wine; in particular, California has made this style of wine, although often with different grapes. The Spanish production of typical Sherry styles is outlined below. Within the EU the names Port and Sherry can only be used for wines from the designated Port and Sherry production regions.

Port wine

Port wine is named after Oporto, a large coastal town in northern Portugal. Therefore *Port* or *Port wine* comes from Portugal, although some other wine-making regions may also produce sweet red wines, possibly using a comparable wine-making process. Port wine is a naturally sweet fortified wine, made by adding fortifying spirit to stop the fermentation approximately halfway to 'dryness' and thereby leaving some glucose and fructose in the wines.

Port wine grapes are grown in a warm climate, and the young wines produced tend to have high phenolic contents, resulting in astringency and making them very suitable for maturation. Port wines are made from grapes grown in a demarcated area that covers the eastern part of the river Douro and its tributaries in the region of the Douro valley in northern Portugal. Although there are numerous varieties recommended or permitted for Port production, there are now five red and four white varieties generally regarded as superior. Many older vineyards are terraced with hand-built slopes, and planted with a mixture of varieties that are harvested and vinified together.

Many of the steps involved in the production of a standard Port wine (outlined in Fig. 1.3) are similar to red table wine vinification. Red grapes are picked, transported to the winery, and crushed and at least a proportion of the stems is removed. The crushed grapes are pumped into a fermentation tank, often with an addition of sulfur dioxide. In this warm climate, the acid level of the grapes may be low, giving a high pH; hence the must may be adjusted to about pH 3.6 using tartaric acid. Fermentation takes place mostly with the naturally occurring yeasts and the fermentation temperature is kept under 28°C. Unlike red wine, fortified red Port wine employs a very short maceration time and so it is very important to 'work' the must before and during the fermentation, for example, by using a robotic leg that simulates traditional treading by foot or more commonly by pumping the must over in the tank. Traditional treading by foot has almost entirely disappeared. The fermentation time on the skins is short; approximately halfway through the fermentation the 'mash' is pressed and the juice is fortified with spirit containing 77% v/v ethanol, to give a final concentration of about 18% v/v ethanol, which prevents further fermentation and results in a naturally sweet fortified wine.

Port wines are normally stored and matured in old oak vats, ranging from about 550 L capacity (pipes) to up to 200 000 L. The young wines are frequently racked as they need to be taken off the yeast. Chemical changes during maturation modify the colour and flavour of the Port wine; the young red Port wine becomes browner, changing from a deep red colour with a purple edge at the rim of the glass, to a brick red or even amber tawny colour. The flavour changes from an intensely fruity, even spirity, character when the wines are very young to a rich fruity ruby wine after three to five years maturation in wood. Wines left an average of ten years or so in wooden containers evolve into tawny wines, having an amber colour and a flavour reminiscent of nuts, raisins and crisp apples.

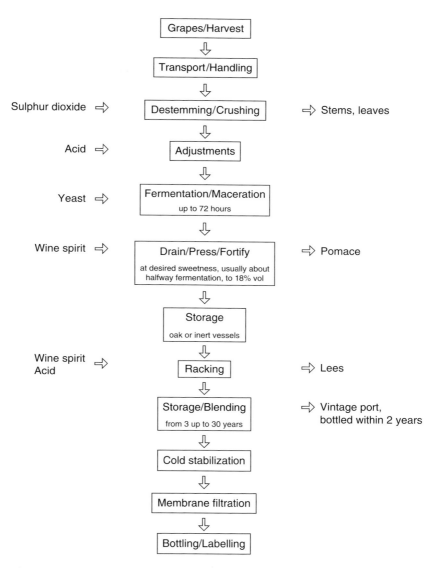

Figure 1.3 Typical processing sequence for the production of fortified Port wines. Some steps are optional (see text).

In contrast with table wines, where wines from each vintage are marketed separately, most Port wines are blended. Annual differences in climate that affect the quality of the grapes and the resulting wines can be corrected by blending. The use of especially made sweetening wines (*jeropiga*) and dry wines allows the final blends to be adjusted to the required sweetness. This allows the shipper to produce a consistent product. The indication of age on most Ports is an average age. Wines sold have an average age of at least three years, with many wine styles being much older.

Vintage Port, being the premium red wine and all from one vintage year, is matured in wood for two years, followed by a considerable period (often several decades) of bottle ageing and so it develops a different character from those wines matured solely in vats or barrels. Vintage Port is sold with a harvest date on the label, and is not treated before bottling; consequently, it will need decanting before drinking to take the wine off the considerable deposit in the bottle.

White Port wines are made in much the same way as the red ones but there is a trend to reduce the skin contact and even to fermenting clarified juice at a low temperature (18-20°C), to obtain a fruity aroma. The wines are aged in old oak vats for a minimum of three years before they are drunk.

Sherry

Sherry is the name given to several related fortified wines made from grapes grown in Jerez de la Frontera, in the province of Cadiz in the south of Spain. The white wine in this region is fairly neutral and lacking in acidity, but it forms an excellent base wine for the delicate flavours produced by the maturation and blending procedures evolved to make Sherry. There are three main types of Sherry, made from the base wine by three different maturation techniques. Sherry can be matured under *flor* (i.e. a layer of yeasts growing on top of the wine), developing into 'Fino' Sherry, or matured without *flor* yeasts, developing into 'Oloroso' Sherry. A combination of *flor* maturation followed by a period of ageing without *flor*, results in 'Amontillado' Sherry.

The climate in Jerez de la Frontera is generally warm and rainfall is moderate. The best vineyards consist of rather chalky soils, known as *Albariza* (very white soil, high in chalk). Palomino Fino is the main grape variety grown. Figure 1.4 outlines the steps involved in the production of base wine, which is fed into the maturation system to produce the Fino (or Amontillado) and Oloroso wine styles. At harvest, the grapes are picked, crushed and pressed. It is at this stage that the types of juice are separated for the intended wine styles. The free run juice and early pressings are the most suitable to make Fino, while juice containing pressed juice is more appropriate for Oloroso or Amontillado. The juice is settled to reduce the solids to about 1% before the fermentation stage. An addition of tartaric acid is made to juices with low acidity. Sulfur dioxide may also be added to the juice. Fermentation to 'dryness' takes place in large cylindrical fermentation tanks at about 25°C. Commercial yeast inocula are not normally used.

The dry wines are classified on the basis of their quality and fortified accordingly. Fino is usually made from the lighter dry wines and Oloroso is usually made from the darker wines, containing more phenols. Fino and Oloroso wines are fortified to 15.5 and 18.5% v/v alcohol, respectively. The wines are matured in seasoned oak casks (500–600 L) in 'bodegas', which are tall, well-ventilated buildings, designed to stay relatively cool.

Fino wines are stored in the coolest part of the bodega and, hopefully, develop *flor*. These wines are kept in vats only 80% full, providing a relatively

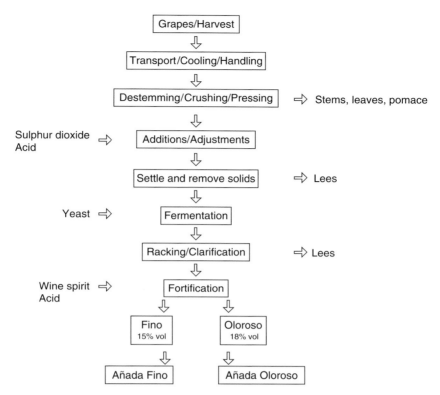

Figure 1.4 Typical schematic processing sequence for the vinification of Sherry. Maturation in butts in the solera system follows from añada (see text for further explanation).

large wine-air interface on which *flor* can develop. The origin of the *flor* yeasts and their taxonomy is still debated, although several are *Saccharomyces* species. The organisms in the *flor* layer consume oxygen, thus protecting the wine from the uptake of oxygen and preventing the oxidative browning of phenolic compounds to which the wine is very susceptible. The *flor* results in numerous biochemical changes, which greatly influence the flavour of the wine.

Young Oloroso wines, which contain more phenolic compounds than Fino wines, are kept in casks 95% full and stored in the warmer part of the bodega. To produce a good Oloroso, the neutral base wine needs to contain sufficient oxidizable phenols and is stored in warmer parts of the bodega, under oxidative conditions. The dark golden colour of the Oloroso can be attributed to the oxidation of phenols. The higher alcohol concentration, usually combined with a higher storage temperature, may increase the extraction of phenols from the wood during maturation of Olorosos and so partly explain the higher concentrations of phenols. Oloroso takes a long time to develop and requires about 7-8 years of maturation.

Some wines mature initially under *flor* and develop all the characteristics of a Fino. However, when the wine starts to lose its *flor*, it is further fortified to about 17.5% v/v alcohol and matured in a second solera system (see below) in casks, about 95% full. The maturation processes then change, with oxidation changing the initially pale yellow wine to amber and dark gold, together with the development of a nutty, complex flavour typical for this wine, called Amontillado.

Large quantities of wines with a consistent quality are obtained by an elaborate fractional blending system, referred to as the 'solera system' (Fig. 1.5), that contains blends of wines of different ages and vintages, and aims to mature wines to achieve a steady supply of comparable and consistent quality. The very complex blending procedure of the solera system involves many transfers of wine. The labour cost is high and it is not easy to make a rapid, significant change to the character of the wine.

The youngest Sherries are stored in the *añada* (wines up to one year old) and wines from the añada are gradually fed into a solera system. The solera system consists of several stages of ageing wines, called *criaderas*. Some of the oldest wine is periodically taken out of the criadera containing the oldest wine (also referred to as the solera) and prepared for bottling and shipping. Wines from the second oldest criadera are used to replace the volume of wine taken from the oldest criadera and basically the topping up of older wines with younger wines continues sequentially throughout the solera system. The volume of wine in the youngest criadera is replenished with a suitable añada wine. Each criadera may consist of a hundred or so butts, and usually the wine drawn from the butts in each criadera is blended before being used to top up the butts in the next criadera stage. There are from three to about 14 criaderas in each solera system, with Fino Sherries generally having more than the other styles. Up to about of 1/3 of the wine can be drawn from the solera system in one year, without affecting the quality of the wine. Fino Sherry needs to be removed regularly from the solera system, to allow the butts to be topped up regularly and so supply nutrients to maintain a healthy growth of *flor*.

Special colouring and sweetening wines are prepared, so that dry Sherries can be blended to the desired colour and level of sweetness before stabilization procedures and bottling. There are two sweetening wines, PX, made from the juice of Pedro Ximenez grapes and Mistela, made from Palomino grapes. The Pedro Ximenez grapes are usually dried, pressed and the juice is fortified to 9% v/v alcohol, and the wine contains 40% sugar. After settling, it is aged in special soleras. Mistela is made from pressed juice of the Palomino grape, which is fortified to 15% v/v alcohol and contains about 15% sugar. It is not usually aged in a solera.

Madeira

In addition to Port and Sherry, Madeira is the third 'classic' fortified wine. The island of Madeira is part of Portugal, lies in the Atlantic about 1000 km from the Portuguese mainland and about 600 km off the coast of North Africa. The

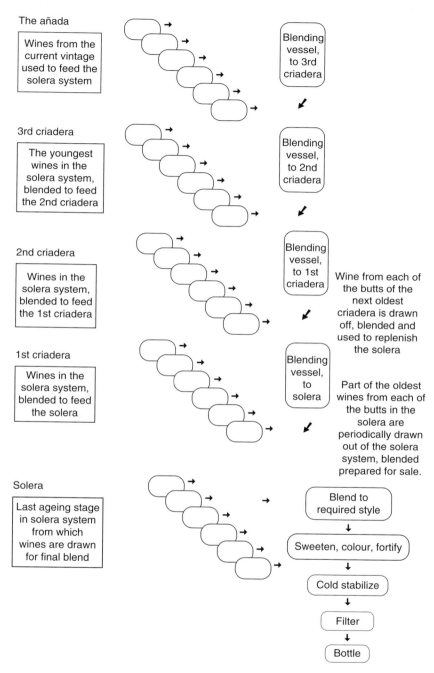

The añada

Wines from the current vintage used to feed the solera system

Blending vessel, to 3rd criadera

3rd criadera

The youngest wines in the solera system, blended to feed the 2nd criadera

Blending vessel, to 2nd criadera

2nd criadera

Wines in the solera system, blended to feed the 1st criadera

Blending vessel, to 1st criadera

Wine from each of the butts of the next oldest criadera is drawn off, blended and used to replenish the solera

1st criadera

Wines in the solera system, blended to feed the solera

Blending vessel, to solera

Part of the oldest wines from each of the butts in the solera are periodically drawn out of the solera system, blended prepared for sale.

Solera

Last ageing stage in solera system from which wines are drawn for final blend

Blend to required style

↓

Sweeten, colour, fortify

↓

Cold stabilize

↓

Filter

↓

Bottle

Figure 1.5 Schematic solera blending system for Sherry ageing, showing Sherry butts in añada, three criaderas and solera stage. The number of stages depends on the style: typically Oloroso, three; Fino, eight and Amontillado, five stages (see text). There can be about 100 butts in each criadera. Up to 1/3 of the wine can be drawn from the solera system each year.

wines are essentially shaped by their maturation, which involves heating them up to 50°C. This heating process is commonly referred to as the *estufagem* and confers a strong and characteristic flavour on the wine.

There are four traditional 'noble' grape varieties (Sercial, Verdelho, Bual and Malmsey) but these account for only some 12% or so of the grapes used for Madeira. However, several others are also grown, and the main grape variety planted is Tinta Negra Mole, in many wine books referred to as Madeira's workhorse. Temperature does not vary much over the year, the winters are relatively warm (average about 14°C) and the summers are relatively cool (average about 23°C). The soil is fertile and light red, consisting of volcanic basalt bedrock. The vineyards tend to be small and on terraces on steep slopes.

The grapes are picked, crushed and pressed. Fermentation, in fermentation tanks, is by the natural yeast population. Several wine-making techniques are in use. Some wines are made by stopping the fermentation by the addition of 95% v/v grape spirit to about 18% v/v alcohol, thus preserving some of the sweetness of the grapes (a process similar to Port wine-making). Other wines are fermented to dryness and fortified after the estufagem.

After fermentation and fortification of the wine, the estufagem process begins. Most wines are matured in this way. The wine is stored in an inert tank (ca. 20 000 L) and heated to ~45°C for three months. Smaller lots of wines are heated in wooden butts (600 L), kept in the warehouse and warmed by the main estufagem. This process tends to take place at between 30 and 40°C. Clear wines are then obtained by fining, and a minimum of 13 months further maturation is needed before the wines can be prepared for shipping.

Many wines are aged considerably longer and after about two years most shippers start preparing a blend to obtain their house style. Ageing typically takes between 13 months and five years. Small lots of wines from exceptional years are called vintage and aged 20 years in wood and two years in-bottle. The wines are orange-amber to deep brown, have a high acidity, and a typical aroma sometimes described as smoky or rancio. Classic grapes (Sercial, Verdelho, Bual and Malmsey) each produce a unique style of wine, and the wines should be made of 85% of the named variety if the grape name appears on the bottle. Sweetening wine is used to sweeten the wine. The wines are sold ready to drink but due to the high alcohol, high acidity and the estufagem process they have a very long shelf-life, remaining in good condition.

1.6 Physiological effects

Wine is primarily drunk for pleasure, consequent upon a mixture of its stimulant activity and its flavour, to which there seems to be early biblical reference: 'and wine that maketh glad the heart of man,' (Psalms 104:15). However, despite the primarily sociable and pleasurable aspects of a moderate consumption of wine, there is increasing evidence of beneficial physiological health aspects, both due to ethanol and some of the phenolic compounds; in particular, reservatrol. Over the last ten years or so more scientific evidence of the health aspects of wine compounds, mostly phenolic compounds, has become available and is briefly discussed below. There are also negative effects, in particular an excess intake of alcohol can lead to both social and health problems and these issues are well documented. However, other attributed negative effects, such as migranes induced by wine, are scientifically less well documented.

1.6.1 Attributed negative effects

Wines are sometimes associated with allergic reactions and the development of migraines for particular people, believed to be triggered by histamine (and some other so-called amines), although some studies suggest that the levels of histamine in wine are generally too low to have a significant biological effect on most humans. After normal vinification the levels of histamine in wine tend to be low, although the concentration is enhanced by prolonged maceration, low fermentation temperatures and the growth of certain bacteria (some belonging to the bacteria inducing malo-lactic fermentation). Migraine is also often attributed to sulfur dioxide in wines, though there appears to be no scientific evidence to support this.

Methyl alcohol (methanol) is definitely toxic, leading to blindness and death, so that drinking 25–100 ml of pure methanol has been reported to be fatal (LD 50 = 350 mg per kg bodyweight). Wines made by normal wine-making procedures have very low concentrations of naturally formed methanol, although some cases of adulteration leading to toxic levels have been reported. Higher alcohols, for example amyl alcohol, known as fusel oils, are a product of the fermentation and their concentrations in wines vary.

The chemistry and physiological effects of wines were investigated by the International Agency for Research on Cancer (IARC) in Lyon, and reported in *Alcohol Drinking* (1988). This report prompted the US Government to issue 'health warnings' on the labels of wine produced in California. The monograph on *Alcohol and Health* by Stuttaford (1997) provides further reading on this topic.

1.6.2 Wine ethyl alcohol (ethanol)

Wine ethyl alcohol (ethanol), the most important physiologically active compound of wine, has stimulating properties when ingested. However, alcohol is essentially a poison that the body detoxifies. The negative effects of excessive alcohol consumption have been well documented. Many social problems, accidents and illnesses are associated with an excessive intake of alcohol, either as a result of chronic long-term overexposure or excessive intake in a relatively short time. Heavy consumption of alcohol causes cardiomyopathy (heart muscle disease) and cirrhosis of the liver. However, many other medical conditions can also, to an extent, be attributed to an excessive intake of alcohol, such as cancer of the mouth and larynx, nerve and muscle wasting, blood disorders, skin problems and increased infertility. Excessive consumption of alcohol during pregnancy may lead to abnormal babies (fetal alcohol syndrome).

Alcoholism is a physical dependence on the consumption of alcohol, so that sudden deprivation may cause withdrawal symptoms; tremors, anxiety, hallucination and delusions. Alcohol dependence takes time to evolve and is not induced after a few binge drinking excesses. Several years of excessive alcohol intake is needed for alcoholism to develop (*Oxford Concise Medical Dictionary*, 1985). Alcoholism impairs intellectual functions, physical skills, memory and judgement, though social skills such as conversation are preserved until a late stage. Even an occasional moderate intake of alcohol affects intellectual functions, physical skills, memory and judgement, hence the need to legislate against the intake of alcohol when driving. In the UK the legal limit for 'drink-driving' has been set at a maximum $80\,mg\,d\,L^{-1}$ (or $17.4\,mmol\,L^{-1}$) for samples taken from blood, breath or urine. Whilst highly dependent upon individual response and related to time elapsed since the last meal, it is generally considered that about 180 ml of wine will cause these limits to be reached in a short time. It is of interest to note that the accepted minimum of a trade measure of wine in the UK is 125 ml.

The effects of alcohol intake on the human body engendered the Temperance movements in the nineteenth century. Especially in the USA, several different organizations emerged, one of which was the Order of the Rechabities derived from the biblical Rechabite house, who denied the use of wine (Jeremiah Chapter 35, verses 5-6). The Temperance movement in the USA led to the Prohibition (1920-1933), which prohibited the production or use of any drink containing alcohol, with devastating effects on the wine industry. It took many years to restore the consumer market and the wine-producing industry.

There are also positive aspects related to a moderate intake of alcohol and some such references are already made in the Bible. Saint Paul said: 'Drink no longer water, but use a little wine for thine often infirmities,' (1 Tim. 5:23). However, there are also some additional warnings against excess, instructing not to be 'a wine bibber' (Tit. 1:7), nor to be 'drunk with wine' (Eph. 5:18). Modern medical research and opinion to date appear to endorse that a moderate consumption of alcohol, in particular red wine, may have positive health benefits compared with total abstinence. A regular moderate consumption of alcohol, in particular wine, is associated with a decrease in mortality. The protective action of wine against some infections of the intestine has long been known and was exploited by wine-drinking cultures, for example the Romans. The antibiotic action of wine is thought to be associated with the presence of certain phenols, particularly in red wines, in addition to any effect of alcohol, and is discussed in the next section.

Alcohol is now also believed to have beneficial anti-cancer effects and has long been thought to have a protective effect against certain heart problems but only for a regular and moderate consumption. There are two mechanisms that seem to decrease the risk of these heart problems associated with a moderate intake of alcohol. The first is an increase in plasma high-density lipoprotein cholesterol (HDL), which is generally associated with a lower risk of heart disease. The second is a reduced activity of the platelets in the blood, which reduces the development of atherosclerosis and coronary thrombosis, thus reducing the onset of angina or infarction. It is stressed that any health benefits ever reported from the intake of alcohol are only applicable in regular moderate alcohol consumption.

1.6.3 Effects of phenols

Numerous studies have shown that red wine may have antibacterial or even antiviral effects; however, there is very little direct scientific proof linking identified compounds with mechanisms of such action. The binding of proteins to phenols is thought to inhibit certain enzymic actions. Although direct proof is limited, phenols (flavonoids and particularly reservatrol) have antioxidant properties and limit the action of free radicals in the body and, therefore, are thought to have anti-carcinogenic effects (Stuttaford, 1997). Supporting evidence comes from research on green tea, consumed in great quantities in the Far East and which has been reported to have beneficial anti-cancer effects due to compounds similar to those in wine.

Epidemiological studies have suggested a moderate consumption of alcoholic beverages, particularly red wine, is associated with a reduction in overall mortality attributed mostly to a reduced risk of coronary heart disease. In particular the so-called French Paradox survey, where the incidence of heart disease was low, despite the high intake of saturated fats and cholesterol, attributed to a daily intake of red wine. Further studies have suggested that plant-derived dietary constituents of food, in particular phenolic compounds in wine, can play an important role in the prevention of disease. However, although the potential role of phytochemicals, compounds having a positive effect on health in our diet, has been discussed and researched, it still remains difficult to get scientific proof of their role on disease prevention, for example heart disease (Visioli *et al.*, 2000). Despite the apparent difficulties, there been numerous studies *in vivo* and *in vitro* to determine physiological effects of specific phenolic compounds in order to unravel effects of these compounds in our diet, and some of the more recent literature reviews have summarized research of phenolic compounds focusing on particular health benefits.

A recent review on the role of wine polyphenols and the promotion of health (Cooper *et al.*, 2004) discusses the effects of wine, in particular its polyphenolic content, on early indicators of coronary heart disease and examines whether it is the wine or the alcohol which affects the onset of heart disease. They concluded that red wine polyphenols have little effect on plasma lipid concentrations, which at elevated concentrations, in particular of the so-called low density lipids, are a risk factor in the onset of heart disease. However, red wine consumption may lower the susceptibility of low density lipids to oxidation, indicating a protection of wine phenols against heart disease. They also found evidence that alcohol has a positive synergistic effect with wine polyphenols on platelet aggregation and adhesion, a heart disease risk factor, and they suggested that this may be relevant after red wine consumption, since the concentration of wine flavonoids in the circulation is usually lower than those found to be effective in platelet aggregation and adhesion. Overall, Cooper et al. (2004) concluded that despite the great research input over the last decade in this area, we are still some way of understanding the health benefits of wine; there are still many questions regarding the functions of the individual compounds and there is little research done on the metabolism and biological activities of phenolic compounds at cellular, molecular and biochemical levels.

The beneficial effects of numerous polyphenolic compounds has been attributed to their antioxidant and free radical scavenging properties, although in many instances direct links between identified compounds and their effects is tentative. A number of studies in vitro support the concept that flavonoids are involved in the decrease of age related disease, however, a review by Halliwell et al. (2005) on the health promoting effects of flavonoids, found little substantive evidence to link the protective effects of flavonoids to their anti-oxidant activity in vivo. They suggested the following reasons for the lack of evidence. Firstly, foods rich in flavonoids may not necessarily give the desired action of the flavonoids in the body, there may be other interactions required to obtain an effect. Secondly flavonoids are complex molecules with multiple potentially biological activities, for example, they are known to inhibit numerous enzymes in the body and interact with cellular drug transport systems. Thirdly, due to our rapid metabolism, aided by colonic bacteria, the plasma concentrations of phenols are usually low, the authors suggested they may be too low to exert any systemic antioxidant action in the body, although this suggestion should be backed up by further research. Fourthly, many flavonoids have cytotoxic effects both in vitro and in vivo, although the relevance of these observations is still poorly understood. These authors concluded that despite the enormous interest in flavonoids and other non-flavonoid phenolic compounds having a potential role in protecting against disease, there were no unambiguous data to show their beneficial effects on maintaining human health, nor data to support any mechanisms of how they may protect health. Antioxidant effects are found in vitro, but consistent evidence in vivo is lacking. They suggested that the biological effects of these compounds and their metabolites may be in the gastrointestinal tract.

A number of chemical groups have been studied for their specific health benefits, for example anthocyanins have been attributed with a wide range of biological activities. In young wines anthocyanins may still be present but in most red wines the anthocyanins will have polymerized to larger molecules. Further information on anthocyanins (Pascual-Teresa & Sanchez-Ballesta, 2008) and on flavonoids (Anderson & Markham, 2006) is documented elsewhere.

Resveratrol

There has been a long history of interest in the potential health effects of resveratrol and according to the review by Aggarwal et al. (2004), its use in the form of grape extracts for human health can be traced back over

2000 years, to a well known Indian herbal preparation *Ayurvedic*, with *Vitis vinifera* grapes as main ingredient, used for heart as well as other disorders. Although sourced from other plants, resveratrol has been used in traditional Chinese and Japanese medicine. To date numerous studies indicate that resveratrol, a non-flavonoid phenolic compound found in red grapes, wine and grape derived products, has a number of pharmacological properties and can protect against heart disease and cancer.

Aggarwal *et al.* (2004) also review the occurrence of resveratrol and related stilbenes in numerous sources. Resveratrol (3,5,4′-trihydroxystilbene) is a phytoalexin naturally produced in plants as a result of stress, such as fungal infection, ultraviolet radiation or injury and is also occurs in grapes. It has been identified in the skins of fresh grapes, with concentrations between 50-100 mg g^{-1} and in wines between 0.2-7.7 mg L^{-1}.

Aggarwal *et al.* (2004) review the role of resveratrol in the prevention and therapy of cancer, both preclinical and clinical studies. Clearly resveratrol has great potential in the prevention and therapy of a wide variety of tumors. It has antiproliferative effects through the induction of cell death (apoptosis) *in vitro* of cells from cell lines of carcinomas of breast, prostrate, colon, pancreas, head and neck and leukemias. Most studies indicated that no cell death was induced in normal cell lines.

Two recent reviews by Ahtar *et al.* (2007) and Bishayee (2009) on cancer prevention and treatment with resveratrol, ranging from rodent studies to clinical trials, focused on the use of on resveratrol, a compound now attributed with health benefits as well as having the potential to be used as a drug in treatment of cancers. The anti-cancer property of resveratrol has been demonstrated in a wide variety of human tumor cells *in vitro*. Bishayee (2009) summarizes the *in vivo* studies and their proposed mechanisms. It seems that resveratrol influences the three stages of carcinogenesis (initiation, promotion and progression) due to its influence on cell division and growth. Resveratrol affects the vascular system by suppressing the growth of new blood vessels, a process involved in the initiation, development and progression of many diseases, including cancer. However, since the growth of blood vessels forms an important part of wound healing, resveratrol delays wound healing.

A recent study on the role of resveratrol on liver cancer cancer (Bishayee & Dihr, 2009) showed that resveratrol can inhibit cell proliferation and there were significant effects of resveratrol on the development of liver cancer *in vivo*. The primary mechanism of resveratrol was attributed in part to the inhibition of cell proliferation and induction of cell death, although further studies are required to confirm the underlying mechanisms. Even low exposure to resveratrol in the diet may be effective in protecting against liver cancer in certain instances. The authors suggested that since resveratrol has very low toxicity, it can potentially be developed as a drug against human liver cancer.

According to a recent study by Kazuhiko *et al.* (2009) even dimers and tetramers of resveratrol isolated from parts of the vines were shown to have cell death effects on heart cells and the authors suggested that one of the tetramers may have use in clinical care for the prevention of acute heart infarcts.

Bibliography

General texts

Anderson, O.M., Markham, K.R. (eds.) (2006) *Flavonoids: Chemistry, Biochemistry and Applications*. CRC Press, Florida.

Barr, A. (1998) *Drink*. Bantam Press, London.

Clark, O. (2009) *Let me tell you about wine*. Pavillion Books, London.

Galet, P. (2000) *General Viticulture*. Oenoplurimédia, Chaintré.

Goolden, J. (1990) *The Taste of Wine*. BBC Books, London.

Johnson, H. (2005) *A life uncorked*. Weidenfeld and Nicholson.

Johnson, H. & Robinson, J. (2001) *World Atlas of Wine*. 5th edn. Mitchell Beazley, London.

Loftus, S. (1985) *Anatomy of the Wine Trade*. Sidgwick and Jackson, London.

Parker, R. (1999) *Wine Buyers Guide*. Dorling Kindersley, London.

Robinson, J. (1986) *Vines, Grapes and Wine*. Mitchell Beazley, London.

Robinson, J. (ed.) (1994) *The Oxford Companion to Wine*. Oxford University Press, Oxford.

Smith, B.C. (ed.) (2007) *Questions of taste: the philosophy of wine*. Oxford University Press, New York.

Regional wine books

Anderson, B. (1992) *The Wines of Italy*. The Italian Trade Centre, London.

Elliott, T. (2010) *The Wines of Madeira*. Trevor Elliott Publishing, Gosport.

Fletcher, W.J.S. (1978) *Port, an Introduction to its History and Delights*. Sotheby Parke Bernet, London.

Livingstone-Learmonth, J. (1992) *The Wines of the Rhône*. 2nd edn. Faber and Faber, London.

Read, J. (1982) *The Wines of Spain*. Faber and Faber, London.

Read, J. (1987) *The Wines of Portugal*. Faber and Faber, London.

Robertson, G. (1982) Port. Faber and Faber, London.

Scarborough, R. (2000) *Rioja and its Wines*. Survival Books, London.

Vandyke Price, P. (1979) *Guide to the Wines of Champagne*. Pitman, London.

General technical

English texts

Amerine, M.A. & Joslin M.A. (1970) *Table Wines: Technology of their Production*. 2nd edn. University of California Press, Berkeley.

Amerine, M.A., Berg, H.W., Kunkee, R.F., *et al.* (1980) *The Technology of Wine Making*. Avi Publishing, Westport, USA.

Amerine, M.A. & Ough, C.S. (1980) *Methods for the Analysis of Wines and Musts*. John Wiley & Sons, Inc., New York.

Amerine, M.A. & Roessler, E.B. (1983) *Wines, their Sensory Evaluation*. 2nd edn. W.H. Freeman & Co., San Francisco and New York.

Forbes, J. (1988) *Technology and Biochemistry of Wines*. Vol. I & II. Gordon & Breech, New York.

Linskens, H.F. & Jackson, J.F. (eds.) (1988) *Wine Analysis*. Springer-Verlag, Berlin.

Peynaud, E. (1984) *Knowing and Making Wine*. John Wiley & Sons, Inc., New York.

Peynaud, E. (1986) *The Taste of Wine*. Macdonald, Orbis, London.
Various (1988) Alcohol Drinking. In: *Chemical Composition of Alcoholic Beverages*. IARC Monograph. Vol. 44. IARC, Lyon.
Waterhouse, A.L. (1995) Wine and heart disease. *Chemistry and Industry*, 1 May, 338.
Webb, A.D. (ed.) (1974) *Chemistry of Wine Making*. Advances in Chemistry Series 137, American Chemical Society, Washington.

French texts

Peynaud, E. (1973) *Le Goût du Vin*. See English translation above.
Ribéreau-Gayon, J., Peynaud, E. & Sudreau, O.P. (1972–1977) *Traite d'Oenologie: Sciences and Techniques du Vin*. Vol. I–IV. Dunod, Paris.
Ribéreau-Gayon, J., Peynaud, E., Ribéreau-Gayon, P. & Sudreau, O.P. (1982) *Sciences et Technologies du Vin. Analyse et Contrôlé des Vins*. Vol. I, 2nd edn. Dunod, Paris.

German texts

Troost, R. (1988) *Technologie des Weines: Handbuch der Lebensmittel Technologie*. 2nd edn. Ulmer, Stuttgart.
Würdig, G. & Woller, R. (eds.) (1988) *Chemie des Weines*, Ulmer, Stuttgart, Germany.

Chapters in books

Bakker, J. (1993) Port, Chemical composition and analysis. In: *Encyclopaedia of Food Science, Food Technology and Nutrition*. (eds. R. Macrae, R.K. Robinson & M.Y. Sadler), pp. 3658–3662. Academic Press, London.
Bakker, J. (1993) Sherry. In: *Encyclopaedia of Food Science, Food Technology and Nutrition*. (eds. R. Macrae, R.K. Robinson & M.Y. Sadler), pp. 4122–4126. Academic Press, London.
Boulton, R. (1993) Wine. In: *McGraw-Hill Yearbook of Science & Technology*. pp. 485–487. McGraw-Hill Inc., New York.
Boulton, R. (1994) Red Wines. In: *Fermented Beverage Production*. (eds. A.G.H. Lea & J.R. Piggott), pp. 121–157. Blackie, Glasgow.
Goswell, R.W. & Kunkee, R.E. (1977) Fortified wine. In: *Economic Microbiology Volume 1, Alcoholic Beverages*. (ed. A.H. Rose), pp. 477–535. Academic Press, London.
Reader, H.P. & Dominguez, M. (1994) Fortified wines: Sherry, Port and Madeira. In: *Fermented Beverage Production*. (eds. A.G.H. Lea & J.R. Piggott), pp. 159–207. Blackie, Glasgow.
Ribéreau-Gayon, P. (1978) Wine flavour. In: *Flavour of Foods and Beverages*. (eds. G. Charalambous & G.E. Inglett), pp. 355–380. Academic Press, New York.
Schreier, P. (1979) Flavour composition, a review. *Critical Reviews in Food Science and Nutrition*, Vol. 12, pp. 59–111. CRC Press, Cleveland, OH.

References

Aggarwal, B.B., Bhardwaj, A., Aggarwal, R.S., Seeram, N.P., Shishodia, S. & Takada, Y. (2004) Role of resveratrol in prevention and therapy of cancer: preclinical and clinical studies. *Anticancer Research*, **24**, 2782–2840.

Athar, M., Back, J.H., Tang, X., Kim, K.H., Kopelovich, L., Bickers, D.R. & Kim, A.L. (2007) Resveratrol: A review of preclinical studies for human cancer prevention. *Toxicology and Applied Pharmacology*, **224**, 274-283.

Bakker, J. & Timberlake, C.F. (1997) Isolation, identification and characterisation of new colour-stable anthocyanins occurring in some red wines. *Journal of Agricultural and Food Chemistry*, **45**, 35-43.

Bartowsky, E.J. & Henschke, P.A. (2004) The 'buttery' attribute of wine – diacetyl – desirability, spoilage and beyond. *International Journal of Food Microbiology*, **96**, 235-252.

Bell, S.J. & Henschke, P.A. (2005) Implication of nitrogen nutrition for grapes, fermentation and wine. *Australian Journal of Grape and Wine Research*, **11**, 242-295.

Bishayee, A. (2009) Cancer Prevention and Treatment with Resveratrol: From Rodent Studies to Clinical Trials. *Cancer Prevention Research*, **2**, 409-418.

Bishayee, A. & Dhir, N. (2009) Resveratrol-mediated chemoprevention of diethy-lnitroseamine-initiated hepatocarcinogenesis: Inhibition of cell proliferation and induction of apoptosis. *Chemico-Biological Interactions*, **179**, 131-144.

Bisson, L.F. & Karpel, J.E. (2010) genetics of yeasts impacting on wine quality. *Annual Review of Food Science and Technology*, **1**, 139-162.

Boido, E., Medina, K., Farina, L., Carrau, F., Versini, G. & Dellacassa, E. (2009) The effect of bacterial strain and aging on the secondary volatile metabolites produced during malolactic fermentaion of tannat red wine. *Journal of Agricultural and Food Chemistry*, **57**, 6271-6278.

Campbell, J.I., Pollnitz, A.P., Sefton, M.A., Herderich, M.J. & Pretorius, I.S. (2006) Factors affecting the influence of oak chips on wine flavour. *Wine Industry Journal*, **21**, 38-42.

Clarke, O. (1986, 1989, 1998 and 2001) *Wine Guide*. Webster/Mitchell Beazley, London.

Cooper, K.A., Chopra, M. & Thurnham, D.I. (2004) Wine polyphenols and promotion of cardiac health: a review. *Nutrition Research Reviews*, **17**, 111-130.

Diban, N., Athes, V., Bes, M. & Souchon, I. (2008) Ethanol and aorma compounds transfer study for partial dealcoholisation of wine using membrane contactor. *Journal of Membrane Science*, **311**, 136-146.

Downey, M.O., Dokoozlian, N.K. & Krstic, M.P. (2006) Cultural practice and environmental impacts on the flavonoid composition of grapes and wine: A review of recent research. *American Journal of Enology and Viticulture*, **57**, 257-268.

Ehsani, M., Fernandez, M.R., Biosca, J.A., Julien, A. & Dequin, S. (2009) Engineering of 2,3-butanediol dehydrogenase to reduce acetoin formation by glycerol-overproducing, low-alcohol Saccharomyces Cerevisiae. *Applied Environmental Biology*, **75**, 3196-3205.

Fleet, G. (2008) Wine yeasts for the future. *Federation of European Microbiological Sciences Yeast Research*, **8**, 978-995.

Halliwell, B., Rafter, J. & Jenner, A. (2005) Health promotion by flavonoids, tocopherols, tocotrienols, and other phenols: direct or indirect effects? Antioxidant or not? *American Journal of Clinical Nutrition*, **81**(suppl), 268-276S.

Jackson, R.S. (2008) *Wine Science, Principles, Practice, Perception*. 3nd edn. Academic Press, San Diego.

Johnson, H. (1989) *Vintage: The Story of Wine*. Simon & Schuster, New York.

Karbowiak, T., Gougeon, R.D., Alinc, J.B., Brachais, L., Debeaufort, F., Voilley, A. & Chassagne, D. (2010) Wine oxidation and the role of cork. *Critical Reviews in Food Science and Nutrition*, **50**, 20-52.

Kazuhiko, S., Kanemaru, K., Sugimoto, C., Suzuki, M., Takeo, T., Motomura, S., Kitahara, H., Niwa, Masatake, Sshima, Y. & Furukawa, K.-I. (2009) Opposite effects of two resveratrol (trans-3,5,4′-trihydroxystilbene)tetramers, vitisin A and hopeaphenol, on apoptosis of myocytes isolated from adult rat heart. *The Journal of Pharmacology and Experimental Therapeutics*, **328**, 90-98.

Kilmartin, P.A. (2009) The oxidation of red and white wines and its impact on wine aroma. *Chemistry in New Zealand*, **73**, 18-22.

Li, H., Guo, A. & Wang, H. (2008) Mechanisms of oxidative browning of wine. *Food Chemistry*, **108**, 1-13.

Loureiro, V. & Malfeito-Ferreira, M. (2003) Spoilage yeasts in the wine industry. *International Journal of Food Microbiology*, **86**, 23-50.

Maicas, S. & Mateo, J.J. (2005) Hydrolysis of terpenyl glycosides in grape juice and other fruit juices: a review. *Applied Microbiology and Biotechnology*, **67**, 332-335.

McGovern, P. (2003) *Ancient wines: the search for the origins of viniculture*. Princetown University Press, New Jersey.

Morata, A., Gomez-Cordoves, M.C., Calderon, F. & Suarez, J.A. (2006) Effects of pH, temperature and SO_2 on the formation pyranoanthocayanins during red wine fermentation with two species of Saccharomyces. *International journal of Food Microbiology*, **106**, 123-129.

Pascual-Teresa, S. & de Sanchez-Ballesta, M.T. (2008) Anthocyanins: from plant to health. *Phytochemistry reviews*, **7**, 281-299.

Perez-Serradilla, J.A. & Luque de Castro, M.D. (2008) Role of lees in wine production: A review. *Food Chemistry*, **111**, 447-456.

Pickering, G. (2000) Low- and reduced alcohol wine- a review. *Journal of Wine Research*, **11** (2), 129-144.

Rentzsch, M., Schwarz, M. & Winterhalter, P. (2007) Pyranoanthocyanins – an overview on structures, occurrence and pathways of formation. *Trends in Food Science and Technology*, **18**, 526-534.

Ribéreau-Gayon, P., Glories, Y., Maujean, A. & Dubourdieu, D. (2006) *Handbook of Enology, Volume I, Wine and Wine Making*. John Wiley & Sons, Ltd, Chichester.

Ribéreau-Gayon, P., Glories, Y., Maujean, A. & Dubourdieu, D. (2006) *Handbook of Enology, Volume II, The Chemistry of Wine, Stabilization and Treatments*. John Wiley & Sons, Ltd, Chichester.

Robinson, J. (1993 and 1995) *Jancis Robinson's Wine Course*. BBC Books, London.

Sacchi, K.L., Bisson, L.F. & Adams, D.O. (2005) A review of the effect of winemaking techniques on phenolic extraction in red wines. *American Journal of Enology and Viticulture*, **56**, 197-206.

Stuttaford, T. (1997) *To Your Good Health, The Wise Drinkers Guide*. Faber and Faber, London.

Swiegers, J.H., Saerens, S.M.G. & Pretorius, I.S. (2008) The development of yeast strains as tools for adjusting the flavor of fermented beverages to market specifications. In: D.H. Frenkel & F. Belanger (eds.), *Biotechnology in Flavour Production*. Chapter 1, pp.1-55. Blackwell Publishing Ltd., Oxford, UK.

du Toit, W.J., Marais, J., Pretorius, I.S. & du Toit, M. (2006) Oxygen in must and wine: A review. *South African Journal of Enology and Viticulture*, **27**, 76-94.

Ugliano, M., Henschke, P.A., Herderich, M.J. & Pretorius, I.S. (2007) Nitrogen management is critical for wine flavour and style. *Wine Industry Journal*, **22**(6), 24-30.

Ugliano, M. & Henschke, P.A. (2009) Yeasts and Wine Flavour. In: *Wine Chemistry and Biochemistry* (eds. M.V. Moreno-Arribas & M.C. Polo), pp. 313–392. Springer ScienceBusiness Media, LLC.

Visioli, F., Borsani, L. & Galli, C. (2000) Diet and prevention of coronary heart disease: the potential role of phytochemicals. *Cardiovascular Research*, **47**, 419–425.

Wildenradt, H.L. & Singleton, V.L. (1974) The production of aldehydes as a result of oxidation of polyphenolic compounds and its relation to wine aging. *American Journal of Enology and Viticulture*, **25**, 119–126.

Chapter 2
Grape Varieties and Growing Regions

2.1 Wine grapes

The wine grape comes from the plants of the botanical genus *Vitis*, and mostly from one important commercial species, *vinifera*, of the vine plant. This genus belongs to the family, Vitaceae, characterized by a climbing habit, and includes other genera such as *Parthenocissus quinquefolia*, otherwise known as Virginia creeper. The genus *Vitis* includes also other species, for example *V. labrusca*, which has some interest for wine-making. The Concorde grape is thought to be a cross between *V. lambrusca* and *V. vinifera*, with its characteristic content of the ester, methyl methranilate that gives Concorde its specific flavour. Other species produce grapes more suitable for eating fresh.

Varieties (and cultivars) of *V. vinifera* selected for specific flavour and agronomic purposes abound. An obvious division is made between grapes with light green skins for making white wines, and those with black skins, generally for red wines, though not exclusively so. Some of the 10 000 known varieties are described as 'noble', largely grapes of French origin. The prevalence of French 'noble' grapes is, to an extent, due to tradition and to the history of France. 'Noble' grapes are those that give wines that are appreciated by wine connoisseurs and wine-drinkers in general for their sensory properties. Table 2.1 shows eight noble 'whites' and seven 'reds' – although this selection is not without some controversy. There are many other viniferous grapes of quality, which are largely of local rather than international significance, such as the Tempranillo grape of Spain and the Nebbiolo grape of Italy. These and possibly others may well become popular in the future, since wine-makers are 'seeking' to test their skills on new varieties, while consumers like to drink new wines.

The main controversy probably arises around two of these 'noble' grape varieties, Müller-Thurgau and Gamay. All these grapes and their vines originate in France and Germany (except Zinfandel); most have international repute, having been successfully transported to the newer wine-producing

Wine Flavour Chemistry, Second Edition. Jokie Bakker and Ronald J. Clarke.
© 2012 Blackwell Publishing Ltd. Published 2012 by Blackwell Publishing Ltd.

Table 2.1 'Noble' varieties of the wine grape.

Green-skinned grapes for white wine	Black-skinned grapes for red wine
Sauvignon Blanc	Cabernet Sauvignon
Sémillon	Pinot Noir
Pinot (Chardonnay)	Merlot
Chenin Blanc	Gamay
Muscat	Grenache
Sylvaner	Syrah (France)/Shiraz (Australia)
Müller-Thurgau	Muscat
Riesling	Zinfandel (US)

countries of the world, such as the USA, New Zealand, South Africa and Australia, and to other longer established European wine producing areas.

A differentiating characteristic between grape varieties is their 'aromatic' property, some grape varieties being quite 'neutral', others being quite 'aromatic'. Most of the odour/flavour components in wine, as will be seen in Chapters 4 and 7, originate during fermentation and ageing, the so-called secondary and tertiary aromas. Some of the 'noble' varieties above have varying contents of terpenoid compounds, with the Muscat varieties having the highest. These compounds confer a 'grapey' or sometimes 'floral' flavour, which can give distinctive wines made from those varieties with sufficiently high terpene content, so that terpenes from the grapes constitute the main part of the so-called primary aroma in the wine. However, although many of these so-called 'noble' grape varieties tend to give wines with a recognizable sensory aroma, in most cases unique impact compounds have not (yet) been identified. The typical chemical composition of a grape variety supplies the 'raw materials' for fermentation and ageing processes (Chapter 7), and even grape varieties with not very distinctive varietal aromas give wines with recognizable flavour characteristics, which in wine appear highly dependent on the grape variety.

These noble 'quality' varieties do not constitute necessarily the largest acreage of vines cultivated in any given country. Some quality grapes are widely planted, but there are other varieties that also fill that position (Table 2.2), which produce viniferous grapes destined for lower quality denominations such as *Vin-de-France* quality or they are used in blending wines. Some wines are specifically produced to be distilled into brandy.

There are several other grape varieties with quality connotation, which are currently mostly of local significance (Table 2.3). The currently widely planted noble grapes were, to an extent, fairly haphazardly selected and in future we may well see other quality grapes gain in popularity with both the wine industry and the consumer. Viognier is an example of a grape traditionally grown in the Rhône region but now becoming more widely planted, and giving white wines with a distinctive varietal aroma.

Table 2.2 Grape varieties with the largest area of vine plantation.

White wine	Red wine
Airén (Spain) (but low yield)	Grenache (world-wide)
Rkatsiteli (Russia)	Merlot (France, now world-wide)
Trebbiano (Italy) (known as Ugni Blanc in France)	Sangiovese (Italy)
Ugni Blanc (France)	Syrah (France)/Shiraz (Australia)
	Carignan (France)

Table 2.3 Some other grape varieties with quality connotation.

For white wine	For red wine
Aligoté (France, Burgundy)	Barbera (Italy)
Marsanne (France, California, Australia)	Carignano (Italy)
Muscadelle (France)	Cariñena (Mazuelo) (Spain)
Pinot Gris (France); Ruländer (Germany)	Cinsaut (France)
Pinot Grigio (Italy)	Malbec (SW France, Argentina)
Pinot Blanc (France); Pinot Bianco (Italy)	Mourvèdre (France, Bandol)
Weisburgunder (Germany)	= Monastrell (SE Spain)
Scheurebe (Germany)	Nebbiolo (Italy, Piedmont)
Sylvaner (Alsace, France, Austria)	Petit Verdot (France)
Viognier (France)	Pinotage (South Africa)
	Tempranillo (Spain)
	Zweigelt (Austria)

Table 2.4 Some grape varieties with a poor quality connotation.

White wine	Red wine
Airén (Spain)	Carignan (France)
Trebbiano (Italy)	
Ugni Blanc (France)	

There are also some grape varieties that have a poor quality connotation (Table 2.4), although they are widely planted (Table 2.2). Presumably their popularity is due to other factors, such as good yield. The quality connotation may not always be well deserved, for example, Carignan can give good wine, especially from old, low yielding bush vines.

2.2   Vine plant characteristics

The vine plant (*Vitis vinifera*) is indigenous to the Northern Hemisphere and grows in temperate regions, though with restrictions based upon soil and actual meso-climate (climate of the vineyard) generally. It is a shallow rooting plant, whilst the aerial parts have some particular botanical characteristics, apart from a climbing habit already noted, requiring training of cultivated species on poles. Information on botanical and genetic aspects can be found elsewhere. Very importantly, the cultivated plant is primarily a self-pollinating plant. The plant needs to reach some maturity before cropping fruit, since a young vine does not bear flowers until it is in its second or third season. The duration of full sunlight is important to achieve fruit bud development and flowering occurs within eight weeks of bud formation.

The fruits of the vine are, strictly speaking, berries. Maturity or ripeness of the grapes is expected to occur between 12 and 22 weeks after flowering, dependent on climatic conditions and grape variety.

After fruits set, the grapes develop, for 40 to 60 days, but remain approximately half their final size, hard, high in acid and green. Dependent on the climate, the next stage is *veraison*, the commencement of ripening period, taking between 30 days in hot regions and 70 days in cool regions. Some varieties are early maturing, others late maturing, needing less or more time to reach maturity when grown under the same circumstances. During veraison the grapes gain their red or yellow colour, gain sugar, lose acid, soften and increase in size. Other changes include the development of flavour (actual or precursor) compounds typical for the grape variety, often referred to as varietal aroma. Samples are normally taken throughout the growing season for chemical analysis of acidity, and sugar content, etc. and a tentative harvest date is set. However, in cool regions complete grape maturity at harvest may not be achieved, so that adjustments may have to be made at the vinification stage. Harvesting is usually accomplished in September or October in France and Germany. A special case is Beaujolais Nouveau, where vinification, bottling and distribution are accomplished by the official release date set on the third Thursday in November each year. In warmer vine growing regions, such as Southern Spain, the harvesting is carried out in August or September to avoid the grapes becoming over-mature.

The leaves of vine plants photosynthesize, so that carbon dioxide taken from the air is converted into carbohydrates and oxygen is expired. Photosynthesis takes place via the so-called Calvin biochemical pathway using UV light as an energy source. The leaves also absorb water as available. The roots also absorb water, but also take up many other essential nutrients, some in trace quantities. The roots absorb nitrogenous components, mainly the nitrate ion (NO_3^{2-}), from the soil, regardless whether the soil is enhanced by organic manure or artificial fertilizers in aqueous solution. They also absorb phosphorus compounds (phosphate ions) and metallic cations, mainly potassium (K^+) but also a wide range of other compounds, some in trace quantities. An adequate input of water through the roots is essential for the effective and healthy growth of the plant.

A recent review by Conde et al. (2007) states that wine quality largely depends on the vineyard and the wine grower. Sugars and acids are produced by the leaves, whilst acids and phenolics are produced in the berry. Some molecules related to aroma and taste, the so-called primary or varietal aromas, are formed during fruit development and are typical for the grape varieties used for wine-making. The characteristics of the grape when picked at commercial maturity for wine-making are crucial for the resulting wine quality.

2.3 Soil, climate and ripeness

2.3.1 Soil

The quality of the wine is thought to depend to a great extent on the vineyard and the skill of the viticulturalist being able to produce good quality grapes. Hence it is not surprising that the nature of the soil in which the vine is grown has always been a part of the mystique of desirable wine flavour and this is usually referred to as the effect of *terroir*. In the review of Conde et al. (2007) the definition is much broader defined, the term *terroir* is used by viticulturalists to define rather widely the geographical and environmental conditions of the vineyard in which the grapes were grown, with many factors contributing, such as composition of soil, climate, topography, even including strains of micro-organisms on the berry skin contributing the fermentation process. A *terroir* offering good growing conditions to a particular grape cultivar helps the plant to produce good quality grapes, forming a good start for a good vintage.

An important aspect of soil quality is water retention and drainage. This capacity of the soil is related to both chemical and physical characteristics. The vine plant does not like waterlogging, hence excessively 'clayey' soils are undesirable. Gravel (defined as mineral particles, >2 mm in diameter), in contrast to clay (particle size around 0.2 mm or less) is necessary for free water draining properties. However, clay has the advantage of attracting and holding humus, containing nitrogen and other nutrients. The vine plant is thought not to need an especially fertile soil; in any event, the plant will not pick up nutrients in excess of its need, even though they may be present. Suitable soils for growing quality grapes are not particularly fertile or deep (see Conde et al., 2007). A recent review on geology and wine suggests that the importance of geology in wine-making may have been over rated (Huggett, 2006). Most of the nourishment vines take from the top 0.6 m, although the plants can take up water from as far down as 2 m, even deeper in drought conditions. In most areas, even where the soil is thin, geological influences on vines will be minimal. The quality of the grapes is only influenced indirectly by factors such as soil composition, geomorphology and water retention. The direct effect of soil on resultant wine flavour is therefore questionable and no scientific proof currently exists. Indirect effects along with the meso-climate (conditions of the vineyard, as for example influenced by the slope and orientation towards the midday sun) will be important. Very light soil will reflect the sunlight, warming the area round the vine. Currently the Old World tends to emphasize the effect of *terroir* (generally the nature of the soil rather than the wider definition), while the New World tends to emphasize climate as having the greatest influence on wine quality and flavour. Further studies are needed to show the true relevance of these two factors. This difference is possibly also reflected in the aims of the wine-makers, Old World wines tend to be made reflecting the style typical for the area, whereas New World wines tend to be made to reflect the best the fruit can give.

2.3.2 Climate

Climate conditions above the soil have an important influence on growth of micro-organisms on grapes, which in some cases cause the grapes to rot. Botrytized grapes (resulting from Noble Rot) have a separate commercial importance, since they can be prized for the production of quality sweet wines, with a very distinctive flavour. Noble Rot is associated with particular grape varieties, the type of grape cluster and meso-climate (i.e. cyclical conditions of fluctuating humidity, humid mornings followed by dry afternoons). Heavy rains and protracted damp periods are more likely to lead to conventional rotting, such as bunch rot, which essentially makes the grapes unfit for wine-making.

Different varieties of grapes will have documented agronomic characteristics, which mean they are favoured for use in particular growing areas and the associated climates. Specific botanical factors can be also important, such as the type of cluster formation of the grapes on the vine, and resistance to various forms of fungal rot and diseases, which, to an extent, depend upon such factors as the thickness of grape skins. Pruning of vines involves removing growth, usually in winter, although lighter summer pruning can also be carried out. It ensures that the otherwise rampant vine is kept in a manageable shape, thus also avoiding so-called over-cropping. The excess production of grapes (over-cropping) of a vine leads to inferior quality grapes and wines, and it is thought to compromise the life span of the vine. Hence, controlling the yield is an important issue in vineyard management, although there does not seem to be firm scientific evidence to indicate the optimum yield for vines. There are a number of different pruning techniques which influence the character of the grape and the resulting wine, even within one varietal type. For example, insufficient pruning in cooler climates will give grapes that take too long to mature; hence a heavier pruning technique would be more appropriate to ensure a ripe grape crop. Summer pruning is often associated with canopy management and helps to ensure a desirable canopy micro-climate, for example, sufficient sunlight on the bunches of grapes.

A most important characteristic of the vine is the rate at which the grapes can reach maturity, i.e., whether the vine is early or later ripening. Late maturing varieties need a warmer climate with a sufficiently long warm autumn and they do not thrive in cool climates. As already mentioned, many grape varieties are currently only of local regional interest but a relatively small number of others, with well known names, are internationally grown and are suitable for vinification in widely different countries. However, often much research is carried out regarding climate, meso-climate and even micro-climate to be created around the vines, in addition to the suitability of the soil. All this ensures that, once the preferred variety is selected and planted, there will be healthy vines yielding healthy good quality grapes.

In Tables 2.5 and 2.6 the second column indicates some of these agronomic characteristics, which can be studied in greater detail in other texts. The third column indicates some botanical features of the grapes, including some general comments on the chemical composition of the grape must, in particular, relation to tannin content and acidity which are both modified during wine-making. Some grape varieties are used on their own in wine production, giving the so-called varietal wines, particularly popular in the New World. Many others are used for blending purposes, and despite the negative connotation of the word, it forms an essential part of good wine-making practice. For example, blending gives the opportunity to add complementary characteristics, adjust sweetness, enhance acidity, or reduce the 'oaky' character of a heavily oaked wine from ageing in barrels.

Chemical composition both of the must and of the resultant wine is the key to a full study of vinification practices and wine flavour, which is dealt with in subsequent chapters. Notably, internationally grown varieties such as Cabernet Sauvignon and Chardonnay can show some marked differences in wine flavour according to the region of vinification (e.g. France compared with Australia or California); these differences are explored in more detail later.

An important characterizing feature of chemical composition is the so-called terpene content of particular grapes, which is relatively unaffected by subsequent vinification. These terpenes are volatile compounds that confer a very distinctive grapey flavour, often referred to as 'aromatic fruity', in contrast to 'ester fruity' flavour aroma which can be generated during fermentation. As discussed in Chapter 4, terpene content varies with different varieties, and is highest in the Muscat variety, but it is also significant in certain 'noble' varieties, notably Riesling and Gewürztraminer, and hence in the wines produced by these grapes. Varieties with very low or zero concentrations of terpenes tend to be more neutral in character but often develop a more generally 'fruity' aroma. These distinctions are also reflected in comments in the third column of the tables. The whole subject is dealt with in later chapters. Terpene content can be used for distinguishing purposes between some varieties. During ageing, changes occur both in the type and amount of terpenes, which has a noticeable influence on the aroma bouquet of the wine (discussed in detail below).

Table 2.5 Some grape variety characteristics for red wines.

Black grapes (red wines)		
Variety (country of origin)	**Agronomic**	**Botanical/composition/ vinification**
Cabernet Sauvignon (France)	Ripens late. Planted in warm climates. Susceptible to fungal disease. International variety (California and Australia, increasing in many wine regions).	Small thick-skinned grapes. High ratio of solids, tannins, to juice, rich in colouring matter. Very suitable for 'oaking'.
Merlot (France)	Early maturing/ripening. Can grow in cool climes, but can suffer frost damage. Likes 'clayey' soil. Susceptible to fungal disease. Most planted variety in Bordeaux. International variety (California, Argentina, South Africa and New Zealand).	Thinner skins than above. Often blended with Cabernet Sauvignon.
Gamay (Gamay Noir à Jus Blanc) (France)	Ripens early. Sensitive to fungal disease, and frost damage. The 'Beaujolais' grape.	Medium sized grapes with tough skins. Low tannin/high acidity. Usually processed by Maceration Carbonique.
Syrah (France) Shiraz (Australia)	Unusual relationship of pruning technique with subsequent taste/flavour of wine. Northern Rhône.	Tannic wine with ageing potential, but loss of acidity past optimal ripening of grapes.
Cabernet Franc (France)	Vine similar in appearance to Cabernet Sauvignon. Ripens earlier. Useful in cooler regions. Internationally planted.	Wines, fruity and less tannic than from Cabernet Sauvignon.
Pinot Noir (France) Spätburgunder (Germany) Pinot Nero (Italy)	Ripens early, not very suitable for very warm regions. 'Fussy' about soil; sensitive to bunch rot; not easily transferable (except to Oregon and some other regions). The 'Red Burgundy' grape.	Ancient French vine, very prone to mutation with slight varietal difference, and even variation between clones of the same cultivar.

Table 2.5 *(Continued)*

Black grapes (red wines)		
Variety (country of origin)	Agronomic	Botanical/composition/ vinification
Nebbiolo (Italy)	Late ripening. Susceptible to powdery mildew. Grown largely in Piedmont.	Wine made is markedly high in tannin and acidity. Grapes need to be ripe (sugar content) to counterbalance. Requires long ageing.
Sangiovese (Italy)	Late ripening. Susceptible to bunch rot. Most planted type in Italy. The 'Chianti' grape.	High acid.
Tempranillo (Spain) (Other Spanish regional names e.g. Cencibel)	Early ripening. Subject to both powdery and downy mildews. Important in Rioja and Navarra. Spain's best quality grape.	High in tannins and acidity. Mid-sized/thick skinned. Usually aged. Not necessarily high in alcohol content.
Zinfandel (USA) (= Primitivo of S Italy)	Uneven ripening. Needs warm climate and long growing season. Most planted variety in California.	Gives red, white and rosé wine.
Grenache (France) (Garnacha (tinto) Spain)	Flourishes in barren soils. Needs careful pruning to keep down the yield. World's most widely planted red wine grape. The 'Southern grape' (France).	Thin skins. High alcohol producer. Generally used for blending purposes, e.g. S Rhône.
Malbec (France)	SW France grape. Fragile in cooler climates. Very popular in Argentina.	Used in blending or as varietal wines (Argentina).
Petit Verdot (France)	Very late ripening limits possible growing regions. Small quantities in Bordeaux. Some interest for Chile, Australia and California.	Acid.
Mourvèdre (France) (= Monestrell) (Spain)	Needs very warm climate to ripen fully. Grape planted in SE France and SE Spain. Old fashioned grape coming into fashion again.	Wines with high alcohol.
Pinotage (South Africa)	Early ripening. High yielding vine, hence requires pruning to maintain quality and yield. South Africa.	South African variety by crossing of Pinot Noir and Cinsaut.

Table 2.6 Some grape variety characteristics for white wines.

Variety (country of origin)	Agronomic	Botanical/composition/vinification
Chardonnay (France)	Ripens early and useful in cool areas. Susceptible to powdery mildew and 'bunch rot'. Famous white, internationally planted. The 'Chablis' grape.	High alcohol, low acidity wine. Low terpene content. Fermentation and maturing in barrels possible. Wine flavour especially dependent upon area of production. Lends itself to different vinification techniques in individual wines.
Sauvignon Blanc (France) (Muscat-Silvaner, Germany)	Particularly sensitive to powdery mildew and black rot; with partial resistance to bunch rot and (downy) mildew. Bordeaux/Loire Valley white grape. Internationally planted.	Small sized grapes. Popular variety for making crisp, dry, aromatic wines all over the world, and distinctive from Chardonnay. Moderate terpene content.
Sémillon (France)	Susceptible to bunch rot and frost, and to Noble Rot. Bordeaux grape, also international (e.g. Australia).	Small clusters of medium sized fruit. Used in fine white wines and also for sweet wines.
Chenin Blanc (France)	Requires considerable sun to ripen. Susceptible to both downy and powdery mildews, bunch rot. Loire Valley, but international.	Tough skinned, medium sized grapes. Used for sweet and dry table wines, sparkling wines.
Riesling (Germany)	Ripens slowly, moderate yield. Slate soil. They are sensitive to powdery mildew. Particularly suited to Germany.	Small to medium sized grapes. Terpene content high. Characteristic flavour on ageing. Varietal wine not much blended.
Muscat Blanc (Muscat à Petit Grains, France)	Grown in Australia. Widely planted variety with many other names. Central and Eastern Europe.	Small round grapes light skinned. High terpene content. Used for sweet and sparkling wines (Italy), but also dry.
Muscat of Alexandria (France)	Less distinguished variety than foregoing. Different names according to country. Used in hotter climates.	High terpene content but most is in unavailable glycosidic form.

Table 2.6 *(Continued)*

Variety (country of origin)	Agronomic	Botanical/composition/ vinification
Muscat Ottonel (France)	Grown in cooler climates, e.g. Alsace.	Lower terpene content than foregoing.
Morio-Muscat (Germany)	Needs a good site for growing and rots easily.	German hybrid of Silvaner and Weissburgunder (Pinot Blanc) High terpene content.
Silvaner (Germany) = Sylvaner (France)	Ripens earlier than Riesling, but later than Müller-Thurgau. Central Europe planting.	
Müller-Thurgau (Germany)	Ripens early, and suitable for cool climates. Subject to both powdery/ downy mildew and bunch rot. Grown in Germany and UK.	Crossing, Riesling and Chasselas (1882). High yielding mid-sized grapes. Used in cheaper German wines (QbA).
Traminer (Germany), Savagnin Rose (France)	Planted in cooler regions. Subject to powdery mildew and bunch rot. N.B. Sauvignin grape of Vin Jeune (Jura).	Modest clusters of small grapes. Tough skins. Terpene content moderate.
Gewürztraminer (Germany)	Planted particularly in Alsace.	Higher terpene content than above. Produces wines of high alcohol content, but may be low acidity.
Aligoté (France)	Burgundy's second wine grape; also grown in Romania and elsewhere.	Acid wine.
Muscadelle (France)	Planted in Bordeaux and SW France.	Largely used for sweet white wines. Related to Muscat.
Viognier (France)	Northern Rhône, but also S France, Australia and California. Can be poor yielding in cool climates.	Fashionable variety. Wines best drunk young: good for blending.
Pinot Gris (France) (= Pinot Grigio, Italy) (= Ruländer, or Grauburgunder, Germany) NB. EU insists on name Tokay-Pinot Gris	Quality grape in Alsace, also grown in Germany (Baden/ Pfalz).	Colour mutation of Pinot Noir. Pink skinned.
Melon de Bourgogne (France)	Grown for Muscadet; and little known elsewhere. Pays Nontais area (R Loire). Withstands cold well.	Dry wine, neither very acid nor strongly flavoured. Often 'Sur lie'. Alleged highly sulfured wines.

(Continued)

Table 2.6 *(Continued)*

Variety (country of origin)	Agronomic	Botanical/composition/vinification
Scheurebe (Germany)	Grown in Germany (Pfalz etc.), also Austria.	Crossing (1915), Unknown variety* and Riesling. Good quality and wine flavour.
Viura (Spain) (= Macebeo)	Grown in Rioja, Spain.	Good acid, but poor ageing capacity.
Grüner Veltliner (Austria)	High yields.	Not very aromatic, ages well in-bottle.

* http://www.vivc.de/datasheet/dataResult.php?data=10818 (last accessed 26.11.2010)

2.3.3 Ripeness

Harvesting the grapes at optimum ripeness is now thought even more important than in the past and research has focused on ensuring that grapes are picked at the level of maturity that gives the most desirable wine flavour characteristics. Traditional physiological ripeness is reached when grapes have accumulated sufficiently high sugar levels, without having lost too much acidity. However, compounds contributing to the flavour of the grape and the resulting wine also are taken into account. Studies on changes of grape composition, with particular interest on compounds thought to influence wine flavour, have been comprehensively reviewed by Conde *et al.* (2007). The development and production of compounds during grape maturation contributing to the sensory properties of the grape and wine are still not well understood. During ripening, the grapes become larger, sweeter and less acid, mostly due to malic acid being metabolized, resulting in a change of the frequently used acid/sugar balance. On a per berry basis tannins and aromatic compounds formed during the early part of berry development decline.

Concurrent there are also important changes in volatile compounds characteristic for the grape variety and contributing to the aroma and regional character of the wine, often referred to as forming part of the wine quality. Best studied are the monoterpenes, C13-norisoprenoids, methoxypyrazines and sulfur compounds. Conde *et al.* (2007) summarize the monoterpene concentrations in three categories: (1) up to $6\,mg\,L^{-1}$ in intensely flavoured grapes (Muscat), (2) between $1 - 4\,mg\,L^{-1}$ in some aromatic varieties (Tramines, Huxel and Riesling) and (3) very low concentrations, generally below perception threshold and thought to contribute only minimally to varietal flavour (Cabernet Sauvignon, Sauignon Blanc, Merlot, Shiraz and Chardonnay). The levels of methoxypyrazines, contributing to the herbaceous aroma of, for example, Cabernet Sauvignon, diminish during maturation. Carotenoids break down during ripening; the process is enhanced by sunlight. Excessive hydrocarbon smells, due to 1,1,6-trimethyl-1,2-dihydronaphtalene, can develop in Riesling wines and have been attributed to extremely high temperatures during grape maturation. This illustrates that although hot climates favour the accumulation of sugars in grapes, it may not always bring out the best in terms of flavour quality.

The increased attention given to the production of wines with the varietal characteristics in terms of optimized flavour has increased the level of ripeness at which the grapes are picked. This means the sugar contents are higher and the fermentation leads to wines with higher concentrations of ethanol (even up to 15% v/v) than in the past. There is research ongoing in selecting yeast strains less efficient in the conversion of sugar into ethanol, enabling the production of wines with lower ethanol concentrations from very sweet and ripe grapes.

2.4 Grape growing regions of the world

2.4.1 World wine production

The main wine producing countries are France, Italy and Spain, and they have well established industries. Quality denominations in these regions date back many years and are determined by area, discussed briefly in Section 2.6. Grapes for wine-making have been selected over centuries, driven by a desire to improve wines to satisfy knowledgeable consumers. Hence it is not surprising that many of the grapes now successfully established in new wine regions in the New World are exported mainly from France and, to a lesser extent, from Germany. Interestingly, not infrequently the grapes traditionally grown in the cooler wine regions of France thrive best in cooler regions elsewhere in the world, although the sensory properties of the wines can be quite different. An example is the successful Sauvignon Blanc wine production in New Zealand, a relatively cool wine region.

However, as can be seen from Table 2.7 the increases in wine production in 2006 since 1996–2000 in some of the New World countries, such as Australia, Chile and New Zealand, are enormous, as are their increases in exports. This no doubt reflects their successful application of research information and technology in both viticulture and wine-making techniques. These countries have no long established histories and traditions in their wine industry, hence they may feel free to experiment in their viticulture and wine-making. In particular Australia has moved traditional thinking about viticulture and wine-making forward. Although their wine industry is not large, their influence is probably world-wide. For example, the fact that grapes grown in hot, dry areas produce top quality white wines contradicts the traditional view on climatic requirements for viticulture and shows that irrigation can have a positive effect on wine quality.

Styles of wine produced have adapted with the changes in fashion, for example, heavily oaked, buttery Chardonnay wines from California have virtually disappeared and have been replaced by lighter, fruity Chardonnay wines. Market

Table 2.7 Wine production and export data in 2006, with calculated increases in production and export since 1996–2000. Only countries with a significant export have been listed.

Ranking in production in 2006	Country	Production in 2006 (1000 hL)	Increase in production since 96-00 (%)	Export in 2006 (1000 hL)	Increase in export since 96-00 (%)
1	France	52127	93	14720	96
2	Italy	52036	96	18390	124
3	Spain	38137	112	14340	163
4	United States	19440	95	3761	163
5	Argentina	15396	114	2934	286
6	Australia	14263	193	7598	364
8	South Africa	9398	120	2717	227
9	Germany	8916	89	3197	137
10	Chile	8448	167	4740	211
11	Portugal	7266	106	2900	136
19	Moldavia	1968	91	1020	77
20	Bulgaria	1757	63	1127	84
21	New Zealand	1332	301	578	388

Data from OIV, 2006.

research and good marketing presumably helps to ensure the wine industry adapts their products to meet the consumer preferences and demands.

The United States is the fourth biggest producer, with many prestigious wineries in California, especially the Napa and Sonoma valleys. However, most of these wines are consumed within the USA and exports tend to come from the large Central Valley. Australia is at the leading edge of technology in wine-making and successfully implements know-how and new technology. The wine industry in South Africa dates from the mid seventeenth century and after the abolition of apartheid the export markets opened, giving impetus to the wine industry. Wines continue to improve and are good value. New Zealand is only a small producer, producing less than 1% of the world wine production. However, current popularity keeps the wine prices high. Detailed accounts on wine-making regions can be found in many popular wine books and websites.

2.4.2 Regions

Many books on wines are devoted to the grape growing and wine production regions of the world in greater or lesser detail, together with assessments of the wines produced in the numerous sub-areas, districts or groups of vineyards, e.g. Châteaux in Bordeaux, France.

Tables 2.8, 2.9, 2.10, 2.11, 2.12, 2.13, 2.14, 2.15, 2.16, 2.17, 2.18, 2.19, 2.20, 2.21 and 2.22 which follow are primarily indicative, showing (1) region, (2) best known wine types, where only the quality wines are mentioned (i.e. AC in French wines),

Table 2.8 Wines in France: area/type Bordeaux.

Region	Best known wine types/red wines[a]	Grape varieties used
Haut-Médoc (High quality clarets)	Haut-Médoc, AC e.g. Château Cantemerle Pauillac, AC e.g. Château Lafite – Rothschild, Latour and Mouton-Rothschild (all 1st growth)[a]; Château Pichon-Longueville Lalande (2nd growth)[a] Listrac, AC e.g. Château Fourcas Holsten Cru Bourgeois) e.g. Château Margaux (1st growth)[a] St Julien, AC e.g. Château Léoville-Barton (2nd growth)[a]	Cabernet Sauvignon Cabernet Sauvignon blended with Cabernet Franc and Merlot
Bourg and Blaye	Côtes de Bourg, AC	Merlot, also blended with Malbec
Graves	Graves, AC Pessac-Léogman, AC e.g. Château Haut-Brion (1st growth)[a]	Cabernet Sauvignon and Merlot
St-Emilion, Pomerol, Fronsac	Pomerol, AC e.g. Château Pétrus St Emilion, AC e.g. Château Ausone Fronsac, AC Castillon Côtes de Bordeaux, AC	Merlot dominant Merlot/Cabernet Franc/Malbec/ Cabernet Sauvignon Cabernet Franc
	White wines	
Graves-Sauternes regions	Barsac, AC (sweet) Graves, AC Dry, aged Sauternes, AC e.g. Château Yquem	Sauvignon Blanc/ Sémillon Sémillon, blended with Sauvignon Blanc
Entre-Deux-Mers	Entre-Deux-Mers, AC Premiôres Cètes de Bordeaux, AC (sweet)	

[a]Growth = cru classé, 1855 classification.

with a brief reference to some typical producers, and (3) main grape varieties used for the particular wine types.

Comments about the flavour of particular wines from wine writers and wine merchants are dealing often with wines made from blends of grapes and

Table 2.9 Wines in France: area/type Burgundy.

Region	Best known wine types/red wines	Main grape varieties used
Côte de Beaune	Aloxe-Corton, AC Beaune, AC Premier Cru[a] Chassagne-Montrachet Premier Cru[a] Volnay AC Premier Cru[a]	Pinot Noir (most red Burgundy by law has 100% of this variety) Pinot Gris Pinot Liebault
Côte de Nuits	Chambertin, AC Grand Cru[a] Gevrey Chambertin, AC Nuits-St-George Premier Cru[a]	Pinot Noir
Côte d'Or	Bourgogne Rouge AC	
Côte Chalonaise	Bourgogne Passe-Tout-Grains, AC	Gamay, 1/3 Pinot Noir
Beaujolais	Beaujolais AC Beaujolais-Villages AC	Gamay Gamay
White wines		
Côte de Beaune	Corton Charlemagne AC Grand Cru[a] Le Montrachet, AC Grand Cru[a]	Chardonnay (a little Pinot Noir) Chardonnay
Côte de Nuits	Musigny, AC	Chardonnay
Côte Chalonnaise	Bouzeron AC	Aligoté
Maconnais	Pouilly-Fuissé, AC	Chardonnay
Chablis	Chablis AC Grand Cru AC and Premier Cru[a], AC	Chardonnay (100% use)
Beaujolais	Beaujolais Blanc AC	Chardonnay

[a]Burgundy classification.

information about blends is not usually labelled. The flavour of the wines will depend upon the vintage year, which is being tasted, in addition to many other factors (age of wine, vinification method, maturation procedure, etc.).

The flavour characteristics of wines from single varieties are of the most interest to the scientist and chemist, since they can be related to the chemical composition of the grapes. Undoubtedly the grape variety has the greatest influence on wine flavour, but the conditions of vinification, including the type of yeast used, will also determine the wine flavour, especially by the formation and modification of the volatile components. Some important factors are agronomic, particularly the yield per vine. It will be evident that certain taste characteristics of the grapes, such as acidity and sweetness, will be determined by the level of maturity at which the grapes are harvested (see above), though the specification for vinification will nearly always be 'ripe' and 'healthy' grapes. Ripeness is usually defined in terms of concentrations of sugar and acid. As grapes ripen, they lose acid and gain sugar. The grapes are ideally

Table 2.10 Wines in France: area/type Champagne.

Region of production: Épernay area		
Grape varieties	Chardonnay Pinot Noir Pinot Meunier	
Wine styles	*Non-vintage*: Basic blend (80% of area output) Typically 3 yrs old. e.g. from Lanson, Père & Fils, Mercier, Pol Roger, Veuve Clciquot – Ponsardin. *Vintage*: Wine of grapes of a particular year, released typically at 6 yrs. e.g. from Pol Roger, Louis Roederer. *Prestige de Luxe*: Top quality, vintage dated. Highly priced. e.g. Bollinger (R.D), Charles Heidsieck (Collection), Krug (Grande Cuvée), Laurent-Perrier (Grand Siècle), Taittinger (Comtes de Champagne),Moët et Chandon (Dom Perignon). *Blanc de Blancs*: Made exclusively from Chardonnay grapes. *Buyer's Own brand*: BOB. Blended to buyer's specification/cost. e.g. Tesco, Waitrose, Sainsbury. *Non-dosage*: No added sugar and therefore bone-dry.	
Wine designation	Brut Zero Brut Sec Demi-sec Doux	Absolutely dry Very dry Medium dry Medium sweet Sweet
Label designations (Two letter code at bottom) Source	CM (co-operative-manipulant) MA (marque d'achéteur) NM (négociant-manipulant) RM (récoltant-manipulant) RM (récoltant-cooperateur)	Comes from a co-operative. Subsidiary brand. Comes from a merchant-handler. Wine 'champenized' by the Champagne house, whose label is used. Comes from a grower who makes his own wine, e.g. Michel Gonet, Albert Le Brun. Comes from a grower selling wine made by a co-operative.

picked when they have reached sufficient sugar to give about 11% alcohol (Chapter 3) while still having sufficient acidity to help to preserve the wine and to give an acid taste. Often sugar/acid ratios are used, but their use is probably very area specific. Little information is known about the flavour of the wine in relation to grape ripeness. The sugar and acid content of the grape at harvest will depend on grape variety and climate.

Table 2.11 Wines in France: area/type Loire.

Region	Best known wine types (white)	Main grape varieties
Upper	Sancerre AC	Sauvignon Blanc
Loire	Pouilly-Fumé, AC	Sauvignon Blanc
Middle	Anjou Blanc Sec, AC	Chenin Blanc
Loire	Touraine, AC	+ Chardonnay. Sauvignon Blanc
Muscadet	Muscadet AC	Melon de Bourgogne
	Best known wine types (red)	
Upper Loire	Sancerre Rouge, AC	Pinot Noir
Middle Loire	Touraine, AC	Variety of grapes
	Chinon, Saumur-Champigny, AC	Cabernet Franc
	Saumur – Champigny AC	Cabernet Franc

Table 2.12 Wines in France: area/type – south western.

Region	Best known wine types (red)	Main grape varieties
Bordeaux fringe country (Bergerac)	Bergerac, AC	Merlot, Cabernet
River Lot area	Cahors, AC	Malbec (70%), Merlot and Tannat
Armagnac region	Madiran, AC	Tannat, Cabernets, Fer
	Best known wine types (white)	
Bergerac	Bergerac Sec, AC	Sémillon, Sauvignon Blanc,
	Monbazillac AC	Sémillon, Sauvignon Blanc, Muscadelle
SW France	Gaillac, AC	Local varieties, Mauzac
	Pacherenc du Vic-Bilh AC	Local mixed varieties

Chemical composition is considered broadly in the next section. In Chapters 3 and 4, the basic taste factors and volatile compounds are dealt with separately in much greater detail.

In Tables 2.8, 2.9, 2.10, 2.11, 2.12, 2.13, 2.14, 2.15 and 2.16 France is recognized as historically the most important area in Europe for wine production, and still occupies a major position in the industry. Not all wine-producing countries have been included; other texts should be consulted for full descriptions of the world's wine-growing regions. A comprehensive reference guide to the wines of the world is Sotheby's *World Wine Encyclopedia* (Stevenson, 1997). Numerous wine writers have written books on wine regions, including listing of wines produced and their producer, some of which have been mentioned in the Bibliography of Chapter 1.

Table 2.13 Wines in France: area/type Rhône.

Region	Best known wine types (red)	Main grape varieties
Northern Rhône	Côte-Rôtie, AC	Syrah/Viognier (white)
	Crozes-Hermitage, AC	Syrah
	Hermitage, AC	
	Best known wine types (white)	
	Condrieu, AC	Viognier
	Clairette de Die, AC (sparkling)	Clairette, Muscat
	Hermitage, AC	Marsanne, Roussanne
	Best known wine types (red)	
Southern Rhône	Châteauneuf-du-Pape, AC	Some 13 different varieties allowed (mainly Grenache, Mourvèdre, Syrah)
	Côtes du Rhône, AC	Grenache
	Lirac, AC	Grenache, Cinsau(l)t and others
	Tavel, AC (rosé wine)	Grenache

Table 2.14 Wines in France: area/type – south east.

Region	Best known wine types (red)	Main grape varieties
South east (Provence)	Bandol, AC	Mourvèdre, with some Grenache, Syrah
(Island of Corsica)	Vin de Corse, AC	Sciaccarellu, Niellucciu
(Languedoc-Roussillon)	Fitou, AC	Carignan, Grenache
	Corbières, AC	Carignan, Cinsaut Grenache
	Côtes de Roussillon, AC	Carignan, mainly
	Banyuls AC (Vin Doux Naturel with added grape juice)	50% Grenache
	Best known wine types (white)	
South east (Provence)	Bellet, AC	Rolle
	Cassis, AC	Marsanne
(Languedoc-Roussillon)	Clairette de Bellegarde, AC	Clairette

2.5 Chemical composition of grapes, must and finished wines

2.5.1 Grapes and must

Plants generally show marked differences in the chemical composition of the fruits of different species within a given genus (e.g. differences between *Coffea arabica* and *Coffea canephora*) but there are rather fewer differences

Table 2.15 Wines in France: area/type Alsace.

Region	Best known wine types (white)	Main grape varieties
Alsace (white)	Types characterized by grape name	
	Alsace AC	Muscat (Muscat Ottonel Muscat Blanc Group)
		Pinot Blanc
		Pinot Gris
		Riesling
		Sylvaner
	Alsace Grand Cru Ac (50 vineyards, in 1992 classification)	Riesling
		Muscat
		Gewürztraminer
	Alsace Vendanges Tardives (VT)	Late picked grapes. Riesling, Gewürztraminer, Pinot Gris, Muscat

Table 2.16 Wines in France: area/type Jura and Savoy.

Region	Best known wine types	Main grape varieties
Jura (red)	Arbois AC	Trousseau
(white)	L'Etoile, AC	Savagnin/
	Arbois, AC	Chardonnay
Vin Jaune	Also Chateau Chalon, AC	Savagnin
Savoy (white)	Rousette de Savoie, AC	Altesse/Chardonnay (Roussette)
	Seyssel, AC	(Roussette)
(red)	Vin de Savoie, AC	Pinot Noir

between varieties of a given species (as in *Coffea arabica*, between Bourbon and Typica) grown under similar conditions. With wines, we are generally considering only one species of grape, *Vitis vinifera, L.*, but there are a large number of varieties within that species, as already noted. Chemical composition differences between some grape varieties have been published, either in total or more likely with reference to particular groups of substances. There can also be differences, however, within a given variety, due to the location of vines (climate, soil, etc.), the different agronomic practices (pruning, yield, etc.) and degree of ripeness at harvest. Of more interest for observing changes in subsequent vinification and maturation is the chemical composition of the grape juice (*must*) immediately prior to fermentation, without any additions. The must for red wine will be very similar in percentage composition terms to that of the originating grapes, since it consists of the crushed grapes (see *Vinification* in Chapter 1). However, the clear juice for white wine production is free of skins, pips, etc. and during the clarification of

Table 2.17 German wines.

Region	Major districts (Bereich)	Best known wine types	Main grape varieties
Ahr	Walporzheim-Ahrtal	Red wines White wines	– Riesling (slow ripening)
Baden (widely separated areas)	Ortenau Kaiserstuhl	Red White generally good quality Rose	Pinot Noir (Spätburgunder) Ruländer (equivalent to Pinot Gris, Gerwürztraminer, Müller-Thurgau Silvaner
Franken	Würzburg	Dry white wines	Silvaner Müller-Thurgau
Mittelrhein	Bacharach	White	Riesling
Mosel	Bernkastel Piesport (village)	White	Riesling
Nahe	Kreuznach Bad Keuznach	White Middle slopes provide best Kabinett Spätlese	Riesling
Rheingau	Johannisberg	White Best quality German wine – Hock	Riesling
Rheinhessen	Bingen Nierstein	White, mixed qualities, e.g. Nierstein Vineyard Grosslage Gutes Domtal, Liebfraumilch	Riesling also Silvaner, Müller-Thurgau, Scheurebe
Pfalz (The Palatinate)	Mittelhaardt Deidesheim (village) (SW) Südliche Weinstrasse	White White (Liebfraumilch)	Riesling Scheurebe Müller-Thurgau
Württemberg		Red and white: small production for local consumption	Trollinger

the white juice most insoluble solids are removed. The soluble solids of the musts in either case are clearly important, as they ultimately affect the subsequent wine flavour obtained.

Water, of course, is the main component of the grape, amounting to some 70-80% by total weight, though in a clear must, the water content is even higher at around 83%. Fresh grapes weigh about 200-225 g per 100 berries. Grapes will contain 15-25% w/w 'sugar', constituting the reducing sugars,

Table 2.18 Italian wines.

Region	Wine town	Best known wine types	Main grape varieties
White			
Piedmont	Turin and Asti	Asti, DOCG (sparkling wine)	Moscato (white)
Tuscany	Sienna	Vernaccia Di San Gimignano, DCOG	Vernaccia
Latium	Rome	Frascati, DOC	Malvasia and Trebbiano
Veneto	Venice	Soave, DOC	Garganega (now using Chardonnay also), Trebbiano
Friuli-Venezia-Guillia		Friuli, some DOC	Pinot Bianco, Riesling Italico Chardonnay and Sauvignon Local (NB not Totcoy) Friulano
Umbria	–	Friulano, DOC (aromatic) Orvieto, DOC (dry wine)	Friulano Greshetto and Malvesia Trebbiano
Trentina Alto Adige (South Tyrol)	Bolzano	Trentino, DOC	Pinot Bianco and Chardonnay Riesling Renano, Riesling Trentino
Red			
Piedmont	Turin Alba	Barberesco, DOCG Barbera, DOC Barolo, DOCG (aged) Dolcetto, DOC (to be drunk young)	Nebbiolo Barbera Nebbiolo Dolcetto
Tuscany	Siena	Brunello di Montalcino, DOCG (4 year ageing)	Sangiovese Sangiovese
	Florence	Chianti DOCG or DOC (range of qualities, Chianti Classico, the best) Also the super – Tuscans	(also 2–5% white grapes) Cabernet Sauvignon (10%)
Umbria	Perugia	Torgiano DOC (Rubesco trade name)	Rubesco Sangiovese, Canaiolo
Emilia-Romagna		Lambrusco, some DOC	
Veneto	Verona	Valpolicella, DOC Breganze, DOC	Local grapes, such as Corvione, Rondinella and Molinara Dried grapes
	Vicenzo		Mostly Merlot
Friuli-Venezia-Guilia	–	Grave del Friuli, DOC	Cabernet Franc and Merlot
Trentino-Alto Adige (South Tyrol)	–	Trentino, DOC	Local grapes such as Lagrein, Teroldego

Table 2.19 Spanish wines.

Region	Best known wine types (red)	Main grape varieties
Central to North Spain		
Rioja	Rioja, DOC	Tempranillo,
Rioja Alta	Styles	Garnacha, Some
Alavesa	Joven, to be drunk young	Graciano and Mazuelo
Baja	Aged – American oak (Crianza, Reserva, Gran Reserva)	(Cariñena)
NW Spain		
Ribera de Duero	Ribera del Duero, DO	Tempranillo also Cabernet/Merlot/ Malbec
Central to Southern Spain	Valdepeñas, DO	Cenibel (Tempranillo)
NE Spain		
Navarra	Navarra, DO	Tempranillo, also Garnacha also Cabernet Sauvignon
Penedés (Catalonia)	Penedés, DO	Cabernet Sauvignon
Canary Islands	Tacorante – Acentejo, DO	Listan Negro (red) and others Listan Blanco (white) and others
Best known wine types (white)		
Central Spain		
La Mancha	La Mancha, DO	Airén
Rioja	Rioja, DO (oak aged)	Viura
NE Spain		
Penedés	Penedés DO Cava, DO	Chardonnay (only) Champagne – type (Viura = Macebeo)
Catalonia	Cava, DO	Champagne – type
Rueda	Rueda DO	Verdejo

glucose and fructose. Glucose is often referred to as 'grape sugar'. Partially dried grapes contain a higher percentage of sugars, say 40%. When fermented, these sugars provide the ethyl alcohol of wine, but they are also the precursors of many of the flavour/aroma compounds as will be seen later. There are very small quantities of other sugar types, but these are non-fermentable (like arabinose) and hence these sugars remain in the wine. There is a small proportion of nitrogenous matter (e.g. protein and free amino acids) at 0.3–1.0%

Table 2.20 Wines from the USA.

Region	Area	Best known vineyard areas	Main producers	Main grape varieties
California	Sacramento San Francisco	Amador County Napa Valley (with subregions) Main Californian area	Mondavi Berringer	Zinfandel (red) Chardonnay (white) Cabernet Sauvignon Merlot, Zinfandel (red)
	Central	San Joaquin Valley (jug wines)	E and J Gallo	–
	San Francisco (Russian River)	Sonoma County (oldest winery, 1857) San Luis Obispo	Sonoma-Cutrer Sonoma Vineyards	Cabernet Sauvignon/ Chardonnay, and late harvest Rieslings
Central coast	San Francisco to Santa Barbara	Saratoga Santa Cruz Salinas Valley	Paul Masson	Zinfandel, Chardonnay
Oregon	Williamette	Williamette Valley	Adelsheim	Pinot Noir

Table 2.21 Wines from Australia and New Zealand.

Region	Best known wine areas	Main producers	Main grape varieties
Australia			
South Australia (Capital: Adelaide)	Coonawarra Southern Vales Clare Watervale	Wynns Henschke Adams	Cabernet Sauvignon (red) Chardonnay (white) Riesling (white)
New South Wales (Capital: Sydney)	Hunter Valley	Upper area dominated by Rosemount (Treasury) Lower area is the traditional quality area location	Shiraz and Merlot (reds) Shiraz (red) Pinot Noir (red) Sémillon (white) Chardonnay
Victoria (Capital: Melbourne)	Murray Valley (N) Goulburn Valley	Lindemans Château Tahbilk	Chardonnay Cabernet, Shiraz
Western Australia (Capital: Perth)	Margaret River Swan Valley	Houghton	Chenin Blanc Cabernet, Chardonnay
New Zealand			
South Island	Marlborough	Brancott Estate	Sauvignon Blanc
North Island	Hawkes Bay		

Table 2.22 Wines from South Africa, Chile and Argentina.

Region	Best known wine areas	Main producers	Main grape varieties
South Africa Cape of Good Hope	Constantia Stellenbosch Paarl Worcester/ Robertson	Bellingham Warwick Estate KWV (cooperative) Kanonkop	Chenin Blanc, formerly Muscat Cabernet Sauvignon Pinotage, Colombard (white), Cinsaut (red)
Chile Central Valley	Maipo Maule	Santa Rita Concha y Toro Terra Noble	Cabernet Sauvignon, Merlot (red) Carmenieres
Argentina	Mendoza San Juan	Peraflor	Malbec Criolla/Grande (also Chica)

and these are of particular importance in wine-making, for example as flavour precursors. However, protein can also give a haze in wines, leaving the wine unacceptable for consumption. The pigments of the grape are mostly located in the epidermal cells of the skin, and are part of the phenolic and anthocyanin compounds, which will be discussed in more detail later. The skin also contains some of the volatile compounds that contribute to the wine aroma, especially the terpenes responsible for so-called 'aromatic fruity – grapey' flavour.

There is even a small quantity present of phospholipids and glycolipids (vegetable oils linked with phosphor group or sugars respectively) based on the unsaturated fatty acids, linoleic and linolenic acids. On breakdown, these are flavour and aroma precursors. The skins also contain carotenoids. Pectins are present in grape juice and can be a source of problems during vinification, since pectins can make wines difficult to filter. Of the non-volatile acids, tartaric acid is almost unique to the grape and it is in part responsible for the special properties of wines; grapes also contain malic acid (the apple acid) and citric acid (characteristic of oranges/lemons). There are also inorganic constituents of the grape, metal cations such as K^+, which will be evident on ashing grapes in a furnace.

The most complicated chemical group of substances present is the phenol-based group, which includes the so-called flavanoids and non-flavanoids. These are discussed fully in Chapter 3, since they are responsible for the basic taste characteristics such as bitterness and astringency in the finished wine. One group of phenols, the anthocyanins, are the red pigment responsible for the grape, must and wine colour, of red wine. The term *tannin*, though somewhat loose, is often used to describe many of these compounds, which also include some volatile phenolic compounds, that contribute to wine aroma. Grapes also contain a surprising quantity of volatile compounds, e.g. terpenes, some of which, as already noted, will appear in the finished wine, though it is important to emphasize that many new volatile compounds, are generated in the vinification processes in great amounts.

The chemical composition of must or grape juice (and also of the wine) is conveniently expressed in grams per litre, (gL^{-1}, clear juice) of each known non-volatile constituent or group of constituents. The volatile substances will only be present in parts per million or less, and therefore their content is expressed as mg, μgL^{-1}, or even ngL^{-1}. The most important non-volatile substances will be the sugars, which should be present in white wine must, generally above the level of $215\,gL^{-1}$ of solution (see Chapter 3). Similarly, for red wine the sugar concentration determined in clarified must needs to be similar as for white wine must, since this sugar content is reflected directly in the eventual ethyl alcohol content after fermentation. This content of sugar can be measured in a number of different ways, mostly simply by the use of a refractometer reading on the degrees BRIX scale (strictly speaking, for sucrose sugar content) together with conversion tables, which actually give the concentration of sugar (calculated as sucrose) present in g sugar/100 g solution (i.e. % w/w).

Non-volatile acids in must will amount to about $5\,gL^{-1}$ tartaric acid, $5\,gL^{-1}$ malic (1-2$\,gL^{-1}$ in southerly regions) acid and $0.3\,gL^{-1}$ citric acid, giving rise to a pH of 3.3-3.4 (Ribéreau Gayon *et al.*, 2006). These amounts will greatly depend on the ripeness of the grapes.

Data on the volatile compounds of must is somewhat rare, but Vernin *et al.* (1993) have investigated the must from Mourvèdre grapes and reported about 100 different compounds, not all identified, including terpenes, importantly the C_6 aldehydes and alcohols but very few esters (aliphatic or benzenoid) that typically characterize finished wines.

Bourzeix (1983), quoted by Jackson (2008), has provided some interesting quantitative data on the total phenolics content and total anthocyanin content separately, on some 14 different types of grapes. The highest in both categories was found in the Colombard variety, at 10 and $9.9\,g\,kg^{-1}$ respectively fresh weight, whilst the variety Grenache showed 3.6 and $1.2\,g\,kg^{-1}$ respectively.

Immediately before vinification, sugar (sucrose) may be added to the must to ensure 10-12% v/v ethyl alcohol content after the fermentation process, by pre-testing the sugar concentration as described above and further in Section 2.6.5. Acidity may also be adjusted at this stage by adding tartaric or malic acid to decrease a high pH, or by adding calcium carbonate to increase a low pH. Sulfur dioxide is usually added at the must stage at around 50 ppm or less (see Chapter 1) to inhibit oxidizing enzymes and encourage a clean fermentation. All such additions are under legal control in most wine-producing countries (e.g. in the EU, see Section 2.6). Undesirable oxidizing enzymes, which cause browning, specifically very undesirable in white wines, appear to be associated with particulate matter. Pectin-hydrolyzing enzymes may therefore be added to reduce this particulate matter. Various clarifying agents used before or after fermentation will be largely removed by filtration and fining procedures before the wine is bottled.

In the must from 'botrytized' grapes (Noble Rot), certain chemical composition differences from 'normal' must will be noted, as discussed in Chapter 7.

2.5.2 Finished wine

Quality control procedures mentioned in Sections 2.5 and 2.6 encompass a number of factors regarding composition, though not all. A typical content range of actual composition of the more significant constituents is shown in Table 2.23.

The substances present in finished wine will include those from in-cask ageing if practiced, though not necessarily from in-bottle ageing unless specifically mentioned. They are divisible into the non-volatile compounds, responsible for the basic tastes of the wine, i.e. bitterness, sweetness, acidity and the mouthfeel of wine, often referred to as astringency (discussed in detail in Chapter 3), the stimulant compounds and the volatile compounds responsible for the aroma components of the total flavour (discussed in detail in Chapter 4). A number of the compounds, often included in the *tannins* group (Table 2.23) are responsible for the colour of wines and they are discussed separately.

In respect of the flavour of the wine, the sweetness (due to residual glucose and fructose but also the glycerol and possibly even the terpenes), the acidity (due to the non-volatile tartaric, malic and citric acids but also the volatile acetic acid) and the bitterness/astringency (due to the 'tannin' substances listed in Table 2.23) all significantly contribute to wine character. Especially important, however, is the volatile compound content, or the so-called aroma, which is divisible into three groups of flavour significance: (1) primary, originating in the grape, mainly terpenes as in Muscat grapes; (2) secondary, generated during fermentation, by the particular enzymes present in the yeasts, essentially for the production of ethyl alcohol; and (3) tertiary, which arise in the maturing stages of the wine, especially when the wine is kept in oak barrels.

2.6 Quality control and classification of wines

An important aspect of wine production is in Classification, Quality Control and Appraisal systems, together with detailed analytical procedures. Such systems exist in all wine-producing countries and originated in France. Quality control and chemical analytical procedures are particularly prominent in the USA (California). Australia has introduced a system ensuring grapes come from a certain area. Classification systems in France, Germany, Italy and Spain primarily focus on the concept of origin (or *terroir*).

2.6.1 France

In France, the top level of classification enabling appropriate bottle labelling for both white and red wines is the Appellation d'Origine Contrôlée (or AC preceded by the name of the particular region or wine-producing area

Table 2.23 Typical range of substance content in wines.

Component		Typical range found (g L^{-1})
Sugars	Glucose	0.2–0.8 (dry)[a], up to 30 (sweet)
	Fructose	1–2 (dry), up to 60 (sweet)
	Arabinose	0.30–1.0
Alcohols	Ethyl alcohol	72–120 (9.1–15.1% v/v)
	Glycerol	5–15
	Butan,2,3-diol	0.3–1.5
	Inositol	0.2–0.7
	Sorbitol	0.1
Acids (non-volatile)	Total	3.5–15.0 (some present as salts with metallic cations)
(volatile)	Acetic	0.5–1.0 (can be higher in 'spoilt' wines)
Metal cations	Potassium	0.5–1.5
	Sodium	0.03–0.05
	Magnesium	0.05–0.15
	Calcium	0.05–0.15
Tannins (Folin Cicolteau test)	Total	*Content in mg L^{-1}*

		Wine type	Range	Average
		White	40–1300	360
		Red	190–3800	2000
		White – dessert	100–1100	350
		Red – dessert	400–3300	900

Component		Typical range found
Other volatile substances	Total	0.8–1.2 (c. 50% consisting of fused oils; 2-methyl propanol, 2-methyl butanol, 2-phenyl ethanol) the other 50% covering about 400 substances, approximately 10 mg – 0.1 µg each)

Data adapted from Peynaud (1986), Ribéreau-Gayon *et al.* (2006), Amerine & Ough (1974).
[a]EU definition of 'dry', <9 g L^{-1} sugar.

applicable). This system is controlled by the 'Institut National de l'Origine et de la Qualité', as initially organized in 1936. It is based upon a number of factors, for which particular rules apply regulating wine production, as follows:

(1) Geographical. Local factors of soil and climate for grape production are specified.

(2) Grape variety. Each appellation for a given area will have one or more grape varieties permitted, according to proven and historic practice, e.g. only Pinot Noir is allowed in most red Burgundies, like Beaune, AC, though Châteauneuf-du-Pape AC may use 13 different red and white grape varieties.

(3) Alcoholic content. There is usually a minimum alcohol content, dependent upon the district but generally 10–12% v/v is required. It can be achieved

by sugar addition to the must before fermentation in the more northerly regions, where in cool years ripeness of the grapes may not be achieved.

(4) Yield of wine. This is usually expressed per area of vineyard used for grape growing. All appellations have maximum allowable yields, usually 45–50 hL of wine per hectare of vineyard, although in some AC areas the permitted yield is higher, though it can be as low as 25 hL ha^{-1}. Dispensations can be granted from time to time. (NB 50 hectolitres = 66 666 bottles (at 750 ml), and 1 hectare = 2.4 acres.)

(5) Vineyard practice. Control of number of vines per hectare, method of pruning and even the method of picking. In southern regions, irrigation is also regulated.

(6) Methods of wine-making. In particular, monitoring methods of controlling acidity are laid down and certain methods, like ion exchange, may not be allowed.

(7) Analysis. An official taste panel that may reject faulty or untypical wines for the particular AC. Chemical analytical methods are discussed in Section 2.6.5.

In addition, there are two lower levels of wine classification: firstly, Vin Délimité de Qualité Supérieure (VDQS) and secondly, Vin de Pays introduced in 1973. The VDQS category, historically used for wines aspiring to AC status, is being phased out. Vin de Pays has a regional definition but the rules are more flexible, although quantity, alcoholic strength, area, etc., are defined. Wines also have to be assessed by a taste panel. Vin de Pays comes in three categories; regional, departmental and locally specific - the smallest areas. All others are Vin de France.

A much earlier system of classification first devised in 1855 and still applicable virtually in the same form and content, is used for some of the finest red and white wines in Bordeaux. This system applies to particular designated vineyards or rather châteaux, which are classed according to their *cru* (literally translated into English as 'growth') in five groups (*premier to 5 ème*) for reds, and three groups for white wines (Sauternes). AC appellations still apply but this classification is meant for an extra dimension of quality appraisal. The same châteaux may, however, also set aside part of their production, which is not of 'cru' standard; these become so-called second wines, with slightly different châteaux names. Certain other châteaux may be classed as *Cru Bourgeois*, a classification outside the classed growths, though they may in the opinion of some wine experts be classed with them.

The wines in Burgundy, apart from AC, etc., are classified in somewhat different manner. The Burgundy vineyards have an AC ranking, such as grand Cru, Premier Cru, Commune, etc. Differences in classification are to a large extent historical, since large holdings like the châteaux in Bordeaux were not broken up during the French Revolution of 1789. In contrast, the Burgundy vineyards are owned by numerous different growers.

2.6.2 Germany

Labels on German wines are generally regarded as being rather confusing. Essentially the German wine law introduced in 1971 uses a quality classification based on ripeness (sugar content) of the grapes. This is not surprising, since in this cool climate the conditions need to be just right to attain ripe grapes. A vineyard classification of the top sites, *Erste Lage* and *Grosses Gewächs*, was published in 2002. There are two Quality Law levels, one defining the size of the area, the other the level of grape maturity.

These are the two Quality Levels: QbA and Prädikatswein:

(1) QbA (Qualitätswein Bestimmter Anbaugebiete)
This is by far the most dominant quality category and includes all Liebfraumilch and basic Grosslage wine blends. It means 'quality wine from designated regions', i.e. one of the following specified regions:
  (a) Ahr (Northerly region near Bonn)
  (b) Mittelrhein (between Bonn and Koblenz)
  (c) Mosel – Saar – Ruwer
  (d) Rheingau (around Mainz, near Wiesbaden)
  (e) Nahe (along River Nahe, tributary of the Rhine)
  (f) Rheinhessen (biggest, fairly central region)
  (g) Pfalz (West of River Necker)
  (h) Hessische Bergstrasse (above Heidelberg)
  (i) Franken (between Frankfurt and Würzburg)
  (j) Württemberg (above Stuttgart)
  (k) Baden (most southerly region, close to Alsace)
  (l) Sachsen (Former E Germany)
  (m) Saale-Untrue (Former E Germany)
Inside these 13 regions, there are three groupings:
      (i) Einzellage. (Single location, about 3000 names.) Refers to production from a single vineyard, and therefore expected to have the highest quality.
     (ii) Grosslage. This name refers to a group of vineyards of similar type, based on one or more villages. Grosslage names (smaller number than above) can refer to a very large area and can refer to poorer quality wines.
    (iii) Bereich (district). An amorphous term refers to a collection of villages, though often using a single village name and vineyard name, e.g. Bernkasteler Schlossberg, Schlossberg (Berg – Castle) is a vineyard in the Bernkastel district. Bernkastel is the whole of the Mittel-Mosel. Similarily, Nierstein (Rheinhessen) and Johannisberg (Rheingau) are the names of both villages and districts. Hochheim (Rheingau) is a wine town with several vineyards. Deidesheim (Pfalz) is also a wine village with several vineyards.

(2) Prädikatswein

This is a higher quality designation, translating as 'quality wine with special attributes' (or predicates). This part of German wine law is especially concerned with minimum grape ripeness levels and therefore 'sweetness'. Prädikatswein wines are not allowed to be chaptalized. Süssreserve may be used for Kabinett and Spätlese. The following wine categories are classified in ascending order of sweetness:

(a) Kabinett. Least ripe, made from ordinary ripe grapes from a normal harvest.

(b) Spätlese (late harvest). From late picked and therefore riper grapes. Can provide either dry to medium dry wine.

(c) Auslese (selected choice). From selected, particularly ripe, late picked grapes, and may be affected by Noble Rot. Usually a medium-sweet wine.

(d) Beerenauslese (BA) (berry selected). Also made from specially selected very ripe grapes, and usually affected by Noble Rot-dependent upon variety. Very sweet wine sold at high price.

(e) Trockenbeerenauslese (TBA) (dried grapes, which were late picked). Made from even riper grapes than BA, affected by Noble Rot, which causes shrivelling. To qualify, there must be 150 Oeschle degrees of sugar content (the German potential alcohol system), equivalent to 21.5% v/v potential alcohol, but not actual. The finished wine will be very sweet and also expensive.

(f) Eiswein. A wine from grapes picked at dawn in winter, so that the grapes are frozen pellets of ice and which may only be made from BA and TBA quality grapes. Removal of ice prior to vinification gives a very sweet, concentrated wine.

Two other qualities should be mentioned:

(1) Landwein. A 'Vin de Pays' is from one of 20 designated areas, and must be Trocken (dry) or Halbtrocken (semi-dry).

(2) Deutscher Tafelwein. No specific vineyard origin. Usually sugared must has been used.

2.6.3 Italy

The DOC Laws are not unlike the French system. However, some of the growers disagree with the restrictions imposed by these laws, and therefore some very good quality wines are in the lowest quality denomination.

Only 10–12% of the Italian wine harvest is strictly regulated, and most is designated as Vino de Tavola, outside the strict regulation. The regulated wines are:

(1) Vino de Tavola. This is applied to table wines which have some reference to place, colour and maybe grape variety but they do not qualify for DOC status. However, some quality wines can be included. Table wines with

Indicazione Geografica Tipica are allowed to use area of production, grape variety and production year on the label.

(2) Denominazione di Origine Controllata (DOC). This is designated for specified wine varieties, grown in delimited zones. Wines are vinified by prescribed methods and aged by prescribed methods to certain standards.

(3) Denominazione di Origine Controllata e Garantita (DOCG). This is a top tier introduced in (1982), with more stringent restrictions on grape types, yields and chemical and sensory analysis than DOC wines.

There are two other specified terms:

(1) Classico - meaning the best parts of the original region.
(2) Superiore - reference to ripeness.

Other label terms are used such as Riserva (generally denoting prolonged ageing), Spumante (sparkling) and Frizzante (highly sparkling).

2.6.4 Spain

Spain has the largest acreage of vineyards in the world but only a small number of demarcated regions, which are called Denominación de Origen (DO). DO is akin to the French system and based on suitable terrain, permitted grape varieties, restricted yields and approved methods of vinification.

It also has an equivalent of Vin de Pays, i.e. 'Vino de la Tierra', which are wines with a local character, 'Vino de Mesa' is Table wine (as opposed to both foregoing); this is the basic category, many wines being blends.

Some other label descriptions are of significance, especially those related to ageing:

- Joven (or sin crianza). Young wine, with minimal ageing.
- Crianza. A wine that has undergone a legal minimum period of ageing in-barrel and vat/bottle.
- Reserva. Wines with a longer legal minimum period of ageing in-barrel and bottle.
- Gran Reserva. Top of range, wines aged for longer than a Reserva.

2.6.5 Australia

Blending is an acceptable practice, in order to get the required fruit characteristics in a wine, blends of wine from different vineyards or even regions are quite common. The appellation control laws introduced are perfectly adapted to encompass blending. In 1993 Australia introduced the term Geographical Indication (GI), and these are classified into:

- South East Australia, the states.
- Zones (which form part of the states).
- Regions (which form part of the zones).
- Sub-regions (which form part of the regions).

The GI claimed on the bottle depends on the origin of the grapes. It is not uncommon in Australia to move the cooled grapes a considerable distance from the vineyard to the winery and grapes from a large area can be used for any particular wine. Hence the claim to be used depends on the area from which the grapes came: if all are sourced from a sub-region that is the claim to be made. Once a GI claim is made on the label, only 85% of the fruit needs to come from the GI area, although more than one area can be claimed on the label, specifying proportions of grapes in the blend.

Some of the zones are large, such as South Eastern Australia, hence grapes could be sourced from any of the vineyards in New South Wales, Victoria, Tasmania, even parts of Queensland and South Australia. A region is much smaller, but must contain at least five independently owned vineyards, and there are criteria on the production, such as size and grape attributes, while a sub-region within the region has more stringent criteria.

The GI system is not dissimilar to the European Appellation systems in use, although it is less restrictive regarding both viticultural practice and wine-making. Both systems aim to protect the use of the regional name, much as many European regional names are protected for wine, for example the on sparkling wines not from the Champagne region in France, the words 'traditional method' have to be used rather than 'Champagne method' to indicate that fermentation took place in the bottle. Currently Australia has about 60 wine regions, with 103 Geographical Indications for wine areas, covering zones, regions and sub-regions.

2.6.6 USA

The USA has introduced the Appellations of Origin, and this can be the name of a:

- state;
- county or counties;
- American Viticultural Area (AVA).

If a name of a state is claimed as the appellation of origin, 100% of the grapes must have been sourced within the state. A county appellation of origin needs to have sourced 75% of the grapes from the claimed county. Up to three counties can be used on the label and the percentage of grapes sourced from each needs to be stated on the label. For an AVA 85% needs to come from the claimed area.

The system of AVAs is regulated by the government, however, it does only approve the area as being distinct from other areas but it does not signify any quality differences between the grapes or wines. Distinguishing features of an area can be geographical, such as soil, climate, elevation, but can also be the result of historical information. There are currently 157 AVAs throughout the USA, most of them (94) are in California.

2.6.7 Quality control systems in the European Union

Reference should be made to Thompson (1987) where he discusses the European system and the system used in the USA in detail. The 1954 International Convention for the Unification of the Methods of Analysis and Appraisal of Wines published in the *Recueil des Méthodes Internationale d'Analyse de Vins* (1962–1973) provided the original compendium of procedures that formed the basis of most of the methods, now used throughout Western Europe. Many of these methods were included in the Annex to EEC Commission Regulation 2676/90 published in the *Official Journal of the European Communities*, 1990, L272, last amended by Commission Regulation No 1293/2005 (OJ No L205, 6.8.2005). This is a detailed document with precise descriptions of basic wine analyses to be used throughout the EU. They are the required methods when it is necessary to verify particulars given on documents relating to wines, for all commercial transactions and QC procedures. The reader should refer to the latest EU publications. All the wine regulations currently enforced in the EC are listed in Statutory Instruments (2009), No 386, Agriculture, England.

Methods are given to determine density and specific gravity at 20°C, alcohol content by volume (also at 20°C), total dry extract (expressed in terms of sucrose), reducing sugars, sucrose, ash, total and volatile acidity, tartaric and citric acids, free and combined sulfur dioxide, sorbic and ascorbic acid, together with the cations sodium, iron, copper and potassium. A method is also included for carbon dioxide content, based on the filtration of an alkaline sample in the presence of carbonic anhydrose from pH 8.6 to 4.0. Test kits for rapid methods for some of these factors have also been developed. Many of these factors feature in the Appellation Contrôlée and other systems already mentioned.

Sulfur dioxide addition is an important aspect of vinification. Thompson (1987) reports that it has been recommended that, after pressing, the juice be analysed for total and free SO_2, pH and titratable acidity, so that the SO_2 can be adjusted according to pH, so that less SO_2 is used at lower pH values. When pectin-hydrolyzing enzymes are added to must, followed by clarification, the free SO_2 is again checked and adjusted as necessary. Sufficient free sulfite must be present to restrict oxidation during processing (compounds formed in fermentation can bind to SO_2 and render it ineffective) and inhibit the development of infecting micro-organisms. Nevertheless, the tendency is to minimize the use of SO_2, for example, malo-lactic fermentation is encouraged in red wines, thus improving biological stability. Just before bottling, the level of

SO_2 may be again adjusted, to about 30 ppm free SO_2, but the relevant national legal limit for total SO_2 must not be exceeded. These national upper limits vary widely; the current EU permitted levels are: 150 mg^{-1} for red wines (160 mg L^{-1} for wines harvested before 2009) and 200 mg L^{-1} for dry whites (210 mg L^{-1} for wines harvested before 2009). Sorbic acid can be used up to 2000 ppm, but it is not very satisfactory in terms of anti-bacterial activity and can give an off-flavour in wine. The relatively low pH of wine inhibits the growth of most micro-organisms but it does not inhibit acetic and lactic acid bacteria. Some of the constituents of wine, notably ethyl alcohol, are also inhibitors of micro-organisms but the modern trend is to keep wine in sound condition by ensuring plant hygiene throughout fermentation and maturation, controlling the temperature during the entire process, together with sterile bottling and filtration.

Quality control is directed to the flavour of the finished wine, the judgement of which primarily resides in the mouths and noses of expert tasters. Modern QC techniques for volatile compound determination are, however, increasingly playing a role, together with multivariate statistical analysis of data to identify features which are desirable or undesirable, or differences between wines. However, it is not yet possible to include such techniques as a major part of quality control management, since these chemical analyses can 'only' identify compounds, including their concentration, up to now they cannot give much information regarding the flavour of wines but see also Chapter 5.

Bibliography

General texts

Clarke, O. (1989, 1998 and 2001) *Wine Guide*. Webster/Mitchell Beazley, London.

Clarke, O. & Rand, M. (2001) *Grapes and Wines*. Webster, London.

Galet, P. (1998) *Grape Varieties and Rootstock Varieties*. Oenoplurimédia, Chaintré.

Grainger, K. & Tattersall, H. (2005) *Wine Production Vine to Bottle*. Blackwell Publishing Ltd., Oxford.

Jackson, R.S. (2008) *Wine Science, Principles, Practice, Perception*. 3rd edn. Academic Press, San Diego.

Johnson, H. & Robinson, J. (2001) *World Atlas of Wine*. 5th edn. Mitchell Beazley, London.

Peynaud, E. (1986) *The Taste of Wine*. Macdonald, Orbis, London.

Ribéreau-Gayon, P., Glories, Y., Maujean, A. & Dubourdieu, D. (2006) *Handbook of Enology, Volume II, The Chemistry of Wine, Stabilization and Treatments*. John Wiley & Sons, Ltd, Chichester.

Robinson, J. (1993, 1995) *Jancis Robinson's Wine Course*. BBC Books, London.

Web general information

L'Alliance des Crus Bourgeois du Médoc (2009) *Crus Bourgeois du Médoc*. http://www.crus-bourgeois.com/documents/selection_officielle_2008.pdf (accessed 4 May 2011).Vitis International Variety Catalogue (2011) http://www.vivc.de/index.php (accessed 4 May 2011).

References

Bourzeix, M., Heredia, N. & Kovač, V. (1983) Richesse de different cépages en composés phenolique totaux et en anthocyanines, *Prog. Agric. Vitic.*, **100**, 421–428 (1983), quoted by Jackson, R.S. (2008) loc. citation, p. 285.

Conde, C., Silva, P., Fontes, N., Dias, A.C.P., Tavares, R.M., Sousa, M.J., Agasse, A., Delrot, S. & Geros, H. (2007) Biochemical changes during grape berry development and fruit wine quality. *Food*, **1**, 1–22. Global Science Books.

Huggett, J.M. (2006) Geology and wine: a review. *Proceedings of the Geologists' Association*, **117**, 239–247.

Stevenson, T. (1997) *Sotheby's World Wine Encyclopedia*. Dorling Kindersley, London.

Thompson, C.C. (1987) Alcoholic beverages and vinegars. In: *Quality Control in the Food Industry*, Vol. 4 (ed. S.M. Herschdoefer), Section 5 'Wine', pp. 57–64. Academic Press, London.

Vernin, G., Pascal-Mousselard, H., Metzger, J. & Parkanyi, C. (1993) Aromas of Mourvèdre wines. In: *Shelf-Life Studies of Foods and Beverages* (ed. G. Charalambous), Developments in Food Science No. 33, pp. 945–974. Elsevier, Amsterdam.

Chapter 3
Basic Taste and Stimulant Components

3.1 Introduction

The stimulant and basic taste components of wines contribute important characteristics to their overall quality. Clearly, the stimulant properties of its main alcoholic component, ethyl alcohol, are well known, whilst acid content determines their so-called acidity in wine tasters' parlance, and final sugar content determines 'dryness' or sweetness. Of the other basic tastes, any bitterness is determined by a range of compounds of differing chemical composition, whilst saltiness, determined by mineral salt content, chiefly sodium chloride, is not important in wines, neither is the taste of umami. Astringency within the mouth as a whole is often an important characteristic, often confused with bitterness, with which it can be associated, since some compounds have a bitter taste as well as an astringent feel.

The compounds contributing to these basic tastes are all non-volatile substances soluble in water or in water/alcohol mixtures. They can be contrasted with the volatile substances responsible for the odour/aroma of the overall flavouring considered in detail in Chapter 4. Ethyl alcohol is a special case, as it is volatile, and in the quantities present in wines it can also contribute to sweetness. Whilst most taste components in their pure separate state are solid substances, others are liquid, essentially non-volatile at room temperature and standard pressure, e.g. glycerol, contributing some sweetness. Some acids such as acetic acid, which are slightly volatile, contribute to both acidity and some odour.

Composition of the non-volatile components is discussed in relation to the must, then to the finished wine; the particular changes occurring on ageing are discussed later in the chapter. A summary on basic taste perception, relevant to sweetness, bitterness and acidity of wine and astringency, particularly relevant for mouthfeel related perception terms, discusses the main research findings to date.

Wine Flavour Chemistry, Second Edition. Jokie Bakker and Ronald J. Clarke.
© 2012 Blackwell Publishing Ltd. Published 2012 by Blackwell Publishing Ltd.

3.2   Basic taste perception

3.2.1   Role of taste

The scientific knowledge of taste recognition and perception has increased significantly over the last decade, in part resulting from the discovery of two families of mammalian taste rerceptors for sugars, amino acids and bitter compounds. These developments are discussed in a number of recent reviews mainly on taste perception (Plattig, 1985; Mombaerts, 2004; Scott, 2004; Scott, 2005; Swiegers *et al.* 2005; Chandrashekar, *et al.*, 2006; Bachmanov & Beauchamp, 2007). Behrens, & Meyerhof (2006) focus on the bitter taste, while Da Conceicao Neta *et al.* (2007) discuss the sour taste. A brief summary of current knowledge is given in the next paragraph, reviewing current knowledge on taste recognition, taste receptors and the mechanisms of taste perception.

Overall our perception of taste and smell contributes greatly to enormous pleasure we get from eating and drinking. Arguably, wine-drinking would not be so popular a passtime if it were not for our sense of taste and smell. People often complain when they perceive that their sense of taste is less acute, typically blaming that the food or drink no longer tastes like it did in the past. However, our sense of taste also has other important functions, and as a chemical contact sense it offers the last chance for us to judge whether we will truly ingest a certain food or drink. Although the human tongue can recognize hundreds of different taste compounds, traditionally we only class them in four different groups: sweet, bitter, sour and salt. Our perception of glutamate or umami is recognized by numerous experts as the fifth taste.

The basic tastes serve an important dietary function. It is thought that taste is primarily used to evaluate the nutritious content of food and drink and to prevent the ingestion of toxic compounds. Taste cells detect sugars and amino acids at very high concentrations, presumably in order to allow us to detect food of high nutritional value. Sweet taste helps to identify energy rich nutrients and it enhances our enjoyment of food, umami helps to identify amino acids, salt helps the intake of required minerals, whilst sour and bitter taste perception help us avoid the intake of potentially dangerous compounds. For example, sweetness of sugar evokes a familiar pleasure reaction in humans, and we attribute these properties to sucrose, rather than the neuronal firing in the brain. It is thought that this relationship between sensory quality, positive hedonic response and behavioral acceptability illustrates how sweet taste detection and perception have evolved to help us recognize a basic source of metabolic energy. Interestingly, even though there are a number of sugars which elicit a sweet taste, we cannot distinguish between them. Presumably, there is no evolutionary pressure on us to do so.

However, there are also taste cells able to detect minute concentrations of compounds that are harmful. Hence taste cells are sensitive to a broad range of compounds, as well as able to detect a wide concentration range. Thus taste helps us enjoy food, as well as serving as a final check that food or drink is palatable and to avoid toxic substances. The bitter taste has evolved to help us prevent the ingestion of a very wide range of toxic chemicals, remarkably all being recognized as bitter. Even though we recognize a wide range of taste stimulants, we only are able to place them in five the taste sensations, however, these five basic tastes help us to recognize, distinguish, select and appreciate the main dietary components.

Although we are able to recognize ethanol, luckily we are not very sensitive to its presence, otherwise our enjoyment and appreciation of wine and other alcohol containing beverages would take on quite a different character, with the perception of the mostly volatile compounds being rather overwhelmed by the relatively high concentration of ethanol. However, the presence of ethanol has a theoretical effect on the volatility of compounds in wines, as discussed in Chapter 4 (Section 4.1), but there is also an effect on sensory perception of aroma compounds as a result of ethanol (Le Berre *et al.*, 2007).

3.2.2 Taste perception mechanism

Taste recognition takes place by specialized epithelial taste receptor cells, which are arranged in a group of 50-100 in so-called taste buds, having an onion-like shape. Taste buds are grouped in three different types of papillea; fungiform papillea are located mostly at the front of the tongue, foliate ones on the sides, and vallate ones at the back. One end of the taste receptor cells are exposed to the oral cavity, and when food is placed in the mouth they interact with the taste stimuli, resulting in the generation of signals transmitted to the brain via branches from three cranial nerves (facial, glossopharyngeal and vagus) and relaying the information in a topographical manner. Taste cells regenerate approximately every ten days, indicating that from a biological point of view taste fulfills a crucial role. As discussed above, the need for our taste receptors to be in optimum condition helps to ensure we select and eat adequate and safe foods.

According to Bachmanov & Beauchamp (2007) it is thought that the wide range of taste information in the brain requires integration of information from other sources, such as hunger and satiety with senses such as taste and vision, thus allowing the generation of a behavioural response to the taste compounds. Taste processing leads to the perception of different aspects of taste, such as judgement of quality, information regarding the hedonics (pleasant or unpleasant food or drink) and persistence. When drinking wine, either as an expert or just for enjoyment, this type of processing invariably takes place, and people form an opinion on the wine, taking into account an array of information, in part taken directly from the wine tasted, but in addition drawn from past experience, ambiance, mood, price, etc.

The emerging picture of taste coding is elegant in its simplicity. Since we can taste five different tastes, it is thought that each taste has specialized taste receptors and current data support this hypothesis. Chandrashekar *et al.* (2006) report that contrary to the long held general belief, it is now clear that distinct cell types expressing unique receptors are tuned to detect each of the five basic tastes: sweet, sour, bitter, salty and umami. Importantly, receptor cells for each of these five basic tastes function as dedicated sensors wired to elicit stereotypic responses, which are transferred to the brain.

Proteins from T1R and T2R families, often referred to as G-Protein Coupled Receptors, are involved in the perception of sweet and umami. The receptor proteins for sweet, bitter and umami are available for tastant compounds to bind to, this resulting very specific binding leads in a number of steps to the stimulation of neutron activation, giving the information to our brain. Bitter receptors are probably encoded by a large group of genes, thus recognizing a wide range of chemicals, which we can label as bitter, although we cannot distinguish between these compounds which is not important from a biological point of view. A family of approximately 30 G-Protein Coupled Receptors are reported to be involved in bitter perception, many appear to be expressed in the same taste receptor cell, further implying that the sensing of bitterness is important, but discrimination between them is not. In wines some phenolic compounds are thought to be able to contribute some degree of bitterness, and in general bitterness in wine is not a desirable sensory property.

Studies to date indicate that the salt and sour tastants are generally ionic (H^+ and Na^+) and act directly through ion channels on the cell membrane of taste receptors, using a different mechanism from the perception of sweet, bitter and umami. However, the physiology of sour perception is still not totally understood. The resulting neuron activation transmits signals to the brain. Excessive sourness is usually avoided, for example, in unripe fruits. However, as part of our food and drink an element of sourness or acidity is appreciated, and of course wine needs adequate acidity in order to be appreciated, as emphasized by numerous sensory terms to describe wine.

3.3   Ethyl alcohol

Ethyl alcohol (ethanol), C_2H_5OH, is by far the most important component of wine, and has long been known for its stimulant, if not intoxicating, properties. Its concentration in wines tends to range between 8 and 15 % v/v, and with the increased interest in picking grapes at optimum maturity, the trend has been for wines to be made with higher alcohol concentrations, especially in New World wines. More emphasis is given to the flavours and flavour precursors in grapes at harvest, crucial to the development of the bouquet of the wine, rather than just the sugar/acid ratio of the grapes.

The measurement of the amount present in a wine is a key aspect of vinification. Since there is no ethyl alcohol in fresh grapes or must, the content of the sugars from which it is derived during fermentation has also been used to predict the final ethyl alcohol content of the wine;– or rather, assessing the potential figure assuming fermentation will be complete. The fermentation of a monosaccharide sugar (like glucose, see Section 3.4) is most simply expressed by Equation 3.1.

$$C_6H_{12}O_6 \rightarrow 2C_2H_5OH + 2CO_2 \qquad (3.1)$$

glucose/fructose (MW 180) $\rightarrow$ ethyl alcohol (2 × MW46 = 92)
+ carbon dioxide (2 × MW44 = 88)   (3.1)

The equation shows that $92/180 \times 100$ (= 51.1%) is the theoretical percentage weight of the original glucose converted into alcohol (or the alcohol yield), with the remainder being carbon dioxide gas. Fructose has the same molecular formula as glucose, and can similarly be converted. In reality, the percentage conversion is lower for several reasons.

In the grape, some of the sugars may be combined with other components, such as pectin and not available for fermentation. The very small quantity of other sugars present, like arabinose, is not fermentable. If added sucrose (MW 342) is used, then conversion into glucose and fructose first occurs (with molecular addition of H_2O) but the theoretical yield on a weight basis will now be $184/342 = 54\%$. In practice, some other alcohols and polyols (in particular glycerol) will be generated in variable quantity; there will be some loss of ethyl alcohol from the fermenting vessels, carried away by the carbon dioxide. The actual overall yield will be considerably less than theoretical (see Equation 3.3). There is also a positive heat of reaction associated with the main reaction, with practical implications, e.g. large commercial scale fermentation vessels have to be designed to allow removal of excess heat (see Chapter 1).

3.3.1 Measurement of ethyl alcohol content in wines

Alcohol (ethanol) content in a wine is normally expressed in percentage terms by volume, i.e. % v/v, or more strictly speaking %/v/v$_s$ which is the ratio of the volume of ethanol present to the *total* volume of the aqueous solution, taken at 20°C. Actual content in commercial wines, which is also now stated on wine labels, will range 9–15% v/v, whilst the maximum possible by fermentation alone is considered to be about 17.5%. The French wine industry sometimes uses Gay-Lussac units (°GL) which, however, correspond with our % v/v numerical values. Whilst both chemical analytical and physical methods are still used to measure content in wines, nowadays, it is simpler and more accurate to use gas-chromatographic techniques. The earliest method, particularly for spirits, was by 'proving'. In British usage, 100° Proof spirit was that spirit which allowed gunpowder wetted with it still to burn; this Proof corresponds to some 57% v/v alcohol content. Higher or lower contents are expressed as degrees, Under or Over-Proof. In US terminology, 200° Proof = 100% v/v ethyl alcohol content, so that the terms, Under and Over, are avoided. Such technical terms are now little used in the wine or spirit industries.

Pure ethyl alcohol (ethanol) is a liquid, with a boiling point at 760 mmHg pressure of 78.3°C, with a specific gravity or density, measured at 20°C, of 0.7893 g ml^{-1} (relative to water at 4°C of 1.0000). The corresponding density of water at 20°C is 0.99823 g ml^{-1}. The numerical value for volume content percentage is always higher than the weight percentage, which is probably why it is favoured for use on wine labels but this value has little real physical meaning. The AOAC (Association of Official Analytical Chemists) has published (AOAC, 1970) a useful table of conversion units for various measures of ethanol content at 20°C over a range of contents relevant to vinification, 0–22.5%, with the corresponding specific gravity, set at 1.00000 for 0% ethanol solution (water). From this table, it can be seen that 10% v/v is equivalent to 8.047% w/w with a measured specific gravity of 0.98650; 15% v/v is equivalent to 12.138% w/w, with a specific gravity of 0.9871. The expression of ethanol content in g per litre solution (g L^{-1}), as with other components of wines to be described in later chapters, is also useful. Thus, 10% v/v = 8.047% w/w = 79.3 g L^{-1}, 0.032 mole fraction; and 15% v/v = 12.138% w/w = 119.0 g L^{-1}, 0.05 mole fraction.

Data for the use of refractometers is available; thus n^D_{20} is 1.3300 for water, and 1.3614 for 100% ethanol.

3.3.2 Measurement of sugar content in musts and wines

The accurate or semi-accurate measurement of the quantity of sugar(s) present initially in grapes and musts is especially important in vinification procedures. As the actual and significant sugars present are the monosaccharides, glucose and fructose, chemical analytical methods are used as standards (i.e. Lane-Eynon method for reducing sugars) for measuring their content. High-performance liquid chromatographic (HPLC) methods are now available to give fully detailed carbohydrate composition data, and are especially suitable in complex aqueous solutions. The chemical nature of carbohydrates or saccharides is discussed in more detail in Section 3.5.

However, in the long-established sugar (sucrose) refining industry, the relationship of specific gravity (or density) of aqueous solutions (conveniently measurable by hydrometers) with their concentration of soluble sugar (usually expressed in % w/w units) was established many years ago. Several different scales of measurement are available, the °Baumé scale, invented in 1784 and still used in Europe and Australia, the °Brix scale, invented in 1854, and still used in the UK and USA, and the °Oeschele scale, used in Germany. These scales were taken up by the wine industry, though the slight difference in specific gravity for glucose against sucrose at the same temperature and soluble solids content can cause confusion. The relationship between specific gravity and sugar content is generally linear at the lower carbohydrate concentrations.

The °Baumé (°Bé) scale was based upon 0°Bé for pure water, and 10°Bé for 10% w/w sodium chloride solution (density, 1.06937 g ml^{-1}), all at 25°C. Alternatively, °Baumé = 145 − 145/(SG.60°F/60°), where SG is the specific gravity of the solution compared with water, both measured at 60°F (15.5°C).

The Oeschele scale (°Oe) again starts at 0°Oe for pure water.

Brix scale

The Brix scale is used to express sugar content in g of sucrose per 100 g of *aqueous solution* and is measured by a hydrometer to determine the specific gravity or density; thus 20°Brix ~ 20% w/w sucrose in solution. A similar

Table 3.1 Relationship of sugar concentrations with solution density and refractive index at 20°C.

| Sugar | Concentration | | Density | | Refractive index |
	% w/w also °Brix[a]	g L⁻¹[b]	g m l⁻¹[c]	g m l⁻¹[d]	n^e_D
Sucrose	16	170	1.0635	1.0649	1.3573
	20	216	1.0810	1.0829	1.3609
	22	239	1.0899	1.0918	1.3672
	24	263	1.0990	1.1009	1.3706
	26	288	1.1082	1.1102	1.3741
D-glucose	16	170	1.0624	1.0643	1.3371
	20	215	1.0797	1.0816	1.3635
	22	238	1.0884	1.0903	1.3668
	24	261	1.0973	1.0992	1.3702
	26	287	1.1063	1.1083	1.3736
D-fructose	16	170	1.0640	1.0659	1.3569
	20	216	1.0816	1.0835	1.3634
	22	239	1.0906	1.0925	1.3667
	24	264	1.0996	1.1015	1.3700
	26	283	1.1089	1.1108	1.3734
Glucose–fructose	16	170	1.0651	1.0670	1.3570
(1:1 mixture)	20	215	1.0806	1.0825	1.3635
	22	238	1.0867	1.0886	1.3668
	24	264	1.0985	1.100	1.3701
	26	287	1.1076	1.1095	1.3735
Water	0	0	0.99823	1.0000	1.3300

[a,b,c,e]Data taken in part from *Handbook of Chemistry and Physics* (ed. Lide, 2001). [b]Calculated figures from density and % w/w concentration. [c]Density of solution relative to water (1.0000 at 4°C), i.e., the true density. [d]Density of solution relative to water (1.0000 at 20°C), i.e., the specific gravity, calculated from footnote [c].

relationship exists with refractive index values, so that a refractometer can be calibrated in Brix units. The relationships with glucose/fructose contents are similar but not identical, as can be seen from the data in Table 3.1. The temperature of measurement is important; the stated basis of calibration of the hydrometer or other device should correspond with the temperature of the solution being measured. Originally, measurements were made at 60°F (15.5°C), but now more usually at 20°C.

The density of sucrose solution is clearly somewhat higher than that for a glucose solution at the same concentration, expressed in % w/w units, but this is not so for fructose, typically found in a wine must from ripe grapes, which has a very similar density.

Jackson (2008) has reproduced the conversion table published by Amerine & Ough (1980) relating Brix-Balling values for the 'sugar content of must' with their Baumé and Oeschele scale equivalents over a concentration range of 0-30°Brix, all at 20°C. The specific gravities in this table are given to five decimal places, and are the same as the corresponding figures in Table 3.1 in Column d, calculated for a density of pure water of 1.00000 at 20°C. °Brix (Balling), which is the same as °Brix, is stated to express the total solubles in the must, but does express the sugar amount in grams of sucrose in 100 g of solution, at least for Brix values above 18°. This relationship is confirmed by inspection of Table 3.1.

Baumé and Oeschele scales

The conversion table of Amerine & Ough (1980) shows 10°Bé to be equivalent to 18°Brix at 20°C, and that the approximate relationship overall is °Brix = 1.8°Baumé.

The table shows 74°Oe to be equivalent to 18°Brix at 20°C; and the footnote says that 'Brix = °Oe/4 − 2.5, approximately'. 74°Oe should therefore be 16°Brix, which is a rather different figure. The exact relationship shows a quadratic equation, so that tables are best suited for obtaining conversions. This scale is, however, of additional interest in that the numbers on the Oeschele scale are taken from the first three decimal places of the specific gravity, e.g. 74°Oe from 1.074/02.

Prediction of alcohol content in the finished wine

Estimation of the so-called potential alcohol content from the known sugar content of the grape must, before fermentation starts, is important in vinification procedures. The relationship between the two values is unfortunately somewhat complex, dependent upon the units used for expressing sucrose (or glucose/fructose) content, e.g. % w/w_s or w/v_s, or $g L^{-1}$ of the solution, and on the % v/v ethanol, whether actually produced, or theoretically calculated. In practice, losses of ethanol to the air will be variable, together with losses from accompanying formation of various polyols, and depend upon the type of yeast used as well as other reasons.

Several equations have been published but the underlying basis is not often described. Thompson (1987) gave Equation 3.2 below, which relates to cider production but is also applicable to wine, so that, for example 200 g L^{-1} or approximately 18.3% w/w glucose/fructose should give 11.8% v/v ethanol, an estimate that is probably high for wine.

$$\text{Prospective ethanol content, % v/v} = (\text{Total reducing sugar content, } g L^{-1})/17 \qquad (3.2)$$

The conversion factors given in the footnotes to the table of Amerine & Ough (Jackson, 2008) are:

(1) % v/v alcohol = 0.52 × °Brix approximately, so that 20°Brix or 200 g sugar per kg of solution should give 10.4% v/v ethanol.
(2) % v/v alcohol = 0.125 × °Oe approximately, so that 83°Oe (= 20°Brix) should give 10.37% v/v alcohol.

(3) The predicted % v/v alcohol is said to be the same as the °Bé figure approximately, e.g. 20°Brix = 11.1°Bé = 11.1% v/v alcohol.

Predicted ethanol content can be obtained from Equation 3.3.

Ethanol content (w/w)

$$= \frac{Y \times \text{weight of sugar (g per kg solution)}}{Y \times \text{weight of sugar} + (1000 - \text{weight of sugar})} \qquad (3.3)$$

where Y is the fractional weight yield of alcohol obtained from the sugar present. For a typical ethanol content of 10.0% v/v, which is equivalent to 8.047% w/w (from the table in Jackson, 2008), and a theoretical yield of 54% for sucrose, then the required sugar content would be 146 g kg^{-1}. At a theoretical yield of 51% for glucose/fructose, then the required 'sugar' content in the must would be 153 g kg^{-1}, or also about 160 g L^{-1}.

Prospective alcohol contents from the the Amerine–Ough table indicate lower yields (37%) in practice, probably on account of polyol formation and effect of acids present in the calculation.

An oft-quoted rule-of-thumb is that each 18 g 'sugar' per litre of solution will give 1% v/v ethanol. Ribéreau-Gayon et al. (2006) give the same figure, with the further comment that the grape must should contain 180, 226, 288 g 'sugar' per litre to produce wines with 10%, 12.6% and 16% v/v alcohol by volume especially. These alcohol contents compare with those using the Amerine–Ough table of 9.0%, 11.2% and 14.0% assuming corresponding Brix figures of 174, 216 and 270.

When pure sucrose is added as in chaptalization (see below), slightly different figures are quoted. Jackson (2008) from Jones & Ough (quoted by Jackson) gives a figure of 17 g to give 10 g alcohol (equivalent to 1.25% v/v). Barr (1988) offers the figure of 7.5 kg sucrose per barrel (225 L) to produce the maximum increase of 2.0% v/v alcohol allowable, i.e. 16.7 g-sugar L^{-1} for 1% v/v alcohol.

3.3.3 Sugar content of grapes and must

As the final alcohol content of a wine is so directly related to the sugar (glucose/fructose) content of the originating grapes and must, variability in the latter in practice needs to be considered.

Their percentage content in fresh grapes that have reached commercial ripeness is 15–25% w/w; but it is even higher in dried or Botrytis affected grapes. As with acidity, the sugar content is determined by the grape variety used and degree of ripeness at picking. Ripeness is determined by the climate and the time at which the grapes are actually picked (e.g. the months of October/November). Times can differ for various reasons. Ribéreau-Gayon (1978) has presented some information for sugar content of musts, together with their acidity in Bordeaux vineyards (1969–1973), but some of the sugar content figures seem rather low (e.g. 170 g L^{-1}), indicating finished wines of relatively low alcohol content. Generally, the warmer the climate, the earlier the grapes will have accumulated the required sugar content for commercial wine-making. In a 'cooler' environment, the grapes are picked in September or October (Bordeaux), whilst in a hotter climate grapes are harvested in September or even August (Sherry, Jerez de la Frontera).

3.3.4 Chaptalization

Chaptalization is a vinification process first suggested in 1801 by Napoleon's Home Secretary, Jean-Antoine Chaptal (Barr, 1988). Sugar (sucrose) is added to the grape must to increase the alcohol content of the final wine, which can often be necessary in 'cool' grape-growing regions to supplement a low glucose/fructose content of the grape at harvest. There are, however, certain legal restraints, so that Burgundies, for example, cannot be chaptalized unless they attain a minimum alcohol content of 10.5–11.0% v/v, and cannot then be increased by more than 2%. This process continues to be controversial, since there are claims of slightly different taste characteristics developed in the fermentation of sucrose, compared with glucose/fructose, naturally present in grapes. The addition of non-grape derived sugars can be detected by measuring the $^{13}C/^{12}C$ ratio of the formed ethanol.

Sucrose can only be added to the must as the pure crystallized, dry solid form. It is obtained either from the beet sugar plant (USA and Europe) or from the cane sugar plant (South America, Caribbean). Sucrose as a crystalline product is commercially available at a number of different levels of purity (defined by between 99.5–99.9%). The highest degree of purity is demanded in sugar for domestic sweetening and bulking use. The slightly lower purity of a sucrose may be acceptable for industrial use, depending upon purpose. The remainder may be invert sugar (0.02–0.06%), mineral ash (0.001–0.06%), organic acids and undesirable moisture (up to 0.01%). Purity may be assessed by measurement of optical rotation as set out, along with other tests, by the International Commission for Uniform Methods of Sugar Analysis. There is no essential difference in the pure sucrose from each of the two sources, but the slight impurities (mineral and/or nitrogenous) represented by the difference in sucrose content from 100% may confer differences detectable by trained experts, especially regarding colour in aqueous solution or the flavour. Any odoriferous compound would need, however, to have an exceedingly low sensory threshold value (ppb). Chaptalization conducted with beet sugar has been claimed (Van Dyke, quoted by Barr, 1988) to introduce a beet sugar note, deriving from the original beetroots which has not been eliminated during its processing or has developed during its processing. The particular flavour compounds in the sugar beetroot are known. This possible effect would depend upon the purity of the beet sugar used. A less pure product might well be used for red wines, where contaminating constituents might not be significant, unlike in white wines. Precise specifications for the beet or cane sugar used are, therefore, important for its use in vinification.

Another comment quoted (Barr, 1988) is that the use of sucrose can confer a 'sweetish glyceriney taste' if overdone. The sucrose has first to be converted into glucose/fructose by the enzyme, invertase, contained in the *Saccharomyces cerevisiae* yeast used in the fermentation, which may at the same time generate a slightly different proportion of polyols. Chaptalization is believed (Jackson, 2008) to enhance the synthesis of higher alcohols and thought to augment the production of glycerol, succinic acid and 2,3-butan-diol and even the synthesis of some aromatic esters can be increased. There appears, however, to be no scientific evidence for its effect.

So-called *Süssreserve* (reserve sugar), used in Germany, is merely unfermented grape juice that is kept in reserve for 'sweetening' rather than alcohol enhancement. The obvious use of purified glucose/fructose from other sources would no doubt be too expensive and illegal in the European Union. Chaptalization reflects the commercial appeal of adequate alcohol content in wines. Barr (1988) suggests the apparently more appropriate use of 15 L of brandy per 225 L barrel to give 2% v/v alcohol rise, rather than 7.5 kg sucrose. However, added alcohol needs to be very neutral so as not to confer flavour characteristics (such as in port!).

3.4 Acidity

Conde *et al*. (2007) reviewed the biochemical changes in grapes during ripening and their effect on fruit and wine quality. The main acids in grapes, tartaric and malic acid, range from 69 to 92% of the organic acid content of grapes and grape leaves, the remainder consisting of citric, succinic, lactic and acetic acid. Sufficient concentrations of acid in the ripe grapes are required in order to be able to make good wine, and if the wine is to be aged, a crucial requirement is sufficient acidity, since acids play an important role in the chemical reactions of compounds forming the so-called wine aroma.

The formation pathway for tartaric acid is outside the oxidative metabolism of sugars (see Conde *et al*., 2007). Its biosynthesis starts with L-ascorbic acid and it is only formed until *véraison* (the moment the grape starts to colour). Malic acid is accumulated in grapes, reaching its peak concentration just before *véraison*, thereafter the malic acid starts to be broken down, whilst sugars are accumulated. There is a negative correlation between temperature and malic acid levels, hence in general, grapes grown in cool regions have higher levels of malic acid than those grown in warmer regions.

Whilst tartaric, malic and citric acids (originating in the grape), with lactic, succinic and acid (traces from grapes, mainly generated during fermentation), are the major contributors to the perceived acidity of wines, several others can be present (discussed in the following paragraphs). These other contributors may be derived from the grape directly or formed by fermentation, or by micro-organisms, including specific compounds produced by botrytized grapes. These are all organic acids, though small quantities of inorganic acids may also be present.

Acidity contributes in many direct and indirect ways to the quality of wine. It is related to the acid taste of the wine (discussed in the following paragraphs); insufficient acidity generally gives flat tasting wines, whilst an excess gives sour tasting wines. Adequate acidity gives wines freshness, and is regarded to balance any residual sugar and the aroma of wines. In particular there is a perceived reduction in sweetness in the presence of a good acidity.

Acidity is also related to the pH of a wine, which is known to affect numerous aspects of wine chemistry, many directly related to the smell, taste or quality. For example, a relatively low pH helps to protect the wine against microbial spoilage and makes the use of sulfur dioxide more effective. It also helps to express the red colour in red wines and reduces the incidence of browning of the phenolic compounds. Also, acids involved during both fermentation and ageing in the formation of esters are responsible for many of the fresh and fruity aromas of wines. As noted in Section 3.9.3, pH influences the redox potential of a wine and is important for wine stability. An 'adequate' acidity is a key prerequisite for wines to age well.

3.4.1 Contents of organic acids

The desired acid content of a wine depends on the residual sugar and wine style, the content in ripe grapes should be within the range of 0.65 to 0.85 g/100 ml (%) (see Conde *et al.*, 2007). If required, the acid content of a must should be adjusted to ensure a balanced wine can be made. Most white wines are preferred to have a higher acid content than red table wines. In wines usually two groups of acids are considered, fixed and volatile acids and total acids is the sum of these two. Volatile acidity refers to acids that can be removed by steam distillation, whereas fixed acids are those that are not that volatile.

Table 3.2 lists the contents of some acids found in wines, together with their trivial and chemical names, origin and content range, mostly quoted from Amerine & Roessler (1983). They are grouped according to their broad classification by chemical class. An even larger number of acids have been tabulated as having been identified in wine listed by Montedoro & Bertuccioli (1986), though only 35 of these secured three references or more to their presence. Ribéreau-Gayon *et al.* (2006) list 16 aliphatic acids (eight found in grapes and another eight after fermentation) and another 11 benzoic-cinnamic acids. They also mention linoleic and linolenic acids, which are of indirect flavour significance.

A few of these acids, in particular formic and acetic, are also volatile in their non-dissociated forms, to which might even be added butyric acid.

The acids that most importantly affect taste are those designated under 'other aliphatic', particularly those also originally present in the grape juice (must).

Lactic acid content can be quite high in wines that have undergone the malo-lactic fermentation process (desirable mainly in red wines). Data suggests contents of 900–2600 mg L^{-1}, though with corresponding reduction in malic acid content (Jackson, 2008). Nykanan & Soumaleinan (1983) report also 60–253 mg L^{-1} in many wines, at 34–205 mg L^{-1} in Italian wines, and at 7–55 mg L^{-1} in Bordeaux wines.

Table 3.2 Contents of the main acids found in wines.

Acid name and group	Origin[a]	Content range mg L^{-1} (ppm)
Aliphatic monobasic		
Methanoic (formic)	F	<60
Ethanoic (acetic)	F	Tr–<600
Butanoic (butyric)	F	<0.5
3-Methyl butanoic (isovaleric)		<0.5
Hexanoic (caproic)	F	1–>3
Octanoic (caprylic)	F	2–>17
Decanoic (capric)	F	0.5–7
Dodecanoic (lauric)	F	>1
Aliphatic monobasic hydroxyl		
2,3,4,5,6-Pentahydroxyhexan-1-oic (gluconic)	G/F (botrytized)	Tr > 30
1-Oxa-2,3,4,5-tetrahydroxyhexan-6-oic (glucaronic)	F	1–140
2-Keto-1,3,4,5-tetrahydroxyhexan-6-oic (galacturonic)	F (botrytized) (pectin breakdown)	10–2000 (0.01–2 g per litre)
Benzenoid monobasic carboxylic (and phenolic)		
2-Furoic	F	<30
4-Hydroxy benzoic (para hydroxybenzoic)	F	<1
3,4-Dihydroxy benzoic (protocatechuic)	F	<5
2,5-Dihydroxy benzoic (gentisic)	F	<5
3-(3,5-Dimethoxy-4-hydroxy-phenyl)prop-2-enoic (3,5-Dimethoxy cinnamic, or sinapic)	F	<5
	G	<50
3,4,5-Trihydroxybenzoic (gallic)	–	–
Ferulic acid (4-hydroxy-3-methoxy cinnamic acid)	G	–
Caftaric (ester of caffeic and tartaric acids)	G	–
Coutaric (ester of p-coumaric and tartaric acids) p-Coumaryl-tartaric	G	
Fertaric (ester of ferulic and tartaric acids)	G	
Combined benzoic-cinnamic	–	100–200 (red wines) 10–20 (white wine)
Other aliphatic		
Butan-2,3-diol-1,4-dioic (tartaric)	G	<1000–75,000 or 1–7.5 g L^{-1}
Butan-2-ol-1,4-dioic (malic)	G	50–5000

(Continued)

Table 3.2 (*Continued*)

Acid name and group	Origin[a]	Content range mg L^{-1} (ppm)
2-Hydroxypropan-1,2,3 tricarboxylic (citric) or 2-Hydroxy-1,2,3-tricarboxy propene	G	130–400 (0.130–0.4 g L^{-1})
Butan-1,4-dioic (succinic)	F	50–750 (0.05–0.75 g L^{-1})
Ethan-1,2-dioic (oxalic)	F	<90
2-Oxo-propanoic (acetyl-methanoic) (pyruvic)	F (ML)	8–50 g L^{-1}
2-Hydroxypropanoic (lactic)		0.2–3 g L^{-1}

Data adapted from Amerine & Rossler (1983) and Ribéreau-Gayon *et al.* (2006). G – from the grape. F – from fermentation. F(ML) incl. malic-lactic transformation.

A number of these acids, such as those based on cinnamic acid, are often largely present in esterified form, e.g. coumaroyl-tartaric acid (coutaric) and caffeoyl tartaric acid (caftaric). Some may be present as glycosidic combinations (e.g. gentistic acid) in the grape, but partly released by hydrolysis (Ribéreau-Gayon *et al.*, 2006).

The presence of the so-called sugar acids gluconic, glucuronic and galacturonic acids in fair quantity is most likely in botrytized wines (Noble Rot), although these acids appear to be produced by acetic acid bacteria growing concurrent with *Botrytis* on the fruit. There is believed to be a correlation of gluconic acid content with quality (Ribéreau-Gayon *et al.*, 2006), although none of these sugar acids affect the odour or taste of the wine. They will be present both free and as glycosides.

Acetic acid is present largely due to micro-organisms, though some may arise in the fermentation produced by yeast. Its maximum content is a subject of legislation, and measured in a particular manner (described in Section 3.4.2).

It is of interest to know how the different grape varieties can have different acid contents, and therefore cause differences in acid taste. Furthermore, there are important factors of climatic conditions during growth up to harvesting, reflected in the level of grape maturity determining acid content and taste.

Ribéreau-Gayon *et al.* (2006) has provided some interesting general data on acid contents, which are those of acids originally present in the grape must (Table 3.3). In 1978, Ribéreau-Gayon reported the data from studies made in 1975 over several harvests. It showed the average acid composition of musts from five red varieties and three white, of vines cultivated in the same soil (but different vineyards). Ribéreau-Gayon found that the tartaric acid content of the white varieties (Sauvignon Blanc, Sémillon and Muscadelle) was lower than that of all red varieties (Merlot, lowest; Cabernet Franc, Cabernet Sauvignon, Malbec and Petit Verdot), as was the total acidity. The relative contents of malic acid in the two groups were similar, as was the potential alcohol content (11–13%). This data was expressed in milliequivalents of each

Table 3.3 Acid content of some grapes/wines.

Tartaric acid	Unripe grapes	Up to 15 g L^{-1}
	Most-northerly vineyards	Often >6 g L^{-1}
	Southerly vineyards	2–3 g L^{-1}
Malic acid	Green grapes (juice)	Up to 25 g L^{-1}
	Mature must – north	4–6.5 g L^{-1}
	Mature must – south	1–2 g L^{-1}
Citric acid		0.5–1 g L^{-1}
Succinic acid	Wines	1 g L^{-1}
Benzoic-cinnamic acid	Red wine	100–200 mg L^{-1}
	White wine	10–20 mg L^{-1}
Gluconic acid	Grape juice (must) affected by Noble Rot	Up to several grams per litre

Adapted from Ribéreau-Gayon *et al.* (2006).

acid. A further study reported on Cabernet Sauvignon for Bordeaux wines of the highest quality over a five-year period (1969–1973) showed a considerable annual variation in total acid content, and indeed of sugar content, though with higher acidity figures than the other study. Data ranging over 29 years for Cabernet Sauvignon grapes samples at reference vineyards in the Bordeaux region at harvest time showed sugar concentrations varying from 170 to 220 g L^{-1}, whilst the total acidity ranged from 7.7–13.7 mEQL^{-1} Ribéreau-Gayon *et al.* (2006).

In practice, the lower acidity of the Merlot variety is used for blending with the basic Cabernet Sauvignon in years in which this grape has insufficient maturity; and the higher acidity of the Malbec and Petit-Verdot varieties are used for blending in years of full maturity (Ribéreau-Gayon, 1978). Pinot Noir grapes are also reported to have a lower level of acidity. The more mature the grapes the lower the acidity but the higher the sugar content.

Maturity, in addition to variety, is therefore an additional factor in determining acidity. The degree of maturity at time of harvest will vary inevitably from season to season even in the same vineyard. Wines made in hotter climates are prone to a lack of acidity (Barr, 1988); it is crucial for grapes not to be left to mature too long on the vine. High lime content of soils produces grapes with high acidity, and is thought to be especially important for sparkling wines (e.g. Champagne, though this is also from a 'cool' growing region). In fermentation, and in malo-lactic fermentation, there will be some changes in acidity from the original must, not least from some additional acid types introduced, so that predictions cannot be firmly made about final wine acidity from must/grape acidity data. It is important to minimize the formation of acetic acid by good hygienic conditions during vinification. There will also be changes in acidity on subsequent storage, as discussed under *Ageing* in Section 3.9.

3.4.2 Measurement of acid content

Acetic acid content is measured by so-called volatile acidity methods, in which a portion of the wine is steam-distilled. For official purposes, the wine has first to be acidified with tartaric acid, to ensure that all the acetic acid (and the few other volatile acids present in much smaller quantities) is in the free state. The acidity can be measured by titration with say N/10 caustic soda, and content expressed in $g L^{-1}$ or millequivalents of any designated acid. Such data is obtained in conjunction with measurement of so-called Total Acidity, i.e. of all the acids present, measured by a similar titration method, though the expression of amount in millequivalents of a particular acid is general. The titration end point is usually set at pH 7, obtained by a pH meter reading, or a bromothymol blue indicator, even though the second acid function of the dioic and trioic acids is not complete, whilst the neutralization of phenolic acids only starts at pH 7.

Significant quantities of acetic acid (or volatile acids) in wines are not favoured for flavour reasons. The threshold value is reported to be $22-54\,mg L^{-1}$ in water, whilst Ribéreau-Gayon (1978) believes $0.6\,g L^{-1}$ H_2SO_4 ($600\,mg L^{-1}$) in wine should be a maximum ($0.72\,g L^{-1}$ acetic acid). There is a maximum permissible legal value in many countries; thus in the EEC/EU, e.g. $0.8\,g L^{-1}$ for white and rosé wines and $0.98\,g L^{-1}$ for red wines (both expressed as H_2SO_4); and California $<1200\,mg L^{-1}$ (red). These figures may be converted into acetic equivalents by multiplying by 1.1/0.9 ($=1.12$). Much lower values are to be expected in marketed wines. Ribéreau-Gayon et al. (2006) state that alcoholic fermentation of grape juice alone leads to $0.2-0.3\,g L^{-1}$ ($200-300\,mg L^{-1}$) (H_2SO_4), so that a good quality well kept wine should not contain much more, as the expected quantity in Table 3.2 indicates.

Whilst the contents of all these organic acids may be assessed individually, the wine industry also uses a Total Acidity figure, which like the Volatile Acid figure is obtainable by titration methods with N/10 caustic soda solution, and the result is expressible in millequivalents of sulfuric acid, or other wine acid of choice. Conversion factors are available for total (titratable) acidity expressed in terms of one another and are given in the table quoted by Jackson (2008) from Amerine & Ough (1980). Thus initial units 'expressed' as sulfuric acid should be multiplied by 1.531 for its tartaric acid equivalent, 1.367 for malic acid and so on. Again, results expressed in millequivalents ($meq L^{-1}$) can be stated more understandably in $g L^{-1}$, but using the known millinormalities of the acid, e.g. 0.049 for sulfuric acid, 0.075 for tartaric acid, so that $100\,meq L^{-1}$ (sulfuric) is equivalent to $4.9\,g L^{-1}$ content. The Total Acidity figure includes that for the volatile acids present, whilst a Fixed Acidity figure can be used only for the content of poorly volatile acids.

The acid composition of must and wines has several other important facets, relating to the stabilization and treatment of wines to prevent deposits on standing, and for other reasons, in particular in relation to wine colour. These aspects, not directly relating to the flavour of wine, are not discussed here but they are, however, very fully covered by Ribéreau-Gayon et al. (2006).

A sufficiently low pH for wines helps to move the equilibrium of the different forms of the anthocyanins towards the red form of the molecule (flavylium ion), thus enhancing the red wine colour. The pH will also affect phenolic changes during maturation (see later sections).

3.4.3 Acid taste

Acid taste on the tongue is clearly related to acid content, though in a complex manner. Da Conceicao Neta et al. (2007) reviewed our current knowledge on sour taste perception and the sour taste properties of acids. They concluded that despite the increase in knowledge, it is still not possible to predict the sour taste intensity based on the composition and concentration of acids and the pH. The physiology of taste perception is still not fully understood and there does not appear to be a simple relationship between sour taste intensity and hydrogen ions. Sour taste is typically associated with acids, and results indicate that in addition to hydrogen ions, anions and/or protonated (undissociated) acid species play a role in determining sour taste of organic acids. One current hypothesis discussed by these authors suggests that sour taste intensity is dependent on the molar concentration of all organic acids that have one or more protonated carboxyl groups plus the concentration of free hydrogen ions. In addition some acids have not just a sour taste, such as succinic acid (see below).

The acidity of a wine or grape juice is also commonly expressed by its pH (or hydrogen ion concentration). Strong inorganic acids, like sulfuric acid, which are fully dissociated into hydrogen ions and anions, show a very low pH in dilute aqueous solution, which would taste very acid. The acids in wines are weak acids, a characteristic that can be assessed from their pK_as, which are tabulated in Table 3.4, along with other physical and relevant characteristics of wine acids.

Thus, from a pK_a of 3 to 4 (including tartaric acid, citric and malic, with the first carboxylic group), the acid functions are partly free, whilst from pK_a 4 to 6 there is little dissociation and formation of hydrogen ions, and even less from 6 upwards. The resultant pH is given by the Henderson–Hasselbach equation (Equation 3.4) for a weak acid HX dissociating into H^+ (actually $H^+.H_2O$) ions and X^- (anions).

$$pH - pK_a = \log [\text{anions}]/[\text{HX}] \tag{3.4}$$

In practice, Equation 3.4 has limitations in preciseness when several weak acids are present. Some of the acids are combined with metallic elements such as potassium and calcium, and some of their salts may be more highly dissociated than the acids themselves. For example, $1.5\,g\,L^{-1}$ of potassium would combine with 5.8 g tartaric acid; Ribéreau-Gayon (2006) reports up to $780\,mg\text{-}K\,L^{-1}$ in wine, and up to $6\,g\,L^{-1}$ tartaric acid in ripe grapes.

The pH of wines may range from 2.8 to 4.0, preferably so that the acidity is relatively high (cf. coffee brews, at 4.9-5.1, orange juice, 3.3 and beer, 4.1-4.7).

Table 3.4 Physical and other characteristics of the main acids found in wines.

Acid chiral orientation[a] polarized light[b]	Molecular Structure	Formula	Weight	Dissociation constant at 25°C Ka	pK[c]	Taste characteristic and threshold	Solubility (g g^{-1}) or % in water at t (°C)	w (°C)
Tartaric (ordinary)(+) R (R*,R*)-2,3-dihydroxybutandioic	HOOC.*CHOH.*CHOH.COOH Crystallized from aqueous solution above 5°C is anhydrous: Below 5°C forms a monohydrate	$C_4H_6O_6$	150	(1) 1.04×10^{-3} (2) 4.5×10^{-5}	3 4.34	Hard acid 10 ppm flavour threshold, pleasant below 5000 ppm (mg L^{-1})	139/100 at 20°C	169–170
Malic (ordinary) L (−), (S)-mono-hydroxybutan-dioic	HOOC.CH₂.CH(OH).COOH	$C_4H_6O_5$	134	(1) 3.98×10^{-4} (2) 7.8×10^{-6}	3.40 5.11	Green Acid	36/100 at 20°C	98–99
Citric, 2-hydroxy-1,2,3-propane-tricarboxylic	HOOC.CH₂.C(OH).(COOH). CH₂COOH	$C_6H_8O_7$	192	(1) 7.4×10^{-4} (2) 1.73×10^{-5} (3) 3.90×10^{-7}	3.13 4.76 6.40	Fresh acid, Pleasant at 0.02–0.08%	59.2% in water	153
Succinic, butan-1, 4-dioic	HOOC.CH₂.CH₂.COOH	$C_4H_6O_4$	118	(1) 6.6×10^{-5} (2) 5.8×10^{-6}	4.18 5.23	Intense acid taste/ some bitter notes	1/13	187–188
Lactic D (+) and L (−) (ordinary), 2-hydroxypropanoic	HOOC.CH(OH).CH₃	$C_3H_6O_3$	90	6.6×10^{-4}	3.18	Mild acid	Soluble in water	16.8
Acetic, ethanoic	CH₃COOH	$C_2H_4O_2$	60	1.73×10^{-5}	4.7	Pungent acid 1g per litre in wine	Miscible	16.7

Compound	Structure	Formula	MW	K_a	pK_a		Solubility	m.p.
Gluconic D (−)	1 COOH 2 HCOH 3 HOCH 4 HCOH 5 HCOH 6 CH$_2$OH	$C_6H_{12}O_7$	196	2.5×10^{-4}	3.6	Mild acid	S in water S in alcohol	131
Glucuronic (β form shows rotation)	−CHO at carbon 1 −COOH at carbon 6	$C_6H_{10}O_7$	194				S in water S in alcohol	165
D-Galacturonic	− ketoacid (C_2)	$C_6H_{10}O_7$	194					159 (α) 168 (β)
Gallic, 3,4,5-trihydroxybenzoic acid		$C_7H_6O_5$	170		1.69		S in water	Sublime at 210°C, 258–265

[a] L and D (or S and R) are used, which apply to the straight chain and ring formula, whilst α and β apply strictly to the ring forms, usually pyranoside. [b] (+) and (−) are used, for dextro- and laevo-rotation of polarized light in solution; in the older literature, d and l (lower case letters) were used. These terms and their numerical values are not related to abstract chemical characteristics of the molecules, but are experimentally assessed in particular circumstances (Appendix I). [c] Dissociation constants $(Ka)_1$, $(Ka)_2$ etc. calculated from $10(I/Ka) = pK$. pKs given in Lide (2002). Similar data is given by Ribéreau-Gayon et al. (2006) but the temperature of measurement is not given.

Using Equation 3.4 above for tartaric acid with a pK_a of 3, in say a wine of pH 3.5, we can calculate that $3.5 - 3.0 = +0.5 = \log$ [anions]/[HX], so that 21% will be present as undissociated acid and 79% present as anions. The acidity of wines, however assessed, is of course, importantly related to their taste on the tongue, which is believed to be primarily due to the hydrogen ions and their concentration. However, pH values are not always strongly correlated with perceived acid taste, as has been found with coffee brews (Woodman, 1985). The sensation of acidity differences overall is more strongly related to titratable acidity, suggesting that the anions, for example those of succinic acid, have some individual effect on flavour. Undissociated molecules that are also volatile (such as acetic acid) also have a rôle in contributing to odour/aroma (see Chapter 4).

Tartaric, malic and citric acids are generally recognized as the most important contributors, perhaps accounting for some 90% of the total acidity, to the taste of quality wines, though the minor rôle of lactic acid should be considered. Acetic acid is, however, highly important, as a potential spoilage agent (i.e. 'vinegary', 'sour'). Conde et al. (2007) discusses the formation, stability and contribution to quality of acids in some detail. Tartaric acid is the strongest acid and contributes to the biological stability, ageing potential and the tart taste of wine. Tartaric acid is gradually lost, during fermentation potassium and calcium form potassium bitartrate and calcium tartate crystals, which precipitate during fermentation and even during maturation. Malic acid gives a green taste to wine, and some is consumed by fermenting yeast, while further losses can occur as a result of malo-lactic fermentation, converting malic into lactic acid.

Much smaller amounts of succinic, lactic, citric and acetic acid are present in wines. Succinic acid, both originating from the grape and derived during fermentation, is quite stable and adds to wine a complex mixture of sour, salty and bitter tastes and its characteristics are attributed to give the special taste typical of fermented beverages. Small amounts of citric acid gives a fresh and slightly acid taste to wine, but like malic acid can easily be converted into lactic acid by some lactic acid bacteria. Most of the lactic acid in wine is formed during the fermentation processes and it gives wine a slight sour taste, reminiscent of yoghurt. Acetic acid, also mostly formed during fermentation, is undesirable above certain levels. The exact contribution to overall flavour of a wine has to be considered in the context of other constituents present; this subject is discussed in Chapter 5.

Acidity is important to wine stability, so that many wines to be aged, like many 'reds' (except Beaujolais Nouveau), need to have adequate acidity, since acids partake in the chemical reactions during maturation and can also precipitate, whilst wines often drunk young can have a lower acidity.

The taste/threshold characteristics of these acids are given in Table 3.4, together with other relevant physical and molecular structure data. Looking at the taste characteristics for malic acid ('green') and lactic acid ('mild') in this table, it becomes clear that in relatively high acid wines the malic-lactic conversion may be desirable. However, in wines lacking acidity, malic-lactic conversion is usually not desirable.

3.5   Sweetness

Residual amounts of the sugars (saccharides), glucose and fructose, in a finished wine determine primarily its perceived sweetness or 'dryness', in wine tasting terminology.

3.5.1   Chemical structure of sugars

Glucose (so-called grape sugar, Gk. *glukus* – sweet, or compound that is sweet; -ose, abounding in) and fructose (fruit sugar, L. *fructus* – fruit) are both reducing sugars or monosaccharides, due to the presence of their aldehydic and keto groups. Their structures and systematic chemical names are shown in Table 3.5. These same sugars are present in a number of other fruits, and, of course, in honey, where the approximate equi-molecular mixture is known as invert sugar (by enzymatic action on sucrose).

Sucrose is a disaccharide (linked glucose-fructose molecules), not originally found in the grape but it sometimes needs to be added in dry form prior to fermentation (chaptalization – see Section 3.3.4) or even afterwards, and therefore, may be a contributor to the sweetness of the final wine. It is, therefore, included in Table 3.5.

The only other aldohexose known to be present in wine is D-galactose but only in small quantities ($0.1\,g\,L^{-1}$). The main aldopentoses in wine are L-arabinose and D-xylose (Gk. *xulon* – wood) at very low levels (0.3–$1.0\,g\,L^{-1}$). They are not fermentable, are more common in red wines than white ones due to their pectic/wood/gum origins and have low sweetness ratings.

3.5.2   Content/sweetness

As already indicated in Table 2.21, the content of residual 'sugar' in a so-called dry wine will be 0.2–$0.8\,g\,L^{-1}$ glucose, and $1.2\,g\,L^{-1}$ of fructose. In a sweet wine, however, these contents may be up to 30 and $60\,g\,L^{-1}$ respectively. Even higher concentrations can be found, for example Sauternes may contain $120\,g\,L^{-1}$ and $175\,g\,L^{-1}$ respectively. The relative sweetness of these sugars in aqueous solution is given in Table 3.6 together with some other relevant physical properties.

For practical purposes, the relative sweetness of sugars is determined relative to a reference compound, usually sucrose. The threshold values for the detection of a compound present are lower than for the recognition of a sweet compound (see also Chapter 4 for aroma compounds).

The contribution of the ethyl alcohol and glycerol to sweetness should be considered in the final wine, though they are generally thought to be negligible and very small. Glycerol concentration is usually between 4–$9\,g\,L^{-1}$ (see a review of Swiegers *et al.*, 2008), with red wines generally having higher concentrations that white wines. Glycerol contributes to sweetness and fullness, although it is not clearly reported in the literature how significant this is. Its sensory threshold for sweetness is determined as $5.2\,g\,L^{-1}$ in white wines. However, the concentrations typical for wines are thought to be too low for a contribution to

Table 3.5 Physical and other characteristics of sugars in grape musts and wines.

Sugar	C. No	Linear structure	Cyclic structures	Molecular		
				Perspective	Formula	Weight
Glucose (grape sugar; dextrose) D (α or β) An aldohexose (gluco-hexose) α-gluco-pyranose in solution (+) dextrorotatory	1 2 3 4 5 6	CHO HCOH HOCH HCOH HCOH CH$_2$OH D-glucose	H — C* — OH HCOH HOCH HCOH HC — O CH$_2$OH α-D-gluco-pyranose (1,5-oxidic)	α-gluco-pyranose β-gluco-pyranose	C$_6$H$_{12}$O$_6$	180

* Asymmetric carbon atom

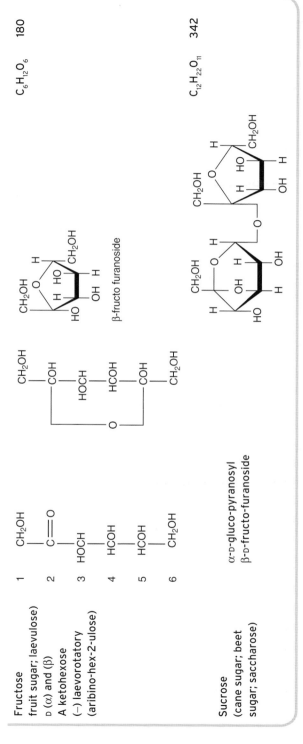

Fructose
fruit sugar; laevulose)
D (α) and (β)
A ketohexose
(−) laevorotatory
(aribino-hex-2-ulose)

$$C_6H_{12}O_6 \qquad 180$$

1   CH₂OH

2   C=O

3   HOCH

4   HCOH

5   HCOH

6   CH₂OH

β-fructo furanoside

Sucrose
(cane sugar; beet
sugar; saccharose)

α-D-gluco-pyranosyl
β-D-fructo-furanoside

$$C_{12}H_{22}O_{11} \qquad 342$$

D (and L) are arbitrary distinctions, referring to the position of the OH group at the C5 atom; whilst α and β refer to the asymmetry of the carbon atom in the C1 position.

Table 3.6 Relative sweetness of different sugars, and other physical properties.

Sugar	Relative sweetness level	Threshold flavour value Water	Solubility in water	Mp (°C)	Optical activity $[\alpha]^t_D$
Glucose	69	0.090[a] (1.63%) 0.065[b] (1.17%)	At 25°C 30.2– 51.2 g/100 g/ soln	Anhydrous α, 146°; β, 150° hydrated (+1 H_2O) 86° 102–104°	pure α, + variable
Fructose	114	0.052[a] (0.94%) 0.02[b] (0.24%)			$\beta \pm 17.5$
Sucrose	100	0.024[a] (0.81%) 0.011[b] (0.36%)	2.07/g water at 20°C Soluble in alcohols	188° (decomposition in aqueous solution) 160° (cooling to vitreous mass)	α + 66.5
Arabinose				160°	
Xylose	67				

[a]Recognition level. [b]Detection level. Adapted from Belitz & Grosch (1987), *Food Chemistry*. Threshold flavour values in moles L^{-1}; in brackets, % concentration.

perceived wine viscosity. A recent sensory study whereby additions of glycerol and ethanol were made to Riesling wines to give a glycerol range from 5.2 to 10.2 g L^{-1} and ethanol range from 11.6 to 13.6 %v/v showed that neither ethanol nor glycerol consistently affected sweetness, acidity, aroma or flavour perception, confirming that the typical glycerol concentration range found in wines may not contribute significantly the perception of the wines. The authors reported a possible weak effect of increasing glycerol on body, although the effect seemed to be wine dependent (Gawell *et al.*, 2007). However, there was a perceived hotness of the wines, correlated with ethanol content of the wines. The production of glycerol in wines is influenced by yeast strain, fermentation temperature, agitation of the must and nitrogen content of the must.

The quantity and the intensity of sweet aqueous solutions depend on the concentration and structure of the sugar and also on other parameters, such as the temperature, the pH and the presence of other compounds (Belitz & Grosch, 1987). Several other foods and beverages also show such interactions. The sourness and bitterness of bilberry and cranberry juice decrease with sweetness, and overall 'pleasantness' also increases. Threshold values of the bitter compounds in grapefruit increased significantly with increasing sucrose concentration. These findings may also be pertinent to the perception of wines, where residual sugar could affect any bitterness imparted by phenolic compounds. More importantly, the interaction between sourness (acidity) and sweetness in wines need a 'good' acidity, in order not to be described as 'cloying' or 'sickly'. Conversely, high acidity wines, without any sweetness, may well be described as 'acid', 'tart' or 'sour', all terms indicating too much acid and without the counterbalance of some sugar. The residual levels of sugar in dry wines is enough to add mouthfeel without a sweet taste, is thought to help the balance of the wine, increase viscosity and may mask a small taste defect in wine (see Conde *et al.*, 2007). These aspects of perceived flavour are explored further in Chapter 5 on wine balance.

3.6 Bitterness, astringency and mouthfeel

Although bitterness and astringency are often confused, especially by the inexpert taster, they are quite distinct, affecting different areas in the mouth and tongue in different ways. Bitterness is one of our basic tastes, as briefly discussed in Section 3.2, whilst astringency is not a taste, more a sensation in the mouth, see for further discussion Section 3.6.4. Mouthfeel generally corresponds with the term 'body', and is a sensation perceived in the mouth as a whole, rather than just the tongue. Even so, like the term 'body', it can be used in different senses by the expert taster, and may relate to the totality of the volatile part of the flavour and even to 'viscosity' of the liquid by others.

3.6.1 Basic chemistry

These organoleptic terms are usually considered together, as their common chemical basis resides in a phenol-derivative group of substances, particularly so-called polyphenols. These compounds are present in many fruits and vegetables and particular kinds are to be found in grapes. Some confusion may be found in their chemical name, polyphenols, differently used by different authors.

Complex systems of chemical classification are inevitable when dealing with such a wide range of compounds. In grapes/wine, and also in oak wood with extractable phenolic compounds from barrel fermentation and ageing, two distinct groups can be first identified, i.e. the non-flavanoid and the flavonoid/flavanoid group. The main phenolic compounds in *Vitis vinifera* grapes are phenolic acids and their conjugates (non-flavanoid), flavan-3-ols (and their polymeric forms of condensed tannins), anthocyanins, flavanols and flavanoids. These groups of compounds are discussed in the following paragraphs. More detailed information can be found in reviews such as Singleton & Noble (1976) and Waterhouse (2002) or text books (Anderson & Markham, 2006). Many of these compounds will not be sensorially important or present in sufficient quantity over threshold values.

Non-flavanoids

These are the small molecular compounds and their conjugates (e.g. esters, glycerides). The two basic unit types are the (1) benzoic acids (carboxyben-zene), of which the most important are the 4-hydroxy acid, 3,4-dihydroxy (proto-catechuic) and the 3,4,5-trihydroxy (syringic), 2-hydroxy (salicylic) and 2,5-dihydroxy (gentisic); (2) cinnamic acids (phenyl propenoic), of which the most important are the 3,4-dihydroxy acid (caffeic), 4-hydroxy (p-coumaric), 3-methoxy-4-hydroxy (ferulic) and the 3,5-dimethoxy-4-hydroxy (sinapic).

Benzoic acids are mainly present in grapes as (1) glycosides in which the components are joined by an ether linkage –O– from two hydroxyl groups (e.g. glycosyl gallic acid), which can be released by acid hydrolysis (in wines, therefore, the free aglycone forms are more prevalent), and (2) esters (e.g. gallic and elagic tannins, 1-galloyl β-glucose), in which the components are joined by a carboxy link, –O–CO– (e.g. carboxyl group of caffeic acid to the hydroxyl group at carbon 1 of the glucose molecule). They are, therefore, true esters, and the components can be released by alkaline hydrolysis (as in saponification).

The cinnamic acids are present in grapes and wines, in small quantities in free form in wines but are largely esterified. Thus, caffeic acid forms caftaric acid (caffeoyl tartaric acid) and coumaric acid forms coutaric acid (coumaroyl tartaric acid). They may also form glycerides (e.g. glucosyl-4-coumaric acid), though rarely so. Cinnamic acids can exist in two isomeric forms; 'trans', generally found in nature, and 'cis'; an equilibrium mixture can be formed under the influence of light.

These phenolic acids and their conjugates have no particular sensorial significance in wines (insufficient amount), except that they can become yellow in a dilute alcohol solution due to oxidation. However, they can be important precursors of volatile phenols, like ethyl phenol, discussed in Chapter 4.

Tyrosol [(4-hydroxyphenyl)ethanol] formed from tyrosine during alcohol fermentation may be included in the group of essentially non-volatile compounds (bp 310°C), also trytophol (benzo-2-pyrro-ethanol).

The coumarins not present in grapes may be present in wines on account of ageing in oak barrels. Coumarins are benzo-pyr-2-ones, present in either free or glycosylated form. The aglycone basic portion (3.I) is either (a) umbelliferone, which has R = H or (b) aesculetin, has R = OH; as (c) scopoletin, R = OCH$_3$. The glycosides are named aesculin at C$_6$ and scopolin at C$_7$.

3.I

Flavan-3-ols

These are important flavanoids, the basic form of which is the flavan molecule, so that the spelling 'flavanoid' is more appropriate (3.II).

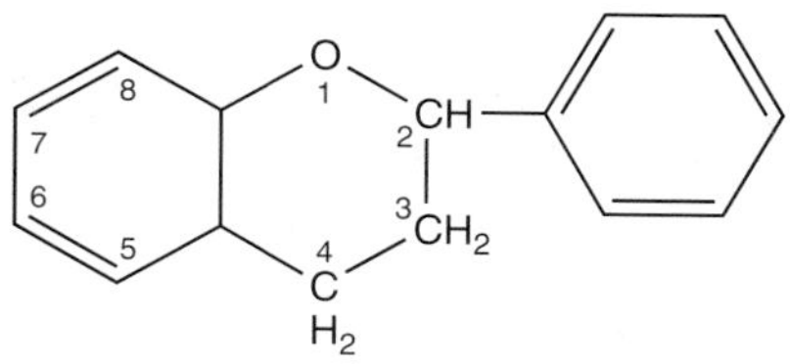

3.II

It is essentially a phenylbenzo-pyran (strictly speaking a dihydropyran) or a phenylchromane. In the numbering above in the benzo-pyran ring, the older system called the 1-position (not substitutable) α, the 2-position, β, and the 3-position γ. The pyran ring can be said to arrive, from a lactonization in the 2-pyrones, or from formation of an internal anhydride of 2-hydroxy-3-(1'-hydroxy-propyl)benzene, when the benzene part becomes benzo- in the molecule (3.III).

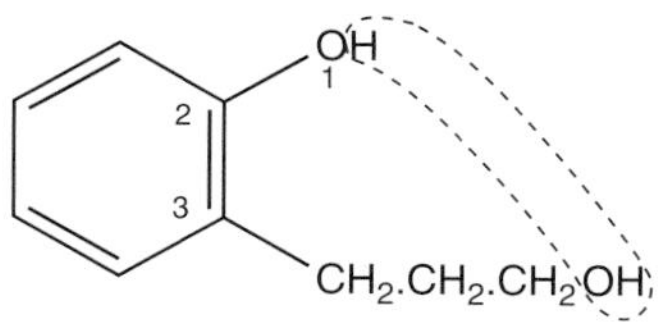

3.III

We then have a group of substances known as the flavan-3-ols, such as the catechins below, as in formula 3.IV, with the new carbon atom numbering [2-(3',4'-dihydroxy-phenyl)-3,4-dihydro-2H-benzo-pyran-3,5,7-triol, $C_{15}H_{14}O_6$. MW = 290].

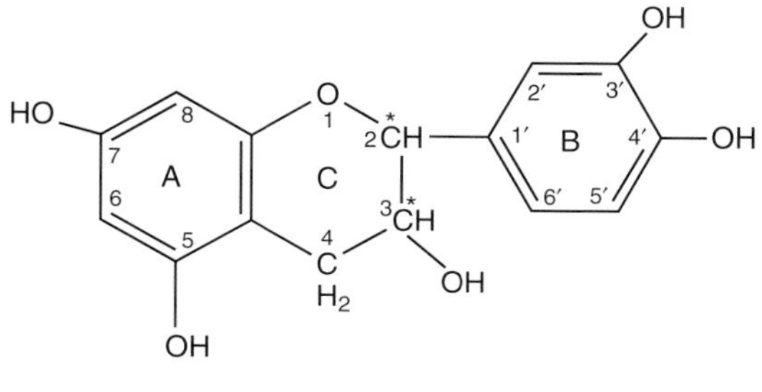

3.IV

The catechin naming comes from the phenyl connection of catechol (*ortho*-hydroxyphenol or 1,2-dihydroxybenzene). The carbon atoms (2 and 3) in the pyran ring allow of stereoisomerism, so that we have catechin (*trans*) and epi-catechin (*cis*); and they are also asymmetric, resulting in optical activity with + and − variants in each. There is also gallocatechin, in which the catechol part is replaced by 3,4,5-trihydroxybenzene (pyrogallol).

These flavan-3-ols are stated not to be glycosylated in the grape or wine (Ribéreau-Gayon *et al.*, 2006). Small amounts of flavan-3,4-diols or leucocyanidins (Gk. *leucos* – white) may be present.

They are, however, particularly important in their polymeric forms (often referred to as procyanidins when including the dimeric form), called 'condensed tannins' in wine literature, as discussed further in Section 3.6.2.

Flavonoids

These compounds are variously referred to as flavonoids, of which the flavonols and flavanonols are present in grapes and wines. The flavonols have the molecular structure shown by 3.V.

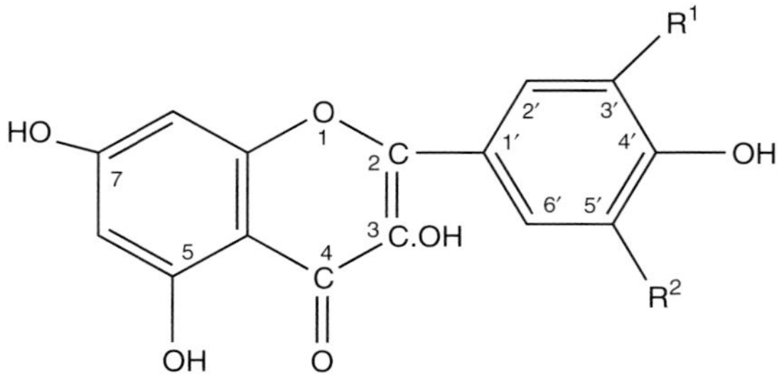

3.V

The pyran ring has now become ketonic, so that 3.V is a 2-phenyl-benzo-pyr-4-one, also known as a flavone or phenylchromone. The known variants in grapes/wines in Table 3.7 below.

Another group of substances, where the pyran ring is dihydropyrone, are known as flavanonols, of which the important representative is dihydro-quercetin or taxifolin (3.VI).

Flavonols are yellow pigments, whilst the flavanonols (Gk. *flavus* – yellow) are much paler. In red wines they are present partly in the aglycone form, as the glycosides in the grapes are hydrolysed in the fermentation. Attachment is at C3 by reaction with the hydroxyl group by an –O– link.

Table 3.7 Substitution patterns and common names of flavonols.

Radical at		Name of the	
R¹	R²	Aglycone	Glycoside
OH	H	Quercetin	quercitin
H	H	Kaempferol	–
OH	OH	Myricetin	myricitin

The glycosides are in the C_3 position.

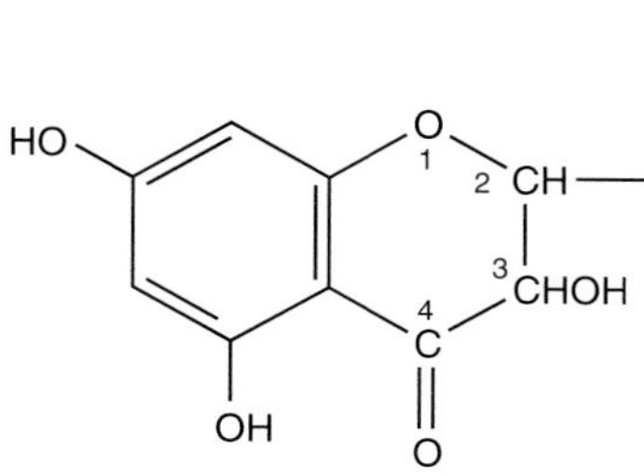

3.VI

Anthocyanins

The other important group of polyphenolics are the anthocyanins (Gk. *anthos* – flower, *cuanos* – dark blue), which are the glycosides of the antho-cyanidins. These compounds are largely responsible for the colour of red grapes, and of many other colours in fruits, vegetables and plants, especially of flowers. They were investigated in the early twentieth century in Germany by Willstätter and later in Oxford, England, by the Nobel prize winner (Lord) Robinson from Australia. Anthocyanins make up a range of red and blue colours.

In red wines, there are some five different compounds, based upon the molecular structure 3.VII, and having different substitutions on the B-ring (Table 3.8).

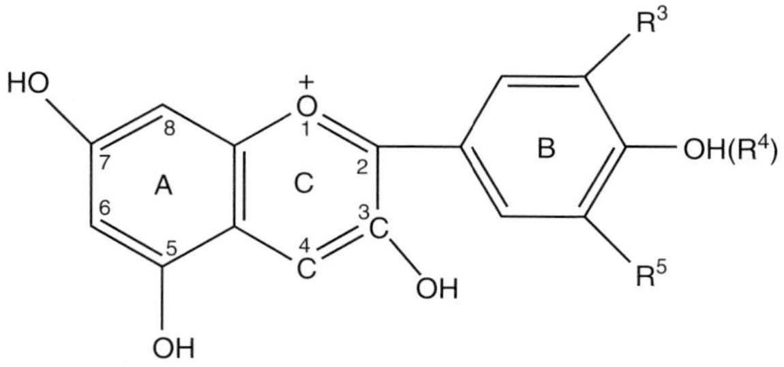

3.VII

Table 3.8   The five different compounds in red wines and the different substitutions on the B-ring.

Radical at			Name of	
R^3	R^4	R^5	Aglycone	Glucoside (glyceride)
OH	OH	H	Cyanidin	cyanidin 3-glycoside
OCH_3	OH	OCH_3	Malvidin	malvidin 3-glycoside
OH	OH	OH	Delphinidin	delphinidin 3-glycoside
OCH_3	OH	H	Peonidin	peonidin 3-glycoside
OH	OH	OCH_3	Petunidin	petunidin 3-glycoside

Differences should be noted from the flavonoids discussed previously, in that the pyran ring is actually pyran, not pyrone or dihydropyran, though other features are substantially the same. The extra conjugation is responsible for the red colour of these compounds.

The saccharide (glucose) is bound at the 1-position in the glucose molecule to the 3-position in the pyran ring, whilst an additional glyceride link is available at the 5-hydroxyl position in the benzo-ring.

The most widespread anthocyanins are the 3-glucosides, and in *Vitis vinifera* all anthocyanins are 3-glucosides. In non-*Vitis vinifera* grapes, 3,5-diglycerides are very common. Anthocyanins do not occur in plants as aglycones, which are very unstable; the 3-glucoside confers both chemical stability and solubility. The 3-glucoside in *Vitis vinifera* can be esterified at C6 of the glucose ring with acetic acid, *p*-coumaric acid or caffeic acid. Malvidin-3-glucoside is the main colouring matter in grapes, from 90% in Grenache to just under 50% in Sangiovese (Ribéreau-Gayon *et al.*, 2006).

The anthocyanin molecule occurs as several different chemical structures, the equilibrium between them being dependent upon pH. The flavylium structure, having a positive charge at position 1 in the C-ring, is the most intensely red form. The delocalized positive charge at C2 and C4, and net negative charges at C6 and C8, make the anthocyanins susceptible to both nucleophilic and electrophilic attack. This accounts for the general reactivity of anthocyanins and the ease of polymerization reactions. Anthocyanins in wines are affected by temperature, light, pH, oxygen, sulfur dioxide, aldehydes, ascorbic acid (minimal amounts in grapes), metals and degradation products of sugars. They are also prone to condensation and polymerization with other phenols. More detailed discussions on anthocyanin chemistry can be found in reviews by Markakis (1982), Brouillard (1983), Timberlake, (1983), Boulton (2001), Brouillard *et al.* (2003), Brouillard (2010) or in textbooks by Ribéreau-Gayon *et al.* (2006), Anderson & Markham (2006) and Jackson (2008).

An important new group of anthocyanins formed in wines and referred to as pyranoanthocyanins, encompassing the group vitisins, which were first reported first reported by Bakker & Timberlake (1997); these compounds are more colour stable and have important properties relevant to wine colour and changes in wine colour during maturation and is discussed in more detail in Section 3.7.

3.6.2 Basic technology

Location of polyphenols in grapes

Polyphenolic compounds are present primarily in the skins, seeds and stems (stalks) of the grapes. The flavanonols (yellow pigments) and anthocyanins (red) reside mainly in the skins, whilst the other flavanoids, flavan-3-ols, -3,4-diols (and their glycerides in small amount), together with their polymers, originate primarily from the seeds and stems. The non-flavanoid phenols reside in the vacuoles of the grape cells. The wine-making procedures determine what parts of these compounds will be transferred to musts and wines.

Although the vinification process will determine the phenolic content in a wine to an extent, the contents and composition of the phenols in the grapes will have a major impact on wine quality. Cultural practice and environmental impacts on phenol in grapes and wines has been reviewed (Downey *et al.*, 2006). Of the many factors influencing the content and composition of phenols in grape varieties, site and season are the most important, and assuming no other variables light and temperature are the most important. Increased temperature will enhance the metabolic processes in the vines and the associated phenol formation. However, if the temperature rises too high (thought to be above 30°C) phenol accumulation stops. Hence, viticultural practice can influence these variables to an extent, for example by ensuring the vines catch most of the light in the vineyard. The authors stressed that not all grape varieties responded in the same way, reflecting that both viticulture and wine-making rely on scientific knowledge, as well as many years of experience.

Use of the term 'tannins' and their classification

The term 'tannin' derives from the capacity of certain naturally found phenolic compounds, historically known to tan animal skins (collagens) and so to form leather, by forging stable combinations with proteins (also polysaccharides). Astringent properties arise for the same reason. The most important early known tannin was so-called tannic acid, actually a polymerized complex of gallic acid, described above, which forms a glycoside with glucose. Emil Fischer in the early nineteenth century synthesized esters (or anhydrides) of 4-hydroxybenzoic acid by reaction of two molecules. The monomer is $HO.C_6H_4.COO.C_6H_4.CO.OH$, which he called a depside (Gk. *depsein* – to tan), together with linear polymers (di-, tri-depsides, etc.) on account of their believed resemblance to the known tannins. They belong to the group of hydrolysable tannins, since their phenol units are separable by hydrolysis (acid/alkali). The second group is the non-hydrolysable tannins (in grapes/wines), based on the flavan-3-ol and polymers, often referred to as procyanidins.

Certain complex phenolic compounds possess astringent properties as a result of the same capacity to react with proteins. The exact mechanism of astringency compounds is still being researched but it is thought to involve binding and precipitation of salivary proteins by phenolic compounds. Again this property is enhanced in their polymeric forms. Tannins for a similar reason are of interest in wine clarification, i.e. protein precipitation. Monomers tend to be more bitter than astringent, even acid/metallic, which characteristics may persist into the lower polymers. Monomers are usually colourless in aqueous/alcoholic solution, unless the molecule possesses suitable chromophores (like C=O groups in the flavonols), whilst polymers will show coloured properties. Anthocyanins are highly coloured compounds, giving red wines their colour.

Tannins are often said to be present in green coffee beans (though they may be in the coffee pulp of the cherry) but this is not strictly so. The corresponding type of compounds present are the chlorogenic acids, based upon the structure of the acyl moiety (hydroxycinnamoyl) (3.VIII), where the most common compound present is where R^2=H, R^3 and R^4=OH, and R^5=H, as in caffeic acid, a cinnamic acid already mentioned in connection with grapes. In coffee, however, this substance (and others related) is esterified with quinic acid, which is a tetrahydroxy-cyclohexanoic acid, not a phenolic acid. They are strictly speaking cyclitols, thus we have monocaffeoyl quinic acid and many others (Clifford, 1985). Some of these monomeric chlorogenic acids (e.g. 2,3-dicaffeoylquinic acid) do have astringent properties but have a bulky molecular structure.

3.VIII

The tannins in green tea are gallates of epi-catechin (ECG) and of epi-gallocatechin (EGCG), and where there is esterification with gallic acid (carboxy group at the 3-position) through to hydroxyl group in the pyran ring. All these substances are present in black tea, though most have been transformed into further more complex compounds under the names of thearubigens and theaflavans (Landau et al., 1977). Again, astringency/bitter factors are involved in these compounds, and also the recently reported medicinal benefits.

Grape tannins

Grape tannins are firmly based upon the particular flavan-3-ols already described, but only in the polymerized forms from dimeric upwards, known as procyanidins. The nature of these polymers has been the subject of considerable studies. The C–C linkage as first shown by Haslam (1981) is through the A-benzene ring of one molecule at C8 to the pyran ring of another at C4 (upper unit). Procyanidin B_1, B_2, B_3 and B_4 are all dimers, their structure consisting of catechin and epi-catechin; B_1 and B_2 are the most abundant in grapes. Other types of linkage can be present in these polymers (see Appendix I). The oligo- and polymeric forms are bitter/astringent, showing increasing astringent activity with increasing molecular weight, especially the so-called condensed tannins. However, increased further polymerization leads to water insolubility and consequently a loss in astringency. The monomeric forms have little astringent capability and are reported also to be acid in taste. Polyphenolic compounds in wines, including monomers of flavanols, flavanonols, tannins and anthocyanins, have had various proposed classifications, of which that of Montedoro & Bertuccioli (1986) is generally accepted, as listed next, with additional comment:

(1) Molecular weight of below 300 (monomers) or 500 with sugar. Responsible for colour (anthocyanins – red; procyanidins – colourless; flavonols – yellow); bitter taste (flavanols and flavonols).

(2) Molecular weights, 500–1500 (2–5 monomer units). Responsible for astringency, body (flavan tannins); colour (flavan tannins – yellow).

(3) Molecular weights 1500–5000 (6–10 monomer units). Responsible for astringency and body, in pseudo-stable solutions (condensed flavan tannins – yellow-red to yellow-brown).

(4) Molecular weight of above 5000 (above ten monomer units). Highly-condensed flavan tannins. Generally water insoluble.

Additional suggestions in respect of colour were made by Ribéreau-Gayon & Glories (1987) quoted by Jackson (2008). Notably, anthocyanins in monomeric forms and in combination with tannin (MW 1000–2000) were thought to be the most influential compounds in determining the colour of red wine. However, the powerful and stable pigments formed by anthocyanins reacting with compounds such as acetaldehyde and pyruvic leads to the group of compounds now known as pyranoanthocynains (Rentzsch *et al.*, 2007), to which the vitisins first reported in wine by Bakker & Timberlake (1997) belong. Undoubtedly, these compounds will contribute to wine colour, as suggested by Bakker & Timberlake (1997) but their contribution to wine colour and possible ways of influencing their formation still awaits to be resolved (discussed in Section 3.7.1). There are no data on the sensory properties of pyranoanthocynains.

Hydrolysable tannins can be separately classified. They include gallo-tannins and elagi-tannins, inner esters based upon gallic and elagic acids respectively, and contain glucose. Elagic acid has a highly complex molecular structure and is essentially a di-lactone of gallic acid. The elagi-tannins are associated with ageing in oak barrels, so-called vescaligin and castalagin being extracted but are capable of hydrolysis to the aglycones of vescalin and castelin. Hydrolysable tannins are stated not to be in grapes themselves, though gallic acid itself from the skins of grapes is always present in wines.

Oxidizability is an important characteristic of phenolic compounds, not least those of tannins in wines, during ageing, as discussed in Section 3.9, and in Appendix I.3.

Quantifying methods

Information as to the actual content of all these polyphenolic compounds in grapes, musts and wines (both young and aged), including monomeric and polymeric forms, is important to flavour/colour assessment. There is clearly a considerable problem in defining all these compounds that can be present, in devising methods of analysing them separately, and in determining content in real weight/volume units. Similarly, determining threshold values (as used for volatile compounds, see Chapter 4) for their sensory properties is a large task.

Phenols (acids, alcohols, esters, flavanoids, etc.) have been traditionally determined by optical density measurements at a wavelength of 760 nm (or other) after reacting a small sample of the wine with a special reagent, in particular the Folin–Ciocalteu reagent (a mixture of phospho-tungstic and -molybdic acids). The unit of measurement, nm, is the nanometre or 10^{-9} m.

In the many papers published by Ribéreau-Gayon and his colleagues, and in review texts based on them, the tannin composition of wines is often expressed in mg L^{-1}, as determined by a particular procedure (the LA method of Ribéreau-Gayon & Stonestreet, 1966, updated by Glories, 1988). This procedure is based upon the original observations of Bate-Smith in 1934 and earlier of the true flavanoid tannin to form a red cyanidin compound in an acid solution with oxidation. For this reason oligo- and polymeric flavan tannins are often referred to as procyanidins. The optical density at 550 nm is determined, with reference to that of a standardized procyanidin preparation. Ribéreau-Gayon *et al*. (2006) comment that the original LA method tended to give higher results than the true concentration and maybe the updated versions of this method also give inflated measurements. Singleton (1974a) reviewed the analysis of phenolic compounds by the Folin–Ciocalteu method and also later in 1988. The most commonly used standard is gallic acid, hence results are expressed in mg L^{-1} gallic acid equivalent (GAE), although conversion factors can be used to express results as catechins, for example. Interference of most substances in wine is low, except for sugars in sweet wines, and occasionally from sulfur dioxide.

Tannin contents of wines are also conveniently expressed by these same workers in terms of various indices, reflecting particular characteristics of the tannins present that are of vinification interest. Thus, a Tannin Gelatin Index reflects the capacity to react with proteins, and so provides an assessment of likely astringency on tasting; values of 20–80 are obtained. A low value indicates a lack of body and may be the reason for 'an impression of flabbiness and bitterness'.

Of various proposed chemical fractionation methods Montedoro & Bertuccioli in 1974 (also reported in 1986), started with a methyl cellulose precipitation to remove higher flavan polymers (greater than trimeric), followed by assays with different reagents on the supernatant liquid analysing different phenolic compounds and groups.

HPLC measurements

With the advent of high-performance liquid chromatography (HPLC) techniques, greater differentiation became possible. Initially identification was mainly by UV/visible detection methods for the determination of tannins not only in wines (Singleton, 1988) but other beverages/fruits, etc. (Ramirez-Martinez & Clifford, 1993). In particular, monomeric anthocyanins can easily be determined using so-called reversed-phase HPLC. The eluant must have a low pH, which is usually achieved by acidifying with an appropriate acid, so that all anthocyanins elute in their coloured flavylium form (Bakker *et al*., 1986). As soon as the anthocyanins have become part of the so-called wine polymers (as a result of polymerization reactions), they can no longer be measured by HPLC, although remaining monomers determined by HPLC, combined with the measure of total wine colour, can give a measure of the degree of polymerization (Bakker *et al.*, 1986). HPLC has evolved enormously over the last two decades. Pascual Teresa *et al.*, (2000) showed the separation and identification of flavanols in wine. Using HPLC together with diode ray scanning and mass spectrometry, four monomers (catechin, epi-catechin, gallocatechin and epi-gallocatechin), eight dimers (B$_1$, B$_2$, B$_3$, B$_4$, B$_7$, GC-4,8-9C, GC-4,8-C and C-4,8-9C) and two trimers (GC-4, 8-C-4,8-C and C$_1$) were identified. The flavanols and prodelphinidins containing gallocatechin were identified in wine for the first time.

An enormous number of papers have been published, using HPLC to analyse water soluble phenolic compounds in wines. In addition to the HPLC and detection techniques available a decade or more ago, so-called hyphenated techniques have developed to powerful tools to aid the identification of anthocyanins and other water soluble phenolic compounds, such as HPLC coupled to mass spectrometry. Aspects of anthocyanins chemistry, including methods for isolation, separation and identification have been reviewed (Castaneda-Ovando *et al.*, 2009). The role of mass spectrometry coupled with a separation method such as HPLC for analyses of phenols in grapes and wines has been reviewed by Flamini (2003). One of the main technical problems to overcome was to ionize these non-volatile compounds prior to mass spectrometry, for which numerous methods have been developed.

Other methods

In addition, nuclear magnetic resonance (NMR) has been particularly fruitful to elucidate the structure of many more complex anthocyanins, for example Bakker *et al.* (1997) proposed a structure for vitisin A and B, new compounds isolated from ageing wines, using NMR. Many complex anthocyanins have been identified using NMR techniques (see Castaneda-Ovando *et al.*, 2009).

Sample preparation is a crucial part of any analytical method. Whereas in some HPLC methods wine samples can be prepared by just a simple filtration before analysing monomeric anthocyanins, prior to placing the wine sample into the HPLC, many new sample preparation techniques are becoming available, giving potentially greater sensitivity and more choice in the analytical technique. Even solventless preparation methods are becoming available, as reviewed by Nerin *et al.* (2009).

Traditional methods, such as those on Folin, are still very useful in wineries, whilst HPLC based methods are possibly more suitable as research tools, or more sophisticated quality control methods. De Beer *et al.* (2004) investigated established methods (Folin-Ciocalteu, reversed phase HPLC, normal phase HPLC and an aldehyde reagent DAC) and compared these with new methods (tannin and polymeric pigment, cyclic voltametry, and an antioxidant assay) for determining the phenolic contents in red and white wines. They found surprisingly good correlations between some of the methods, such as Folin and reversed phase HPLC. This would be expected for young wines, with low concentrations of polymeric compounds, which are not usually measured by reversed phase HPLC (Bakker *et al.*, 1986).

However, De Beer *et al.* (2004) reported that correlations between the analytical methods and sensory properties such as astringency or colour were very weak, indicating that the instrumental analytical methods may not measure just the compounds conferring such sensory properties to wines. A recent investigation by Mercurio & Smith (2008) into sensory properties related to astringency used the sensory term drying, defined as 'feeling a lack of lubrication or desiccation' and found good correlations between sensory panel scores for 'drying' for wines and two analytical methods for wine tannin analyses (methyl cellulose precipitatable tannin and Adams-Harbertson tannin assay), although no explanation could be given why this should be so.

In scientific assessment of sensory impressions, the use of panels and of threshold data is valuable to evaluate the sensory effect of particular compounds, as is particularly described for volatile compounds in Chapters 4 and 5. Some threshold data is available for wines, but values can be very dependent upon the other substances (e.g. sugar – acid) present; absolute values determined in pure water for a given compound may not be helpful, but are nevertheless of interest. Meilgaard (1982) in his massive investigations into the flavour of beer, in his so-called Flavour Index system (also discussed in Chapters 4 and 5) used difference threshold values, considered valid for estimation within an existing content of the same compound in the beer. Interestingly, the reference substance used for astringency in beer is quercetin (difference of $80\,mg\,L^{-1}$ in 0–$10\,mg\,L^{-1}$ existing amount), and isohumulone for bitterness (7–15, 0–30 respectively).

Analyses in grapes and during wine-making

An HPLC method was described by Price (1995), quoted and used by Fischer *et al.* (2000) in their investigations comparing phenolic compound extraction levels over three variants of 'cap management' during fermentation for the production of red wines. They determined the qualities present in the final wines of each of the following characterizing compounds or groups of compounds for taste/colour or other analytical significance: (1) gallic acid, (2) catechin, (3) epi-catechin, (4) caftaric acid, (5) quercetin 3-glycoside, (6) quercetin, (7) malvidin 3-glucoside, (8) monomeric anthocyanins, (9) polymeric anthocyanins and (10) polymeric phenols.

The extraction of flavanoid and other phenolic compounds in vinification depends as always for wine constituents, firstly on the amount available in the grape, itself varying with variety/cultivar and botanical location (skin, seed, juice, etc.), maturity and agronomic conditions (e.g. grape clustering). Secondly, extraction depends on the procedures of pre-fermentation and fermentation, including such factors as temperature and cap management during red wine vinification. The latter aspect features particularly in the work of Fischer *et al.* (2000), where comparison is made in data from traditional *punch down* of floating skins, and newer methods of mechanical punch down, mechanical pump over, and spiral stir-tank use. Comparisons are also made of results using thermovinification, followed by traditional vat fermentation of the juice. In general, up to 60% of the phenolic compounds from grapes will be extracted in red wine vinification, whilst the Fischer data shows marked differences depending on the methods and grape varieties studied. The phenolic content of the wine increases during the early stages of fermentation, whilst the must is in contact with the seeds and skins but subsequently begins to fall, as phenols bind and precipitate out with proteins and other plant materials. Extraction of phenols and coloured anthocyanins during various stages of wine-making is discussed in more detail in Chapter 1. There are further falls in content on fining and maturation. As discussed under ageing (Section 3.9 on changes during maturation), there are additional changes in the concentration and structure of the phenols.

Quantitative information is available. By the older analytical methods, Ribéreau-Gayon (1978) gives data for tannin content in skins, presumably determined by an LA method already described. His figures range from 0.36 to 0.76 mg per 200 grapes (weighing 190–366 g) in Sauvignon Cabernet of different years of harvest, and at two different vineyards – these would equate to approximately 2 g (2000 mg) per kg of grapes. Bourzeix *et al.* (1983), quoted by Jackson (2008), have determined phenolic content (excluding anthocyanins, separately given) in mg per kg of grapes, roughly corresponding to that in the must for red wine grapes. Amongst other varieties cited, there is a figure of 6124 for Cabernet Sauvignon, 7722 for Pinot Noir and 4163 mg per kg grapes for Malbec. Total phenol contents in dry white wines have been given as 50–250 mg L^{-1}, but ten times those amounts in red wines.

Singleton (1992b) quotes figures using the Folin–Ciocalteu method, and also expressed as mg GAE/berry (grape) or mg-GAE/g berry. His figures for white wines are average 0.650 mg per berry, with a range 0.421–1.083; for red wines, 1.777 mg per berry, with a range 0.900–3.478. He also reports that the total phenolic compound content is 5% in the juice, 50% in red grape skins or 25% in white grape skins, and the remaining 45–70% is in the seeds.

Flavanoids (catechins and procyanidins) clearly characterize red wines, where they constitute more than 89% of total phenol content (Jackson, 2008), whilst in white wines they make up less than 20%. Ribéreau-Gayon *et al.* (2006) give figures of 1–4 g L^{-1} for red wines; whilst in white wines, content ranges from 100 if the must settles properly, to 300–400 mg L^{-1} if maturation continues in the presence of lees (e.g. Muscadet-sur-lie). They may also be partly extracted by bonding as complexes with polysaccharides present in the grape.

Singleton (1992b) reports, for red grapes, 300 mg kg^{-1} catechins, of which 82% is in the seeds and 18% in skins. Dimers, and up to tetramers, average 4.94 mg kg^{-1} with 73% in the seeds and 27% in the skins.

The flavonol derivatives (e.g. quercetin), though present in the grape in the glycosylated form, are present largely in the aglycone form, Ribéreau-Gayon *et al.* (2006) quoting contents of up to 100 in red and only 1–3 mg L^{-1} in white.

Table 3.9 Phenolic compound contents after traditional red wine vinification with manual punch down (Pinot Noir fermentation).

Phenolic compound	Content (mg L^{-1})	Taste comments
Gallic acid (3,4,5-trihydroxy-benzoic acid)	6.6	sl. bitter/acid
Catechin (flavan-3-ol *trans)*	21.8	"
Epi-catechin (flavan-3-ol, *cis)*	13.0	"
Polymeric phenols (procyanidins)	36.4	astringent/bitter
Caftaric acid (caffeoyl-tartaric acid)	8.8	bitter
Quercetin-3-glycoside	<2.0	sl. bitter
Quercetin (flavonol)	0.4	sl. bitter
Malvidin-3-glycoside	52.8	colouring matter
Monomeric anthocyanins	61.5	"
Polymeric anthocyanins	0.0	"
Total = 89 mg L^{-1}, excluding all anthocyanins and 114.3 mg L^{-1} anthocyanins		

Data from Fischer *et al.* (2000). Reproduced with permission of John Wiley & Sons, Ltd. Taste comments by the Editors.

In wines not aged in oak, the main non-flavanoid compounds (i.e. derivatives of hydroxy-cinnamic and hydroxy-benzoic acids) commonly occurring as glycerides of various alcohols, are present at about 100–200 mg L^{-1} (GAE, or gallic acid equivalents) in red and 10–20 mg L^{-1} in white. Singleton (1992) quotes up to 260 mg-GAE L^{-1} in red wine and up to 175 mg L^{-1} in white wine.

With HPLC techniques it is possible to characterize individual compounds (or set of compounds) and determine their contents directly in mg per litre of wine (with authenticated reference samples) as fully tabulated by Fischer *et al.* (2000) for the red wines produced in their series of experimental trials. Thus in Table 3.9, data has been taken and presented separately for a Pinot Noir wine, by traditional fermentation in a 200 L tank with punch down in the tanks four times a day of the cap to enhance extraction. The other variant procedures used industrial tanks (5000 L capacity). In all cases, fermentation on the skins was allowed for five days, after which contact time with skins was stopped by pressing. There was no separate cooling or fining, but malo-lactic fermentation, SO$_2$ addition and filtration were practiced before bottling. Analyses took place after 6–10 months.

The finished wine had a pH of 3.8, titratable acidity of 4.9 g L^{-1}, alcohol content 12.5% v/v, sugar content 0.5 g L^{-1} and free/total SO$_2$ content 32/69 mg L^{-1}.

It seems that the quantities of components as determined by this method, when totalled and individually expressed, are substantially lower than those quoted above for red wines and grapes, though they may not encompass all those compounds determined by chemical methods. The polymeric forms were not analysed by the HPLC method used. The compounds chosen are, however, reasonably representative. The quantities of the same compounds produced by the other process variants are much higher, sometimes of the order of 600%. Sections 3.6.3 and 3.6.4 contain further information on contents.

Even so, the relationship of analytical content with sensory properties is not clear-cut. It is noted, for example, that interactions and the presence/absence of other components (aldehydes, SO_2, polysaccharides and indeed, anthocyanins) can modify sensory impressions in terms of balance, harshness and smoothness. For this reason, wine-making claims many craft elements, despite the wealth of accurate analytical techniques and data now available to the wine maker.

3.6.3   Bitter constituents

White wines

The main constituents producing any bitterness appear to be the small amounts of monomeric flavan-3-ols and -3,4-diols (catechins and leuco-cyanidins); $35\,mg\,L^{-1}$ is considered undesirable but not usually reached in white wines. The flavanonols are slightly bitter; the flavanone naringin is present in certain kinds of grapes such as Riesling (as also in bitter oranges, in the peel) and is bitter, though its concentration is considered to be insufficient to affect taste. The non-flavanoid caftaric acid at concentrations above $4\,mg\,L^{-1}$ will give a notice- ably bitter taste, as would with the *red* wines produced by Fischer *et al.* (2000) just described above. At a typical level of about $20\text{–}30\,mg\,L^{-1}$ in both white and red wines, tyrosol (*p*-hydroxyphenyl-ethanol) can cause bitterness.

Red wines

As with white wines, the flavan-3-ols and derivatives will cause bitterness, especially as their quantity is so much higher and above threshold levels. The higher polymers (especially those between 2–5 monomeric units) will tend to be more astringent than bitter in solution. Anthocyanins and their combination with tannins are said not to be very astringent (Ribéreau-Gayon *et al.*, 2006), but have a marked bitterness especially in young wines, when their structure is still being modified. Bitterness is a characteristic of too much extract from skins (as contrasted with that from seeds/stalk), together with an herbaceous element to the flavour if the grapes are insufficiently ripe.

3.6.4   Astringency

Astringency forms part of the mouthfeel of many drinks, such as tea, coffee and wine. It is an essential sensory characteristic of red wine and adds a certain amount of 'bite' to the wine, often described as a dry and puckery sensation. Astringency in red wine is primarily due to flavanoid phenolic compounds, which are natural constituents of the grapes from which wines are made. When a high concentration of such compounds is extracted during wine-making, the wine is considered too harsh and too astringent, and will need considerable time to mature to lose some of its astringency and become more mellow. Conversely, red wines lacking astringency are often considered rather flat.

Astringency is not detected by special taste buds or the olfactory epithelium but, rather, it is a dry and puckery sensation perceived in the mouth. The perception of astringency is not restricted to any particular area in the mouth or the tongue and it can take some time to develop to its full intensity. It also requires time for the astringent sensation to appear and disappear from the mouth. The mechanism of astringency perception is not well understood, while the exact nature of the compounds eliciting an astringent sensation is also still a subject of research. A comprehensive review by Bajec & Pickering (2008) concludes that the wealth of papers suggests that astringency is a complex, multifaceted sensation, complicated by a number of variables and there is no clear hypotheses regarding a mechanism for astringency perception. It is suggested that distinct astringent compounds may have different pathways in eliciting astringency. We still lack clear definitions regarding the precise oral sensations contributing to astringency, which makes it difficult to compare some of the published sensory data. It would appear to be possible to train panels to assess red wines on specified sub qualities of astringency (Gawel et al., 2001) and the suggested astringency wheel with terms to describe the sensory sensations (Gawel et al., 2000) may help future investigations of astringency.

One hypothesis regarding the mechanism and compounds involved was proposed by Bate-Smith in 1973, who suggested that astringent compounds act on the proteins of the saliva. He proposed that polyphenolic compounds form complexes with salivary proteins and/or mucopolysaccharides, either precipitating them or causing sufficient conformational changes, so that proteins lose their lubricating power and make the mouthfeel rough and dry. Specifically the precipitation of proline-rich proteins contributes to astringency perception (Clifford, 1993). However, the direct action of polyphenols on oral tissues is also possible.

Astringency is usually attributed to high molecular weight phenolic compounds, but even small wine flavanoids, (+)-catechin and (−)-epicatechin elicit astringency in red wines and, like most of the astringent compounds, the two monomers were also bitter (Kallithraka et al., 1997a). However, these monomeric phenolic compounds are not chemically defined as astringent. Interaction between astringency and acidity in wine also occurs, as shown by the increase in astringency as the pH of the wine decreased (Kallithraka et al., 1997b), and further discussed by Bajec & Pickering (2008).

3.6.5 Mouthfeel

'Mouthfeel' is a term given to a sensation perceived in the mouth as a whole and often corresponds with the term 'body' (though this may include a combined effect with the volatile substances as well). Body is often correlated with viscosity or 'thickness' of the liquid taken by mouth. The hydrophilic proline-rich protein–phenol complex referred to above may be a good example of a contribution to mouthfeel, although this hypothesis is still a matter of debate (see Bajec & Pickering, 2008).

A sensory study on mouthfeel of red and white wines showed that mouthfeel attributes could not be closely related to the phenolic composition and structure, but the presence of anthocyanins in wines increased the perceived astringency (Oberholster *et al.*, 2009), and the authors suggested that the presence of anthocyanins may partly explain the differences perceived between the mouthfeel properties of a white and red wine, although further studies are needed to determine the effect of anthocyanin polymerization products, such as the vitisins discussed below, on the sensory properties, in particular related to mouthfeel and astringency attributes.

3.7 Colouring matter

Whilst colour is not a direct contributor to wine flavour, it is a part of the overall perceived organoleptic impression. Compounds contributing to colour are all phenolic compounds, and many complex reaction products are formed during wine-making and maturation. Over the last decade or so there has been significant progress in our understanding of these reactions. Of course, these reactions may also influence the sensory properties related to bitterness, astringency and mouthfeel of wines, discussed in Section 3.6. The actual colouring matter in wine is best discussed separately, for red and then white wines since their chemical origin and significance is rather different in the two cases.

3.7.1 Colour of red wines

The colouring matter of red wines derives from the presence of anthocyanins in the form of a glycoside of the flavanoid, anthocyanidin, already described in the introduction of Section 3.6.1. As also stated there, anthocyanins can exist as both mono- and diglycerides, but in *Vitis vinifera* only monoglycerides exist. They may be further complexed by the bonding of acetic acid, coumaric or caffeic acid to the glucose component. Their chemical and water solubility is enhanced by their glucosidic structure but acylation of the sugars makes them less water-soluble.

The properties of, and amount of, each of the five types of anthocyanin varies widely among the cultivar/varieties of grapes and according to growing conditions, as discussed in Section 3.6.2. The proportion markedly influences hue and colour stability. The colour is determined by the number of hydroxyl groups in the B-phenyl ring of the anthocyanin and ranges from blue to orange to red. Resistance to oxidation by whatever mechanism is greater where there are no ortho or adjacent hydroxyl groups in this ring, as with malvidin and peonidin. It is malvidin (3-glucoside) with two methoxy groups and only one hydroxyl group in the B-phenyl ring that is the most dominant in red wines and is largely responsible for the colour of young wines. Red anthocyanins are in the form of flavylium positive ion, but other colourless forms are also present; the distribution between coloured and colourless forms depends on the pH, with lower pH values expressing more colour. The whole chemistry of their existence and change on fermentation and ageing is, in fact, highly complex, involving also agglomerate formation with other tannins.

A mini review (Brouilard *et al.* 2003) summarized these developments in anthocyanin chemistry in wines. Whereas anthocyanins in flowers are often extremely complicated structures with many acylations in order to confer stability, the anthocyanins in grapes and very young red wines are extremely simple, and the anthocyanins disappear rapidly during maturation to be replaced by new pigments, which possess great longevity in colour. This so-called ageing of red wines generally improves the sensory attributes of red wines, the formed red colour is more stable than the anthocyanins, and has a different colour quality. These authors gave an overview of current knowledge of the reactions pertinent to these colour changes, and suggested that recent research indicated possible beneficial effects of red wine pigments on health.

The colour changes in red wines due to condensation reactions between anthocyanins and other phenolic compounds naturally occurring in wines are well documented, and until recently these were the reactions thought to exert the greatest influence on the colour changes in red wines during maturation, including one involving the condensation of anthocyanin and procyanidin by the intermediary of acetaldehyde, as discussed by Freitas & Mateus (2006). Thus polymerization reactions can occur direct between anthocyanins and procyanindins, for example catechin, forming typically a bond on the 4 position of the anthocyanin and 6 or 8 position of the catechin. Another direct condensation suggested is between the 4 position of a procyanidin and the 8 position of the anthocyanin. There are also polymeric structures formed due to acetaldehyde mediated condensations between anthocyanins and flavanol (Timberlake & Bridle, 1976; Bakker *et al.*, 1993), whereby the aldehyde moiety forms a bridge between the anthocyanin most likely at the 8 position and a catechin molecule. Acetaldehyde is very reactive, and can be formed in maturing wines as part of alcohol oxidation, as discussed in Section 3.9.1. Possibly other aldehydes can also participate in this reaction, although such products have not been found in wines.

Several reaction mechanisms have been proposed, but it was not until the introduction of very powerful analytical techniques (discussed in Section 3.6.2) that can be successfully applied to phenolic compounds, including the antho-cyanins and the reaction products, that more information regarding these highly complex reactions has become available. Until recently, the main changes in anthocyanin composition and the resulting colour changes in red wines were believed to be only due to polymerization reactions.

However, publications have reported the formation of new malvidin-derived pigments, indicating that anthocyanins undergo even more complex changes in wines than hitherto had been studied in polymerization reactions. Bakker *et al.* (1997) reported the isolation and identification of such anthocyanin-based pigments, named vitisin A and vitisin B, formed in ageing red fortified wines. These compounds are more coloured in wine than the anthocyanins they are derived from and are resistant to bleaching by sulfur dioxide (Bakker & Timberlake, 1997). Vitisin A is formed by reaction of the main anthocyanin in wine, malvidin 3-glucoside, with pyruvic acid, a compound naturally present in wine (Romero & Bakker, 2000).

More information on this type of anthocyanin reactions has emerged, including the reaction mechanisms, and it is now known that anthocyanins in wines participate in many reactions, not just in polymerization reactions with phenols and acetaldehyde discussed above, but also with other compounds present in wines, giving an array of more colour stable coloured compounds ranging from orange to blue. Anthocyanins easily react in wines, although the type of reaction is pH dependent. Anthocyanins can undergo oxidation and electrophilic substitution, and also nucleophilic addition of the C-ring (see 3.VII) when the molecule is positively charged (reviewed by Fulcrand et al., 2006). Anthocyanins react with small compounds naturally present in wines, such as pyruvic and phenolic acids, acetaldehyde, caffeic acid and p-vinylphenol and form a range of different compounds belonging to pyranoanthocyanins.

The simplest pyranoanthocyanin is vitisin B, where acetaldehyde reacts with an anthocyanin and causes ring closure between the 5-hydroxyl group and the C on the 4 position, thus forming a pyran ring. The pyran ring will protect the anthocyanin from nucleophilic attack, and hinders the formation of colourless carbinol base and protects against bleaching by sulfur dioxide. Anthocyanins react in this way with small compounds in wines and form numerous compounds (vinylphenol adduct, vinylcatechol adduct, vinylguiacol adduct, vitisin A, vitisin B and others), all having in common the pyran ring formed, thus protecting the colour and stability of the anthocyanin molecule, reviewed by Cheynier et al. (2006), Fulcrand et al. (2006), Freitas & Mateus (2006) and Rentzsch et al. (2007).

Pyranoanthocyanin-flavanols have also been detected, whereby a flavanol is attached to the pyran ring of the pyranoanthocyanin. It has been detected in model solutions and later isolated from wines (see a review by Freitas & Mateus, 2006). Portisin, so named since it was isolated from Port wine, is a pyranoanthocyanin linked to flavanol through a vinyl group and is bluer than the above described anthocyanin derived compounds (see Freitas & Mateus, 2006). In fact Rentsch et al. (2007) suggest that considering the pathways of formation, additional structures of pyranoanthocyanins can be expected to be discovered in the near future.

Of interest from a sensory point of view is the contribution pyrano-anthocyanins make to the colour of wine. Many are slightly more orange than malvidin 3-glucoside (the main anthocyanin in wines), although portisin has a rather blueish hue. There is still a debate ongoing regarding the contribution of the anthcyanin derived compounds to wine colour and estimates vary. The compounds are more stable to pH change than malvidin 3-glucoside, and therefore a greater part of the molecules present will express colour at wine pH, but they are less strongly coloured and present only at low concentrations (see Rentsch et al., 2007). Also these authors stressed that two other factors contribute to wine colour: copigmentation, a physical phenomenon (reviewed by Boulton, 2001), and genuine of polymeric pigments formed in wine.

There are also studies on the possible enhanced formation of pyrano-anthocyanins during wine-making, by selecting yeast strains producing high levels of pyruvic acid and acetaldehyde (Morata *et al.*, 2003), looking at the effect of yeast strain on vinylphenolic pyroanthocyanins (Morata *et al.*, 2007), or the effect of pH, temperature and sulfur dioxide during fermentation on vitisin production (Morata *et al.*, 2006). Sulfur dioxide strongly binds to aldehyde and hence reduces vitisin formation. Other parameters tended to depend on yeast strain. It would seem that careful selection of yeast strain and controlling the fermentation conditions can significantly increase the formation of pyranoanthocyanins.

The changes accompanying wine anthocyanin and phenol reactions during wine ageing are a general reduction in astringency, often described as a softening of the tannins. It would be interesting to determine whether any of the changes can be linked to changes in sensory properties, and whether there are specific anthocyanin derived products having an effect on the astringency related descriptors.

For more detailed information, the reader is referred to Jackson (2008) and the references therein, many to Somers and colleagues, and of course to Ribéreau-Gayon *et al.* (2006). For a review covering the chemistry of wine pigments, see Freitas & Matteus (2010). They concluded that the colour in red wine depends on the formation of anthocyanin derived compounds, as briefly discussed above, but also on the physico-chemical phenomena, such as copigmentation and selfassociation, and such phenomena could be the first step of the chemical reactions. An explanation is given of the subtle changes that occur, leading to wine progressively taking on a brickish hue, due to the brownish hue of most polymers, although it is by no means clear what contribution each of the many anthocyanin derived pigments and polymers make to wine colour. Wine-making conditions, as discussed in Chapter 1, may also influence the pigment formation, such as the metabolites (acetaldehyde, pyruvic acid) formed by yeasts during fermentation. The important influence of the sulfur dioxide present should also be noted, since a compound complex of an anthocyanin with sulfur dioxide is uncoloured, although this particular reaction is reversible on loss of sulfur dioxide. It also binds reversibly to acetaldehyde, influencing the acetaldehyde concentration in wine, and hence the pigment formation reactions.

3.7.2 Colour of white wines

There is much less data and explanation on the colour of white wines, but the content of phenolic compounds, potentially colouring the wine, are much lower than in red wine. Most consists of readily soluble non-flavanoid material such as caffeoyl tartaric and related derivatives. Flavanoid phenols are extricated slowly, and they are only found in significant amounts in juice macerated with pomace, primarily as catechins and catechin-gallate polymers.

The mechanisms involved in oxidative browning in wines have been reviewed (Macheix *et al.*, 1991; Toit *et al.*, 2006; Li *et al.*, 2008; Kilmartin, 2009) and some of the information is relevant to the colour of white wine. Overall, the mechanisms relevant for white wines are still incompletely understood. Browning can be enzymic, mostly started during fermentation, and non-enzymic, most relevant during wine storage. Sulfur dioxide has antioxidant properties, and protects wines against browning reactions, discussed in Section 3.8.1. Although it is generally considered a good thing to protect white wines against oxygen, some wines may then become susceptible to *pinking*, i.e. developing a slightly pink colour when accidentally being exposed to oxygen, such as during bottling. The underlying chemistry is still not understood.

The chemistry involved in browning reactions is similar to reactions in red wines. The oxidation of o-diphenols to o-quinones can lead to the formation of coloured polymeric pigments with high molecular weight, a process which can be regenerated in coupled oxidation with other compounds. Once the process has started, it continues readily, since the dimers can be regenerated, and renewably oxidized, accelerating the reaction. In the presence of transition metals hydrogen peroxide is produced and ethanol is oxidized to acetaldehyde. The hydrogen peroxide in the presence of sufficient iron can produce free hydroxyl radicals, which oxidize ethanol and tartaric acid (both abundant in wine) to acetaldehyde and glyoxylic acid. Both these compounds can participate in reactions with flavanols, giving yellow, brown or colourless polymers in wines.

Browning of white wines is thought to be a result of three different types of reactions, all involving phenolic compounds. The first type of browning reactions involves flavanols, such as catechins and dimeric procyanindins, and coupled oxidative reactions give brown polymers. Such coupled oxidation reactions may also involve the hydroxycinnamic acids, also there seems to be little correlation between their content and white wine browning. Sensitivity of the flavanols to browning reactions have been shown to vary, and wine-making techniques such as pressing method, skin contact and cultivar all affect the procyanindin concentration and the wine's sensitivity to browning. The second type of reactions involves the oxidation of tartaric acid to glyoxalic acid, described above. The glyoxalic acid forms a bridge in the form of a carboxymethine link between the flavanol units, in the presence of copper or iron ions. Tartaric acid is added in some countries to lower the pH, and it can contain traces of glyoxalic acid, which can influence the colour of white wine in the presence of sufficient copper ions. The third type of reaction involves acetaldehyde, present as a fermentation product or formed as part of the coupled oxidation of phenols described above for red wines. Acetaldehyde can form ethyl bridges between flavanol molecules, such as catechin. Mixed polymers are formed, all contributing to the colour of wines.

The question still remains to what extent the various reactions occur in white, and what the relative effects are on the colour and the sensory attributes of the wine, such as astringency, although white wines are usually much less astringent that red wines, due to their lower phenolic contents.

Since many white wines are sold to be drunk within a year of bottling, many of the browning reactions are more likely to occur during bulk storage. Assuming the wines are bottled avoiding oxidation, changes in-bottle are expected to be small over the short term. Kallithraka *et al.* (2009) reported that young white wines could be stored for at least nine months in-bottle without any significant changes in colour and they suggested that the changes in catechin and cinnamic acid composition were minimal, hence the wines maintained their antioxidant properties, attributed to have a beneficial effect on human health.

3.8   Other constituents

Two other well known constituents of wine, most conveniently considered here, are of significance to wine flavour. These compounds are gases in their pure form, but can also exist in solution, both in dispersion and in bubble form in aqueous liquids; they are, of course, sulfur dioxide and carbon dioxide. Another well known gas, oxygen, is also relevant here due to its chemically reactive potential at different stages of vinification, even in a finished wine whilst it is being consumed from an opened bottle.

3.8.1   Sulfur dioxide

Sulfur dioxide (SO_2) has a long history of use. Burning sulfur is thought to have been used by the Romans, and reference to its use dates back to the fifteenth century. It is therefore not surprising that even today it is still one of the most useful chemical compounds that can be used in the wine industry. Solutions of sulfur dioxide are useful in sterilizing wine equipment, although care has to be taken to avoid damage by use of high concentrations. In wine-making, additions of SO_2 at various stages can have a beneficial effect.

Firstly, SO_2 has strong anti-microbial activity, and so when added just prior to fermentation it will inhibit the growth of indigenous yeasts and contaminating bacteria, especially lactic acid and acetic acid bacteria. Fermenting yeasts of *Saccharomyces cerevisiae* strains are not particularly sensitive to SO_2, and even produce SO_2 themselves (Rankine & Pocock, 1969). Risk of spoilage by bacteria is also much lower if SO_2 is used during wine storage and just before bottling. Secondly, sulfur dioxide also reduces the incidence of browning of phenolic compounds. The enzyme laccase, also a cause of browning of phenolic compounds but derived from *Botrytis cinerea*, has its activity reduced by SO_2.

Thirdly, sulfur dioxide acts as an antioxidant, suppressing the chemical oxidation of phenolic compounds, thus effectively reducing browning reactions. Especially in white wines, a pale yellow colour can be maintained by preventing browning actions. A further benefit is that wines have a fresher flavour by reacting with acetaldehyde and preventing the loss of varietal odours. The latter action is attributed to suppressing the formation of so-called quinones (from phenolic compounds), which would oxidize the flavour compounds. However, although there are benefits from sparingly using sulfur dioxide in wine-making, its use is limited and there are undesirable effects. It is essential, therefore, to use good wine-making practices, as described in Chapter 1, to reduce the amounts of sulfur dioxide used to a minimum.

Basic chemistry

Sulfur dioxide (SO_2) is a colourless gas (bp $-10°C$) with a characteristic pungent, choking odour (even concentrations above 20 ppm in the air have a marked effect). It is very soluble in water, so that, at 20°C, 11.6 g will be taken up by 100 g water at normal atmospheric pressure (i.e. a water solubility equivalent to about 50 times its own volume of gas), but solubility decreases as the temperature is increased.

The aqueous solution behaves as if it were an acid, known as sulfurous acid H_2SO_3, dissociable into firstly H^+ and HSO_3^-, representing a fairly strong acid (H^+ concentration equivalent to methyl red on filtration, and secondly, $HSO_3^- \leftrightarrow H^+ + SO_3^{--}$ with a dissociation constant of about 10^{-7}). The solution is, however, oxidizable to sulfuric acid (H_2SO_4), especially in the presence of suitable catalysts. The substance is therefore a powerful reducing agent, represented by Equation 3.5.

$$SO_3^{2-} + H_2O \rightarrow SO_4^{2-} + 2H^+ + 2 \text{ electrons} \tag{3.5}$$

Simultaneously, the acidity of the aqueous solution is increased.

In solution, sulfur dioxide will be present in several forms, i.e. active molecular SO_2, ionic SO_2, undissociated H_2SO_3 consisting of free SO_2, unstable combined forms with carboxyl substances, and a more stable aldehyde–SO_2 complex (hydroxysulfonates). Only the free SO_2, chiefly molecular SO_2, will be effective in anti-microbial action – a concentration of 2 mg L^{-1} molecular SO_2 should be sufficient to protect a wine during storage. Analysis expresses content in terms of both free and total acidity. Equilibria between the various forms of SO_2 depend upon the pH, with a lower pH increasing the amount of molecular SO_2.

Technical use

In practice SO_2 is added in the form of potassium metabisulfite ($K_2S_2O_5$) or potassium bisulfite ($KHSO_3$), or as the gas dissolved in water, which means it can be pumped in automatically.

The initial amount used before fermentation will be related to the intensity of microbic population on the grapes expected in vineyard practice, by use or otherwise of chemical sprays (pesticide/fungicides) against micro-organisms, such as the odium fungus and mildew. In use, the amount is controlled by various legislations in practice (see Chapter 2, Section 2.6 '*Quality Control*'); and by and large, its use is minimized by the wine maker.

Sulfur dioxide has known physiological effects on humans, so that there is a WHO/FAO recommendation on maximum daily intake from all sources of 0.7 mg per kg body weight. Jackson (2008) quotes 400 mg (free and bound) SO_2 consumption for several weeks as 'having no adverse effects on some people', while most commercial wines contain less than 100 mg L^{-1}. Large doses are toxic.

Taste effects

Sulfur dioxide has an effect of its own on taste/flavour, though this is likely to result from the free SO_2 rather than the sulfite/bisulfite and hydrogen ions contributing to acidity. Amerine & Roesler (1983) say that most individuals will begin to detect a 'burned-match' distinctive odour at about 15-40 mg L^{-1}. Spanish investigators have published data (Aldave, 1993) showing some marked compositional differences, in respect of a number of other volatile compounds, between those young wines vinified with and without sulfur dioxide. This data will be discussed later, in Chapter 4.

Many wines in the past have been 'over-sulfured', though it would be claimed that the excess sulfur would disappear on long ageing and indeed was regarded as necessary for this process. 'Sulfurous' wines are still detected by wine experts and condemned on that account. The 'sulfury prickle' sometimes detected in wines, especially young and fresh white wines, can often be dispersed by swirling the wine vigorously in the glass for a minute or two.

3.8.2 Carbon dioxide

Carbon dioxide, CO_2, is a colourless gas with a faintly pungent odour and gives a tingling sensation in the nose when inhaled. It is well-known as 'dry ice', but not as a liquid. However, it is now important as a supercritical fluid for various extraction purposes (e.g. decaffeination of green coffee beans). It is a heavy gas, with considerable known quantitative data under various conditions. Under atmospheric pressure, water dissolves in its own volume of carbon dioxide at room temperature (i.e. less than sulfur dioxide), corresponding to about 2 g CO_2 per litre of water, though solutions can easily be supersaturated. Solubility is greater at higher pressures.

The dispersion of carbon dioxide as 'bubbles' in an aqueous solution rising to the top is very familiar, in sparkling wines and carbonated soft drinks.

In aqueous solutions, equilibria are established between carbon dioxide molecules, water, carbonic acid and its ions (Equation 3.6) and have been determined (at 18°C) as given in Equation 3.7, which shows that carbonic acid as an acid is very weak, both on its first and second dissociation constants. From the ratio (1) in Equation 3.7 it can be seen that only 0.7% of dissolved carbon dioxide is in the carbonic acid form. Ratio (4) is the apparent first dissociation constant.

$$CO_2 + H_2O \leftrightarrow H_2CO_3 : H_2CO_3 \leftrightarrow H^+ + HCO_3^- : HCO_3^- \leftrightarrow H^+ + CO_3^{2-} \qquad (3.6)$$

$$
\begin{aligned}
&(1)\ [H_2CO_3]/[CO_2] = 7 \times 10^{-3} : \\
&(2)\ [H^+][HCO_3^-]/[H_2CO_3] = 5 \times 10^{-5} : \\
&(3)\ [H^+][CO_3^{2-}]/[HCO_3^-] = 6 \times 10^{-11} : \\
&(4)\ [H^+][HCO_3^-]/[CO_2][H_2CO_3] = 3.5 \times 10^{-7}
\end{aligned}
\qquad (3.7)
$$

Formation and handling of CO_2

Carbon dioxide, of course, arises in wine, as a by-product of the alcoholic fermentation (Equation 3.8).

$$C_6H_{12}O_6 \rightarrow 2C_2H_5OH + 2CO_2$$

Glucose / fructose $(MW\ 180) \rightarrow$ Carbon dioxide $(MW\ 44)$ (3.8)

Thus, 1 g of glucose/fructose will produce, theoretically, 88/180 g = 0.49 g (49% yield) of carbon dioxide, which occupies 0.49 × 22.4/44 = 0.249 L = 249 ml, or around 260 ml, at normal temperature and pressure [1 mole-gram of CO_2 (44 g) occupies 22.4 L at standard temperature, 0°C, and pressure (STP)].

The greater proportion of this volume will escape in gaseous form to the atmosphere, and will carry off with it some of the ethyl alcohol and volatile aroma substances as they are generated during fermentation. Furthermore, part of the heat generated will be lost with the gas. If the CO_2 gas could not escape, the fermentation would be inhibited. Carbon dioxide gas is a toxic hazard, so that build-up in the atmosphere around the fermentation vessels should be avoided, thus it has to be kept below 1% v/v, while about 3% is a reason for evacuating the premises. Cleaning the inside of fermentation vessels after fermentation is particularly hazardous since the heavy CO_2 can linger in the vessel for some time. Winery workers need to ensure that the CO_2 content in vessels has fallen to safe levels. Conversely, the generation of carbon dioxide has a useful rôle in de-aerating and providing a non-oxidizing atmosphere within the fermenting mass and the space above.

With a typical 212 g glucose/fructose content per litre of must initially, we can expect the generation of up to 212 × 88/180 g = 103 g of carbon dioxide, of which only about 2 g (2000 mg) will remain after fermentation (i.e. about the solubility level), though substantially lower for still wines at the time of bottling.

Sensory factors

Peynaud (1986) describes the taste importance of carbon dioxide as of its prickly sensation on the mucous surface of the mouth as it escapes, though in a wine it is only perceptible at above 500 mg L^{-1}. Dissolved CO_2 has a slightly acidulous nature. Ribéreau-Gayon et al. (2006) suggest a maximum of 300 mg L^{-1} for good red wine.

The presence of low levels of carbon dioxide helps to protect the wine against oxidation. Especially, 'fruity' light white wines made to be drunk young are often bottled with high levels, up to 500 mg L^{-1}, of carbon dioxide to protect their pale yellow colour and even enhance fresh aroma. The vaporizing carbon dioxide during drinking is thought to carry some of the 'fruity' smelling aroma compounds (esters), thus possibly enhancing their perception.

Carbon dioxide is of especial interest in sparkling wines, and especially in Champagne, where it is further generated in-bottle by a second fermentation. The production of gas will be determined by the amount of sugar added with the yeasts to the bottle. The subsequent release of the carbon dioxide gas on opening a bottle is part of the attractiveness of these wines. Jackson (2008) describes this release from a 750 ml bottle as a drop in pressure from about 600 kPa (6 atmospheres ~ 86 psig) to atmospheric (101.3 kPa or 14.3 psig) with a decrease in dissolved CO_2 amount from about 12 to 2 g L^{-1}, and the gradual liberation of some almost 5 L of carbon dioxide gas. This gas is not, however, released immediately and 'there is insufficient energy for bubble formation, and most enters a metastable state'. The description of the mode of release of bubbles by nucleation (homogeneous or heterogeneous) or from stabilized micro-bubbles provides an interesting insight into the observed phenomenon of bubble rising and effervescence.

The characteristics of the bubbles are also considered relevant to the sensory quality of the wine (Jackson, 2008). The formation of durable, continuous strains of small bubbles is an important quality attribute, although factors regulating this quality are still not fully understood. Another important quality aspect is the formation of a ring of bubbles around the ring of the glass. The bubbles are also thought to contribute to perception since several volatile compounds adsorb onto the surface of the bubble, and are released where the bubble breaks, in the air or in the mouth. Factors affecting carbon dioxide in the bottle are temperature, sugar and ethanol content, and for the correct bubble formation in the glass, a suitable shape clean glass is pertinent.

3.8.3　Oxygen

Contact with air, and therefore with oxygen, by all foods and beverages during manufacture and subsequent storage, is an important factor in their changing chemical composition, even though slight, and therefore perceived flavour. This subject is well covered in the book *Shelf-Life Studies* (ed. Charalambous, 1993) with a wide range of different authors/subjects, including a substantial chapter on packaged beer (Kaminura & Kaneda, 1993), dealing with all aspects of its potential spoilage, with some similarities and relevance to occurrences in wine. The several chapters on wine, discussed elsewhere in this particular book, only deal with the storage of bottled wine, where the rôle of oxygen is minimal.

The changes in flavour (taste/aroma) that occur in wine are usually regarded as beneficial when there is a slow and periodic consumption of small amounts of oxygen over a period in the maturation of red wine, but usually detrimental to white wine. Some other foodstuffs, like cheese, develop flavour characteristics over a period in so-called ageing, with flavour implications. There are, however, no clear guidelines as to what levels of contact give optimum quality for wines, despite there having been a long interest in this topic. Pasteur already determined in the 1860s that oxygen plays a crucial role in the maturation of wine.

Basic chemistry

Oxygen is a colourless gas, when pure is without odour, and has a density of 1.327 g L^{-1} at 21°C and 1 atm- pressure. Its density is, expectedly, higher/lower at lower/higher temperatures, e.g. 1.430 g L^{-1} at 0°C. It is only sparingly soluble in pure water. Accurate data (Kirk-Othmer, 1985, see Appendix II) is available over a range of temperatures, e.g. 49 ml at 0°C, 31 ml at 20°C, and 26 ml at 30°C, with volumes being measured at standard temperature (0°C) and pressure (1 atm) (STP). Solubility will also vary with pressure, according to Henry's Law, but this factor is of limited interest in vinification. It appears to be considerably less soluble in ethyl alcohol and aqueous alcohol solution; but similarly in vegetable oils, e.g. 27 ml L^{-1} at room temperature (Hoffman, 1989). The figures available for wine from Ribéreau-Gayon *et al.* (2006) indicate that between 10.5 and 5.6 mg of oxygen per litre from 5–35°C are required to saturate wine, which correspond to approximately 7.3 and 4.0 ml of oxygen (measured at 0°C) per litre of wine. In wine-making practice, weight units are generally used. Furthermore, figures for wine are only immediately valid, due to developing chemical reaction.

Oxygen is a very reactive substance for many of the chemical compounds in wine, especially phenolic compounds, lipid substances, aroma substances, amino acids, sulfur dioxide and, not least, alcohol itself. The situation is highly complex and not fully understood, with many different protective mechanisms around, so that each substance under so-called oxidation is best considered separately. Some understanding in terms of modern electronic theories of valency and chemical reaction is necessary.

The particular unusual configuration of the oxygen molecule accounts for its reactive properties (see Appendix I, Section I.3.3). The oxygen molecule in its ground state is known as triplet oxygen ($^3O_2$), which can cause some oxidative reactions, but other reactions may need a whole range of activated oxygen molecules, for effective action, often by a chain reaction of radicals. These are singlet oxygen ($^1O_2$), superoxide (O^{2-}), hydroperoxyl radical (^-OOH), hydrogen peroxide (H_2O_2) or hydroxyl radical ($^·OH$). In addition, catalysts such as metals (Fe/Cu), enzymes (phenoloxidase, oxidases, lipoxygenase, etc.), may be needed, without which different reactions may not proceed. Chain reactions can be terminated by antioxidants. Subsequent breakdown reactions of oxidized long-chain lipid substances are important, as they may result in the formation of odoriferous small molecules, as shown for wine (see Section 3.7 and Appendix I.3).

Reactions with oxygen are now considered part of an overall group of reactions, referred to as oxidation, even though in some cases, no actual oxygen may be needed. Substances are said to be oxidized when they fix oxygen, lose hydrogen, or lose one or more electrons.

Reduction is the reverse of these reactions, and in practice there is always a balance between the two, so-called oxidizing–reducing systems, with their condition being interpreted by redox potential measurements, discussed below.

Oxygen content in wines

Grapes contain very low levels of dissolved oxygen; the content of oxygen in a finished bottled wine is also low, but not negligible. The amount of oxygen present in the must during fermenting and maturation in vats/barrels will vary during the different stages of vinification, dependent upon pick-up and absorption by oxidation. The overall quantities involved will help determine the final flavour characteristics of the wine.

Although measurement of oxygen content in wines is difficult, Ribéreau-Gayon *et al.* (2006) describe the use of polarographic analysis, using specialized electrodes in a battery. The quantity of dissolved oxygen is directly proportional to the intensity of the electric current as a result of movement of electrons, by so-called redox potential, which is further discussed in Section 3.8.

Ribéreau-Gayon and his colleagues at the University of Bordeaux have published their quantitative findings on the pick-up or ingress of oxygen in each of the various steps involved in the maturation of red wine in barrels. Vivas (1997) in a doctoral thesis, quoted by Ribéreau-Gayon *et al.* (2006), provided data to show that:

(1)   During general handling a total of some $18\,mg\,L^{-1}$ is picked up and dissolved in the wine.
(2)   During treatments, a total of $22\,mg\,L^{-1}$.
(3)   During racking $(0-5\,mg\,L^{-1})$ and topping up of barrels through the bung $(0.25\,mg\,L^{-1})$.

These figures cover an extended period of time and individual steps show no more than $7\,mg\,L^{-1}$ pick-up at a time (i.e. saturation solubility).

A review of Toit *et al.* (2006) quotes individual figures of oxygen uptake at various handling stages: pumping $(2\,mg\,L^{-1})$, transfer from tank to tank (up to $6\,mg\,L^{-1}$) filtration $(4-7\,mg\,L^{-1})$, racking $(3-5\,mg\,L^{-1})$, centrifugation (up to $8\,mg\,L^{-1}$) bottling $(0.5-3\,mg\,L^{-1})$ and barrel ageing $(20-45\,mg\,L^{-1}$ per year). Dependent on the time in-barrel, the uptake during barrel ageing is potentially much larger than any other uptake. However, the additional amounts involved in the seepage of oxygen through the wood of the barrels and their effect has been a subject of controversy, with comments from Barr (1988), Robinson (1995) and Singleton (1974b). The latter tends to discount the significance of this particular uptake source on any flavour changes. Vivas (1997) provides figures, on a $mg\,L^{-1}$ year rate basis, for different types of barrels and bungs, which range from 10 to $45\,mg\,L^{-1}$. Lower humidity, tight grain and thinner staves all allow more oxygen uptake into the wine. Leakage can occur through gaps in the staves and there can be differences of permeability of the oak wood used.

However, conditions can vary very considerably over a range of wines made. Jackson (2008) comments from Singleton (1974b) that well made barrels, tightly bunged, full of wine (low ullage) and turned so that the bung is kept wet, are unlikely to let much air (oxygen) through into the wine, compared with racking, sampling and topping up. Vivas (1987) additionally comments on the difficulty of determining separately the likely absorption of oxygen by the elagi tannins released from and in the wood. Total pick-up is said to be up to $250\,mg\,L^{-1}$ compared with lower amounts (ca. $100\,mg\,L^{-1}$) by ageing in large closed vats. These amounts expressed in millimoles (i.e. 7.8 and 3.12) per litre may be used for calculation in stoichiometric equations of oxidative change of compounds present. These figures seem rather high compared with the amount desirable (40 mg L⁻¹), quoted elsewhere by Jackson (2008) and Ribéreau-Gayon *et al.* (2006), although not entirely out of line with figures quoted in the section below for amounts of oxygen red wines may take up whilst improving during maturation. Absorption will also be determined in practice by the amount of movement and agitation occurring, since mass trans-fer equations involve exposed surface area and film-coefficients of diffusion.

The pick-up amount of oxygen during pre-fermentation stages of wine-making has also been investigated. Thus, during the crushing of grapes, Jackson (2008) mentions a figure of 9 mg or 6 ml per litre of must. Some operations during fermentation may introduce further oxygen, such as in pumping over when used, estimated at 10 mg L⁻¹.

Effect of oxygen on wine

The chemical/flavour effect of the reported amounts of oxygen in the must/wine at different stages of wine-making will now be discussed.

During crushing and early maceration of the grapes, one effect is the development of odoriferous volatile substances. The formation of various long-chain aldeyhydes, like hexanal and the unsaturated hexenals, at this stage was first described by Drawert in 1974, and then of 2-nonenal and 4-heptenal by Grosch & Semmelroch in 1985, from linoleic/linolenic acids in the lipids of the grapes. An explanation of their occurrence is based upon an auto-oxidation mechanism, as described in detail in Appendix I and further discussed in Chapters 4 and 7. This particular type of oxidation is thought to be assisted by the lipoxygenase enzyme present in the must, which is no longer active during fermentation and subsequently. Chapter 4 describes these particular flavours in more detail; they are not necessarily desirable as they may be due to the use of unripe grapes.

At the crushing stage for white wine production, there is the prospect of the desirable oxidation of the caftaric acid present, which is otherwise responsible for browning of these wines. This oxidation by the available oxygen can produce highly polymerized and precipitable substances, which will then be filtered out after the fermentation. Jackson (2008) discusses the investigations of Singleton (1987), whilst Ribéreau-Gayon mentions the work of Cheynier *et al.* during 1994 in this area. Again an enzyme, polyphenol oxidase, is cited as a catalyst for the oxidation, as also discussed in Appendix I, though sulfur dioxide will depress its effect.

During maceration following crushing, the final type/quality of the wine is importantly determined by the extraction of phenolic substances into the must, their subsequent oxidation and the removal of precipitable substances.

There are some important oxidation effects during the fermentation itself, on the functioning of yeasts, synthesis of fusel oils and of esters, and the rôle of spoilage bacteria and reduction of hydrogen sulfide odours, which are discussed in more detail in Chapter 7.

During storage, red wines are thought to benefit from a limited amount of oxygen uptake and the amount benefitting the development of red wine depends on the phenol content (see Toit *et al.*, 2006). In fact a saturation of wine gives typically 6 mg L^{-1} oxygen, and red wines normally improve with up to ten saturations (60 mg L^{-1}) with some wines improving even up to 25 saturations (150 mg L^{-1}). White wines are generally considered oxidized after about ten saturations (60 mg L^{-1}) (see Toit *et al.*, 2006; figures quoted from Boulton, 2001). The changes that occur during maturation, both by oxygen or other chemical means, with particular respect of flavour are separately discussed under *Maturation* in the next section.

The chemistry of the effect of oxygen on various phenolic substances is highly complex, and is still undergoing interpretation. Some of the latest findings are discussed in Appendix I. A recent review by Li *et al.* (2008) summarizes the current knowledge on the chemistry of oxidation in wines, whilst Freitas & Matteus (2010) discuss the role of oxygen in the changes of wine pigments.

3.9   Changes in maturation

The term maturation or ageing covers the storage and handling of wines after fermentation/racking up to and after bottling. The process is generally considered to improve the quality of the wine, especially of red wines made from better quality grapes such as Cabernet Sauvignon, Tempranillo (in Spain) and Nebbiolo (in Italy). It is also true for white wines made from high quality grapes such as Chardonnay and Riesling. Effects can persist for decades. Most wines, however, will only improve after a few months, and then deteriorate slowly, but irreversibly. For wines to improve with maturation they need to have a sufficiently high phenolic compound and acid content.

There are two types of maturation practiced in the wine industry: firstly, by the producer or wine-maker 'in-barrel (cask)' or 'in-vat', especially used for red wines for one or two years (or even longer) as described in Chapter 1; secondly, by both the producer and the consumer, called 'in-bottle' ageing as a result of the choice of time of drinking after bottling. The total desirable times proposed by wine experts are as long as 10–15 years for some red wines, and 0–10 years for some white wines for optimum quality. Even longer periods are allowed for prestige wines, such as the 'Premier Crus' of Bordeaux.

Both of these types of ageing cause changes in chemical composition and therefore flavour and colour, the nature, need and desirability of which are hotly disputed by wine writers, trade experts and academic researchers. The kinds of changes and new chemical compounds formed to be discussed in this section are those of the non-volatile compounds present, affecting basic taste characteristics, particularly astringency. The more important changes are perhaps with volatile components, affecting aroma, and therefore, so-called 'bouquet' which are described in Chapter 4. The overall flavour effect requires, of course, consideration of both, as dealt with in Chapter 5.

The two types of ageing are further differentiated by their chemical type. Thus, *in-vat* maturation is essentially about the effect of absorbed oxygen, i.e. oxidation, as particularly with *in-barrel* ageing. The latter has the important added feature of extractives from the wooden barrels used. *In-bottle* maturation is of an entirely different character in that no oxygen is involved, nor extractives possible – nevertheless important chemical equilibria changes can occur.

3.9.1 'In-barrel' ageing

Vats

Ageing in-cask, or wooden barrels, has many historical and commercial implications, which influence its current usage. Oak barrels have been used for many years, if not centuries when barrels were also used for transportation – though a distinction has to be made between new barrels and old, and between the types of oak used. The two main types of French oak are firstly, the pedunculate oaks (*Quercus robur*) growing widely in the Limousin district, and secondly, the sessile oaks (*Quercus sessilis*) more associated with the Vosges region. They differ in chemical composition and therefore, in extractable substances. The former have a high extractable polyphenol content and relatively low concentrations of odoriferous compounds, whilst the latter have the reverse. In the USA, the dominant species is *Quercus alba*, with a low phenol content but high content of volatile substances, in particular the 'oak lactones'. This type of oak is also widely used in Spain for Rioja wines in preference to French oaks.

The method of construction of barrels is affected by the type of oak used. Chemical composition or rather extractability of substances is also importantly influenced by the seasoning method adopted for the wood, i.e. natural or kiln drying and whether the staves were sawn or split. The construction of wine barrels, an activity known as cooperage, involves the assembly of wood staves by bending and requires two stages of heating. The second stage is called 'toasting', itself divisible into three different possible regimes. The most intense 'heavy toast' involves a surface temperature of about 230°C for about 15 minutes, or 'charring'. It is not surprising, therefore, that further compositional changes take place, again with the amount and type of extractives, both volatile and non-volatile and dependent upon the type of oak. In use, it is also essential that old barrels be kept in good hygienic conditions, often carried out by sulfuring, itself having further possible implications for flavour in subsequent wine storage.

It is the skill of the wine maker to ensure that the wine is aged in an oak barrel which complements the flavour of the wine and is a matter of subjective preference. However, the barrels are rather variable in their properties, depending on the choice of oak and the numerous parameters of the production of the barrels. Skilled taste panels can differentiate between the type of oak barrels when the same wine has been stored in different barrels, but the oak characteristics are more difficult to recognize for a range of different wines. All these aspects have been described by Jackson (2008) in detail and by Ribéreau-Gayon *et al.* (2006) based on considerable experimental data. Singleton (1974b) has also reviewed the many aspects of wooden containers as a factor in wine maturation.

Extraction from barrels

The nature of these extractives from the oak used is the key to understanding their potential effect on the wines being aged, together with the degree and type of oxidation that can take place. All these extractives are essentially phenolic substances, some of which are already present in the wine, as described in Section 3.6. Those additionally extracted compounds from oak and present in the aged wine are again the non-flavanoid phenols, the hydrolysable tannins and their derived products. A separate group of non-flavanoid phenolic substances, however, are derived from the breakdown of the lignin in the oak, and are therefore specific to wines aged in wood, but primarily affect the aroma characteristics and are separately discussed in Chapter 4. Ribéreau-Gayon *et al.* (2006) report some interesting studies of laboratory work determining the amounts by model alcoholic solutions.

The main involatile non-flavanoid compounds extracted are the above described coumarins (Section 3.6.1), which are harsh tasting acids, whilst their glycosides are bitter. Simpler phenolic compounds like gallic acid will be extracted, as will be the polymers of gallic and ellagic acids, with or without glycosation, which are bitter substances. These are, for example, the ellagi-tannins (hydrosoluble by acid to release glucose, as in wine), and the gallo-tannins (esters only hydrolysable in an alkaline medium, like saponification). Monomeric phenolic acids like ellagic acid may therefore be present in aged wine. However, their extraction depends upon several factors, not least the time/temperature of ageing and type of oak, and including the pH/alcohol content of the wine. Although ellagi-tannins could extract readily into wine, they tend to hydrolyse during toasting, and those extracted tend to hydrolyse in the wine, or bind to proteins or polysaccharides, followed by precipitation. The low concentrations of ellagi-tannins together with their low astringency means that they have a low impact on the sensory properties of the wine.

It is stated (Jackson, 2008) that wine dissolves about 30% of the tannins in the innermost few millimetres of the oak staves of the barrel but this is sufficient to influence the sensory attributes of the wine. The amount extracted will depend on the age of the barrel and the number of times a barrel is used to impart oak flavour onto the wine, and choices are at the discretion of the wine maker. It is possible to shave the inside of the barrel, and remove about 4 mm of wood to expose wood that has not been in contact with wine, however, this exposed new wood does not have the same characteristics as the original, in part because the toasting procedure when constructing the barrel affects the composition of extractable compounds on only a thin layer of the wood.

Small amounts of sugars can accumulate in the wine due to the hydrolysis of hemicellulose but this is thought to be insufficient in quantity to affect the perception of wine. Oak also absorbs components from wine. The greatest loss is water and alcohol, both through adsorption (estimated 5–6 L in a 225 L barrel by Chattonnet, 1994; quoted by Jackson, 2008), as well as evaporation of alcohol and ethanol. The relative amounts of ethanol and water evaporation depends on humidity, for example under dry conditions more water than ethanol evaporates.

Oxidation in barrels

Oxidation, as already discussed, produces sensory changes (including those of aroma, Chapter 4, which are perhaps more important). Quantitative data, however, on a whole range of wines, relating taste and colour to actual non-volatile compound composition is scarce. An important task of the wine-maker during the wine-making process is to protect the must or wine from oxidation, however, during the storage of wines in barrels a controlled amount of oxidation is desirable. The general nature of these effects is well documented in Jackson (2008) and Ribéreau-Gayon et al. (2006) with special reference to red wines and Peynaud (1986), with knowledgeable opinions from Barr (1988), whilst the chemistry is reviewed (Toit et al., 2006; Li et al., 2008; Kilmartin, 2009) and briefly described above.

It is generally considered that acidity will decline. Slow de-acidification can result from the isomerization of the natural '+' to '−' optical rotatory form of tartaric acid, which is less soluble. Ribéreau-Gayon (1978) presented data on total acidity and also sulfate content after storage in new oak barrels and in used ones, which figures were both clearly higher in the latter (4.02 g L^{-1}, H_2SO_4:1.26, K_2SO_4). The pH was correspondingly lower with higher acid content. These changes, however, he attributes to the necessary sterilization of used barrels by SO_2, which progressively oxidizes to sulfuric acid (H_2SO_4). In respect of storage in a stainless steel vessel, compared with a new barrel, there is little difference in content, i.e. 3.63 g L^{-1}, H_2SO_4 in a new barrel, and 3.43 in stainless steel (the figures for sulfate content are 0.86 and 0.92 respectively), whilst the pH remains at 3.35.

There is an increase in the volatile acid content (measured as g L^{-1} H_2SO_4), though the increased acid will be largely acetic acid, stated due to oxidation/bacterial ingress of Acetobacter, from 0.46 in a stainless vessel to 0.62 in a new wooden barrel, corresponding to the increase in total acidity noted. Any overall increases in acidity found are said to be probably due to bacterial action. However, another formation route may be purely chemical, which would explain a higher volatile acidity of some old mature wines, without these wines showing signs of spoilage. Wildenradt & Singleton (1974) showed the production of aldehydes from ethanol during oxidative wine storage. Possibly further oxidation of ethanol would lead to the formation of ethanoic (acetic) acid, though this must remain speculative.

Most studies have centred on the effect of oxygen on the phenolic substances, particularly to tannins of red wines, which are key components determining bitterness/astringency, as already described, and on the anthocyanin and anthocyanin–tannin complexes, which determine colour. Red wines contain, as also described, a wide range and content of all the different tannins that can be present, with different responses to the effect of oxygen, during ageing. Good ageing potential is, however, associated with adequate phenolic compound content. It is concluded that 'the softening of flavour' is due to certain transformations by oxidation of these tannins, largely of a structural nature involving more polymerization. This subject is more fully considered in Appendix I. Polymerization can produce phenolic substances that are precipitated, and removed on subsequent clarification, and thus is a kind of cleansing operation. Changes in the tannins can be assessed in respect of the different indices (HCl, etc.) on structure/capacity.

The quantity of tannins (from an index figure) present in a red wine stored in a new oak barrel was found (Ribéreau-Gayon, 1978) to be 5% higher than that from a used barrel, or when stored in stainless steel. However, the increase of tannin content is regarded as relatively insignificant. The colour intensity change was, however, markedly greater with wine from new barrels.

Oxidation is likely to be lower overall during storage and ageing in stainless steel vessels, especially if closed, as often now used for white wine vinification (Chapter 1). In addition, there will be no extractives to consider. In white wines, with lesser quantities of phenolic compounds available in a protective rôle against possible oxidation of 'fruity' aromas and the development of flatness, even those made in a 'modern high-tech' environment with low ambient oxygen and low temperatures are more prone to browning oxidation than those made in a more traditional way. Browning in white wine is particularly associated with the oxidation of caftaric acid (or the caffeic acid released by hydrolysis in an acid medium). The mechanism of oxidation has been previously discussed, as has the pick-up of oxygen during crushing and maceration of the grapes. The result, however, is the formation of precipitable polymers that can be removed in the clarification process prior to fermentation.

3.9.2 'In-bottle' ageing

Ageing in bottles is assumed to take place without further ingress of oxygen in a well-corked bottle and of course without extractives from the container. Changes in phenolic content and the structure of tannins still occur, but Ribéreau-Gayon *et al.* (2006) point out that 'a wine that tastes hard and astringent at the time of bottling will generally retain that character, even after several years'.

It is, however, the changes in the volatile content affecting bouquet and aroma that are of greater flavour interest, which is separately discussed in Chapter 4.

3.9.3 Oxidation–reduction (redox) potential

Ribéreau-Gayon *et al.* (2006) and his colleagues at the University of Bordeaux have applied the concepts of redox potential to the oxidation–reduction phenomena occurring in wine-making. These concepts and their application are somewhat complex and difficult to follow, and the terminology can be confusing. Some of the fundamentals of this subject are first explained below, and in the Appendix, I.3.4, taken from basic but advanced textbooks, such as that of Shriver *et al.* (1994).

General

Oxidation reactions are always accompanied by reduction reactions, so that a total reaction is conceptualized in terms of constituent redox half-reactions. Thus, an important redox couple (O_2, H^+/H_2O) takes part in the forward action, which is expressed in the reduction mode, with a so-called standard reduction potential of +1.23 V.

$$O_2 \text{ (g)} + 4H^+ \text{ (aq.)} + 4e^- \rightarrow 2H_2O \text{ (aq.)} \tag{3.9}$$

The term, standard reduction potential, E_0, is here based upon the substance in the oxidized form, being expressed before that in the reduced form (state), e.g. also Fe^{3+}/Fe^{2+}, $E_0 = +0.76$ V, but Zn^{2+}/Zn has $E_0 = -0.76$ V. These potentials are assessed against a standard reference redox couple (H^+/H_2), which is given an arbitrary value of zero, when hydrogen is in the standard state as a gas (g) at atmospheric pressure (1 bar), and the hydrogen ions or protons are in aqueous solution (aq.) at a pH of 0 or nearby a 1 molar solution of a strong acid. The E_0 of interest can be measured in a galvanic cell with an inert platinum electrode together with a reference hydrogen electrode, though in practice various other designs of precisely known potential are used. The actual potential (E) that is measured depends upon the concentrations of the reactants in the cell, as will be explained below.

In organic solutions, like wines, the substances actually oxidized may be represented molecularly, as AH_2, to form the corresponding redox couple ($A^{\cdot}/AH_2$) in the half-reaction, $AH_2 \rightarrow A^{\cdot} + 2H^+ + e^-$ (in the oxidizing mode), or $A^{\cdot} + 2H^+ + e^- \rightarrow AH_2$ (in the reducing mode), with an E_0' value dependent upon the nature of AH_2, similarly in relation to the standard hydrogen electrode. By subtracting the equations and the potentials of the two half-reactions in the reduction mode, a representation is given of the reaction that can actually occur, and of its likelihood (positive E_0 value difference). Unfortunately, there is little data on E values for half-reactions of phenolic substances, especially for all those in wines, though the couple (quinhydrone/dihydroquinone) has a reported E_0 of 0.59 V. Furthermore, the exact oxidizing mechanism may not be precisely known in all cases, i.e. whether it is directly through molecular oxygen, or through a peroxide intermediate and/or catalysed by iron or copper ions.

Nernst equation

Actual potentials, E_H, whether of a whole reaction or of the component half-reactions, will be higher when the concentrations of the reactants are in a non-equilibrium condition, as for example will result from the dissolution of oxygen gas into the medium. The Nernst equation is derived from thermo-dynamic considerations of the free energy changes involved, when expressed for each half-reaction couple (Equation 3.10).

$$E_H = E_0 - \frac{RT}{nF}\ln\frac{\text{[reduced]}}{\text{[oxidized]}} \quad \text{or} \quad = E_0 + \frac{RT}{nF}\ln\frac{\text{[oxidized]}}{\text{[reduced]}} \tag{3.10}$$

where R is the gas constant ($8.31\,\text{J}\,\text{mol}^{-1}\,\text{K}^{-1}$), T is the temperature (degrees Kelvin), F is the Faraday constant, at $96\,500$ coulombs, and n the number of unit electrons involved. The term [oxidized] is the molar concentration of the substances in the oxidized condition, and similarly for reduced condition of the redox half-reaction couple. E_H also depends upon the actual pH of the solution, since protons may be involved, thus

$$E_H = E_0 (=0) - RT/nF \ln(\text{hydrogen pressure} = 1\,\text{bar})/(H^+) \tag{3.11}$$

Since $\log_{10} = 2.303\ln$ (or $\log_e$), then at 25°C, the Nernst equation for a half-reaction becomes Equation (3.12).

$$E_H = E_0 - 0.059\log\frac{\text{[reduced]}}{\text{[oxidized]}} - 0.059\,\text{pH Volts} \tag{3.12}$$

The actual measurement of a redox potential in wines still presents problems on account of the complexity of the oxidation processes taking place, and in the design of suitable electrodes, both standard and reference.

Redox potentials in wine

Ribéreau-Gayon et al. (2006) describes the work of Vivas & Zamora (1992), in which an operating equation was derived from the Nernst equations, and which can be used with a specially designed combined electrode for a galvanic cell (Equation 3.13):

$$E_H = E_0 + A\text{pH} + B\log[O_2] \tag{3.13}$$

The potentials are most conveniently expressed in millivolts. A is a constant with the value of 59, pH is not a great variable in wine, usually about 3.5, while B is an empirical constant, the value of which reflects the presence of the oxidizable substances in a wine and is typically 150.9. The potential is markedly determined by the dissolved oxygen content of the wine $[O_2]$ in millimole units, which can range up to $7.5\,\text{mg}\,\text{L}^{-1}$ or 235 mM at 25°C. Content can be separately determined by a polarographic device, originally designed by Clark (1950). E_0 is described not as a standard reduction potential, but as a 'normal' potential that would be produced with equimolar quantities of the oxidized/reduced species (i.e. log 1/1 = 0). Initially, a measuring electrode used platinum for the collection of electrons from the half-reaction, already discussed, $O_2 + 4H^+ + 4e^- \rightarrow 2H_2O$, and a reference electrode, based upon the half-reaction, $Ag + Cl^- \rightarrow AgCl + e^-$ using silver/silver chloride, but with a very constant potential in relation to the hydrogen electrode. When this unit was placed in distilled water with differing amounts of dissolved oxygen, though very low (of the order of $1 mg L^{-1}$), the results were consistent with the Nernst equation. The constant B was $59/4 = 14.8$ (millivolt basis) and the E_0 of $1.178 V$ was close to that expected of the oxidizing half-couple at $1.23 V$. The modified combined electrode design, based upon the exchange of electrons through the silver chloride, gave a similar result for distilled water, and its use for wine was justified for differing dissolved oxygen contents. Data was given for the measured E_H at differing O_2 contents, from 0.1 to $5 mg L^{-1}$ in a red wine, though the actual E_0 did not appear to be given.

During wine-making, however, as already described, the oxygen content of the aqueous liquid is highly variable, due to pick-ups from the air at the various stages, together with subsequent periods of absorption by reduction–oxidizing actions. Ribéreau-Gayon *et al.* (2006) describe the use of the sophisticated combined electrode device to measure potential in a cell, showing the corresponding changes in E_H and any changes in E_0. In fact, the technique also enables the measurement of oxygen content in wines at any time. The data presented in general was not diagnostic for particular oxidizable phenolic compounds and their rôle, nor indicated the desirability or otherwise of any specific levels or changes of E_H. However, other data described in Appendix I shows the comparative rates of oxidation of different specific types of compound found in wines, by potential measurements on known quantities of each in model solutions initially saturated with oxygen gas, over a period of time. The non-phenolic substances showed very low rates of oxygen consumption, but in descending order, flavanoids (not fully characterized), anthocyanins, catechins, oligomeric procyanidins and polymeric procyanidins showed higher rates. E_0 values, however, also increased, with increased level of polymerization, reflecting perhaps inherent oxidizability (see oxidation of procyanidins in Appendix I).

The measured E_H in wines ranges from about $500 mV$, when fully saturated with oxygen, to lower than $50 mV$ in the total absence of oxygen and therefore pick-up. Possible differences in the value of B were not mentioned, but might be expected from the wide variety of phenolic compounds and their amount, and so with their different oxidizabilities, E_0 might similarly differ. E_H and E_0 are always positive, with the former reflecting the tendency of the reaction to proceed, on account primarily of the dissolved oxygen available at any given time. The latter figure reflects the equilibrium condition for the particular reactants (or redox couples), and should be a constant for any given wine at the same temperature.

Redox potentials during vinification

Over a period of about 100 days, during the vinification of a red wine, up to the early ageing stages, there were the expected rises and falls of E_H, especially during the racking stages as shown in a graphical plot with time in days on the *x*-axis. Further typical data presented in the form of tables is also given for a red wine, subsequently stored in different kinds of container (oak barrels to plastic vessels), and both types of data are similarly given for a white wine.

Some investigations were also carried out on E_H values prior to fermentation. The pick-up of oxygen and oxidation reactions, during the crushing of grapes, already noted above, is reflected in E_H, which can temporarily reach levels of over 400 mV in red wines and 350 mV in white, compared with stabilized average levels of 300 and 200 mV, respectively, after the main vinification operations, including racking. During fermentation, for both a white and a red wine, graphical plots of E_H with time showed a rapid fall, and therefore absorption by chemical reaction was under way. The need for oxygen or otherwise during fermentation is discussed more fully in Chapter 7. Interestingly, the rate of fall is very similar for both the red and white wines made. In the same red wine, following peak values of E_H at malo-lactic fermentation, the slope of the lines (dE_H/dt) after each subsequent pick-up of oxygen were similar, reflecting a particular composition of oxidizable catechins, procyanidins and anthocyanins. It is possible that precise measurements of these rises and falls determined over a range of different wines could be correlated with 'quality' factors in a wine.

These measurement techniques were also used to show the effect of oxidation metal catalysts during maturation, such as $Fe^{3+} + e^- \rightarrow Fe^{2+}$ (reduction mode) and $Cu^0 \rightarrow Cu^{2+} + 2e^-$ and $Cu^+ \rightarrow Cu^{2+} + e^-$ (oxidation modes). There was a marked difference in oxidation characteristics between red and white wines, probably due to the particular types of phenolic compounds to be found.

Ribéreau-Gayon *et al.* (2006) also relate the development of reductive bouquet (see Chapter 4) in both red and white wines, which develops after a short period of bottle ageing, that is when all the dissolved oxygen has reacted. This condition corresponds with a measured redox potential of below about 200 mV. These desirable reductive reactions will increase with decreasing potentials and vice versa. The tightness of corks and other means of closure in bottles is therefore important, and may be determined by potential measurements; thus, figures of 162, 168 and 320 mV are quoted for wines in bottles with glass stoppers, corks and an ineffective screw cap, respectively. The fact that E_H can still be low in the apparent complete absence of oxygen suggests that complete absence is actually impossible in practice, or that some other oxidizing agent is also present in wines, albeit at a low level. These potential measurements might be of interest in assessing the controversial effects of exposing wine after pouring from a bottle to the air in respect of degree and rate of deterioration.

Bibliography

General texts

Anderson, O.M. & Markham, K.R. (eds) (2006) *Flavonoids: Chemistry, Biochemistry and Applications*. CRC Press, Florida.

Barr, A. (1988) *Wine Snobbery*. Faber and Faber, London.

Belitz, H.-D. & Grosch, W. (1987) *Food Chemistry*. Springer-Verlag, Berlin.

Jackson, R.S. (1994, 2000, 2008) *Wine Science, Principles, Practice, Perception*. 2nd edn., 3rd edn., Academic Press, San Diego.

Montedoro, G. & Bertuccioli, M. (1986) The flavour of wines, vermouth and fortified wines. In: *Food Flavours, Part B, The Flavour of Beverages* (eds. I.D. Morton & A.J. Macleod), Developments in Food Science, 3B, Elsevier, Amsterdam.

Peynaud, E. (1986) *The Taste of Wine*. Macdonald, Orbis, London.

Ribéreau-Gayon, P., Glories, Y., Maujean, A. & Dubourdieu, D. (2006) *Handbook of Enology, Volume II, The Chemistry of Wine, Stabilization and Treatments*. John Wiley & Sons, Ltd, Chichester.

Shriver, D.F., Atkins, P.W. & Langford, C.H. (1994) *Inorganic Chemistry*. Oxford University Press, Oxford.

References

Aldave, L. (1993) The shelf-life of young white wines. In: *Shelf-Life Studies of Foods and Beverages* (ed. G. Charalambous), Developments in Food Science, 33, pp. 923–944, Elsevier, Amsterdam.

Amerine, M.A. & Ough, C.S. (1980) *Methods for Analysis of Musts and Wines*. John Wiley & Sons, Inc., New York, quoted by Jackson, R.S. (2008) loc. cit., pp. 213 and 215.

Amerine, M.A. & Roessler, E.B. (1983) Wines: Their Sensory Evaluation, quoted by Montedoro, G. & Bertuccioli, M. (1986) loc. cit., p. 187.

Anon. *Official Methods of Analysis*, 11th Edition, AOAC, quoted by Jackson, R.S. (1984) loc. cit., p. 214.

Bachmanov, A.A. & Beauchamp, G.K. (2007) Taste receptor genes. *Annual Reviews in Nutrition*, **27**, 389–414.

Bajec, M.R. & Pickering, G.J. (2008) Astringency: mechanisms and perception. *Critical reviews in Food Science and Nutrition*, **48**, 1–18.

Bakker, J., Preston, N.W. & Timberlake, C.F. (1986) Ageing of anthocyanins in red wines: comparison of HPLC and Spectral methods. *American Journal of Enology and Viticulture*, **37**, 121–6.

Bakker, J., Picinelli, A. & Bridle, P. (1993) Model wine solutions: Colour and composition changing during ageing. *Vitis*, **32**, 111–118.

Bakker, J. & Timberlake, C.F. (1997) The isolation and characterisation of new colour stable anthocyanins occurring in some red wines. *Journal of Agriculture and Food Chemistry*, **45**, 35–43.

Bakker, J., Bridle, P. & Honda, *et al.* (1997) Isolation and identification of a new anthocyanin occurring in some red wines. *Phytochemistry*, **44**, 1375–1382.

Barr, A. (1988) *Wine Snobbery*. Faber and Faber, London.

Behrens, M. & Meyerhof, W. (2006) Bitter taste receptors and human bitter taste perception. *Cellular and Molecular Life Science*, **63**, 1501–1509.

Boulton, R. (1983) The in-vivo expression of anthocyanin colour in plants. *Phytochemistry*, **22**, 1311–1323.

Boulton, R. (2001) The copigmentation of anthocyanins and its role in the color of red wine: a critical review. *American Journal of Enology and Viticulture*, **52**, 67–87.

Brouillard, R., Chassaing, S. & Fougerousse, A. (2003) Why are grape/fresh wine anthocyanins so simple and why is it that red wine color lasts so long? *Phytochemistry*, **64**, 1179–1186.

Brouillard, R., Chassaing, S., Isorez, G., Kueny-Stotz, M. & Figueiredo, P. (2010) The visible flavonoids or anthocyanins: from research to applications. *Recent Advances in Polyphenol Research*, 1–22.

Castaneda-Ovando, A., Pacheco-Hernandez, de L., Paez-Hernandez, E., Rodriguez, J.A. & Galan-Vidal, C.A. (2009) Chemical studies of anthocyanins: A review. *Food Chemistry*, **113**, 859–871.

Chandrashekar, J., Hoon, M.A., Ryba, N.J.P. & Zuker, C.S. (2006) The receptors and cells for mammalian taste. *Nature*, **444**, 288–294.

Cheynier, V., Duenas-Paton, M., Salas, E., Maury, J.M., Sarni-Manchado, P. & Fulcrand, H. (2006) Structure and properties of wine pigments and tannins. *American Journal of Enology and Viticulture*, **57**, 298–305.

Clifford, M.N. (1985) Chlorogenic acids. In: *Coffee Chemistry* (eds. R.J. Clarke & R. Macrae), Vol. 1. Elsevier Applied Science, London.

Clifford, M.N. (1993) Phenol-protein interactions and their possible significance for astringency. In: *Interactions of Food Components* (eds. C.E. Birch & M.J. Lindley), pp. 143–164. Elsevier Applied Science, London.

Clifford, M.N. & Ramirez, J.R. (1991) Tannins in the sun-dried pulp from wet processing of arabica coffee beans. In: *Proceedings of the 14th ASIC Colloquium (San Francisco)*, 1991, pp. 230–236. ASIC, Paris.

Da Conceicao Neta, E.D., Johanningsmeier, S.D. & McFeeters, R.F. (2007) The chemistry and physiology of sour taste – a review. *Journal of Food Science*, **72**(2), 33–38.

De Beer, D., Harbertson, J.F., Kilmartin, P.A., Roginsky, V., Barsukova, T., Adams, D.O. & Waterhouse, A.L. (2004) Phenolics: A comparison of diverse analytical methods. *American Journal of Enology and Viticulture*, **54**, 389–400.

Downey, M.O., Dokoozlian, N.K. & Krstic, M.P. (2006) Cultural practice and environmental impacts on the flavonoid composition of grapes and wine: A review of recent research. *American Journal of Enology and Viticulture*, **57**, 257–268.

Fischer, U., Strasser, M. & Gutzler, K. (2000) Impact of fermentation technology on the phenolic and volatile composition of German red wines. *International Journal of Food Science and Technology*, **35**(i), 81–94.

Flamini, R. (2003) Mass spectrometry in grape and wine chemistry. Part I: polyphenols. *Mass Spectrometry Reviews*, **22**, 218–250.

Freitas, V.A.P., de & Mateus, N. (2006) Chemical transformations of anthocyanins yielding a variety of colours (review). *Environmental Chemical letters*, **4**, 175–183.

Freitas, V.A.P., de & Mateus, N. (2010) Updating wine pigments. In: Recent Advances in Polyphenol Research (ed. C. Santos-Buelga, M.T. Escribano-Bailon & V.Lattanzio), Vol. 2, pp 59–80. John Wiley & Sons, Ltd, UK.

Fulcrand, H., Duenas, M., Salas, E. & Cheynier, V. (2006) Phenolic reactions during winemaking and ageing. *American Journal of Enology and Viticulture*, **57**, 289–297.

Gawel, R., Oberholster, A. & Francis, I.L. (2000) A mouth-feel wheel: terminology for communicating mouth feel properties of red wine. *Australian Journal of grape and Wine Research*, **6**, 203–207.

Gawel. R., Iland, P.G. & Francis, I.L. (2001) Characterising the astringency of red wine: a case study. *Food Quality and Preference*, **12**, 83–94.

Gawel, R., van Sluyter, S. & Waters, E.J. (2007) The effects of ethanol and glycerol on the body and other sensory characteristics of Riesling wines. *Australian Journal of grape and Wine Research*, **13**, 38–45.

Haslam, E. (1981) Vegetable tannins. In: *The Biochemistry of Plants* (ed. E.E. Coin), Vol. 17, pp. 527–536. Academic Press, New York.

Hoffman, G. (1989) *The Chemistry and Technology of Edible Oils and Fats and their Products*, Academic Press, London.

Jackson, R.S. (2008) *Wine Science, Principles, Practice, Perception*. 3nd edn., Academic Press, San Diego.

Kallithraka, S., Bakker, J. & Clifford, M.N. (1997a) Evaluation of bitterness and astringency of (+)-catechin and (−)-epicatechin in red wine and in model solution. *Journal of Sensory Studies*, **12**, 25–37.

Kallithraka, S., Bakker, J. & Clifford, M.N. (1997b) Red wine and model wine astringency as affected by malic and lactic acid. *Journal of Food Science*, **62**, 416–420.

Kallithraka, S., Salacha, M.I. & Tzourou, I. (2009) Changes in phenolic composition and antioxidant activity of white wine during bottle storage: Accelerated browning test versus bottle storage. *Food Chemistry*, **113**, 500–505.

Kaminura, M. & Kaneda, H. (1993) The shelflife of beer. In: *Shelf-Life Studies of Foods and Beverages* (ed. G. Charalambous), Developments in Food Science No 33. Elsevier, Amsterdam.

Kilmartin, P.A. (2009) The oxidation of red and white wines and its impact on wine aroma. *Chemistry in New Zealand*, **73**, 18-22.

Landau, J.M. & Young, C.S. (1997) The effect of tea on health. *Chemistry and Industry*, 994-996.

Le Berre, E., Atanasova, B., Langlois, D., Etievant, P. & Thamas-Danguin, T. (2007) Impact of ethanol on the perception of wine odourant mixtures. *Food Quality and Preference*, **18**, 901-908.

Macheix, J.J., Sapis, J.C. & Fleuriet, A. (1991) Phenolic compounds and polyphenol-oxidase in relation to browning in grapes and wines. *Critical Reviews in Food Science and Nutrition*, **30**, 441-486.

Meilgaard, M.C. (1982) Prediction of flavour differences between beers from their chemical composition. *Journal of Agriculture and Food Chemistry*, **30**, 1009-1017.

Mercurio, M.D. & Smith, P.A. (2008) Tannin qualtification in red grapes and wine: Comparison of polysaccharide- and protein-based tannin precipitation techniques and their ability to model wine astringency. *Journal of Agriculture and Food Chemistry*, **56**, 5528-5537.

Mombaerts, P. (2004) Genes and ligands for odorant, vomeronasal and taste perceptors. *Nature reviews: Neuroscience*, **5**, 263-278.

Montedoro, G. & Bertuccioli, M. (1986) The flavour of wines, vermouth and fortified wines. In: *Food Flavours, Part B, The Flavour of Beverages* (eds. I.D. Morton & A.J. Macleod), Developments in Food Science, 3B. Elsevier, Amsterdam.

Morata, A., Gomes-Cordoves, M.C., Colombo, B. & Suarez, J.A. (2003) Pyruvic acid and acetaldehyde production by different strains of Saccharomyces cerevisiae: Relationship with Vitisin A and B formation in red wines. *Journal of Agricultural and Food Chemistry*, **51**, 7402-7409.

Morata, A., Gomes-Cordoves, M.C., Calderon, F. & Suarez, J.A. (2006) Effect of pH, temperature, and SO_2 on the formation of pyranoanthocyanins during red wine fermentation with two species of *Saccharomyces*. *International Journal of Food Microbiology*, **106**, 123-129.

Morata, A., Gonzalez, C. & Suarez-Lepe, J.A. (2007) Formation of vinylphenolic pyranoanthocaynins by selected yeasts fermenting red grape musts supplemented with hydroxycinnamic acids. *International Journal of Food Microbiology*, **116**, 144-152.

Nerin, C., Salafranca, J., Aznar, M. & Batlle, R. (2009) Critical review on recent developments in solventless techiques for extraction of analytes. *Analytical and Bioanalytical Chemistry*, **393**, 809-833.

Nykanan, L. & Soumaileinan, M. (1983) Composition of wines, quoted by the IARC Monograph, *Alcohol Drinking* (1988) Vol. 4, pp. 79-86. International Agency for Research into Cancer, Lyon.

Oberholster, A., Francis, I.L., Iland, P.G. & Waters, E.J. (2009) Mouthfeel of white wines with or without pomace contact and added anthocyanins. *Australian Journal of Grape and Wine Research*, **15**, 59-69.

Peynaud, E. (1986) *The Taste of Wine*. Macdonald, Orbis, London.

Plattig, K.H. (1985) The sense of taste. In: *Sensory Analysis of Foods* (ed. J.R. Piggott), pp. 1-22. Elsevier Applied Science, Barking.

Rankine, B.C. & Pocock, K.F. (1969) Influence of yeast strain on binding of SO_2 in wines and its formation during fermentation. *Journal of Science Food and Agriculture*, **20**, 104-109.

Rentzsch, M., Schwarz, M. & Winterhalter, P. (2007) Pyranoanthocyanins – an overview on structures, occurrence and pathways of formation. *Trends in Food Science and Technology*, **18**, 526-534.

Ribéreau-Gayon, P. (1978) Wine flavour. In: *The Flavour of Foods and Beverages* (eds G. Charalambous & G.E. Inglett), pp. 355-380. Academic Press, New York.

Ribéreau-Gayon, P., Glories, Y., Maujean, A. & Dubourdieu, D. (2006) *Handbook of Enology, Volume II, The Chemistry of Wine, Stabilization and Treatments*. John Wiley & Sons, Ltd, Chichester.

Romero, C. & Bakker, J. (2000) Effect of storage temperature and pyruvate on kinetics of anthocyanin degradation, Vitisin A derivative formation, and colour characteristics of model solutions. *Journal of Agriculture and Food Chemistry*, **48**, 2135-2141.

Scott, K. (2004) The sweet and bitter of mammalian taste. *Current Opinion in Neurobiology*, **14**, 423-427.

Scott, K. (2005) Taste recognition: food for thought. *Neuron*, **48**, 455-464.

Singleton, V.L. (1974a) Analytical fractionation of the phenolic substances of grapes and wine and some practical uses of such analyses. In: *Chemistry of Winemaking*. (Advances in Chemistry series number 137), pp. 184-211.

Singleton, V.L. (1974b) Some aspects of the wooden container as a factor in wine maturation. In: *Chemistry of Winemaking*. (Advances in Chemistry series number 137), pp. 254-277. American Chemical Society, Washington.

Singleton, V.L. (1987) Observations with phenols and related reactions in musts, wines and model systems. *American Journal of Enology and Viticulture*, **38**, 69-77.

Singleton, V.L. (1988) Wine phenols. In: *Modern Methods of Plant Analysis, Wine Analysis* (eds. H.F. Linskens & J.F. Jackson), pp. 173-218. Springer-Verlag, Berlin.

Singleton, V.L. (1992a) Wine and Enology: status and outlook. *American Journal of Enology and Viticulture*, **43**, 344-354.

Singleton, V.L. (1992b) Tannins and the qualities of wines. In: *Plant Polyphenols* (eds. R.W. Hemingway & P.E. Laks), pp. 859-880. Plenum Press, New York.

Singleton, V.L. & Noble, A.C. (1976) Wine flavour and phenolic substances. In: *Phenolic, Sulphur and Nitrogen Compounds in Food Flavours* (ACS symposium Series number 26), pp. 47-70. American Chemical Society, Washington.

Swiegers, J.H., Chambers, P.J. & Pretorius, I.S. (2005) Olfaction and Taste: Human perception, physiology and genetics. *Australian Journal of Grape and Wine Research*, **11**, 109-113.

Swiegers, J.H., Saerens, S.M.G. & Pretorius, I.S. (2008) The development of yeast strains as tools to adjust the flavour of fermented beverages to market specifications. In: *Biotechnology in Flavour Production* (eds. D.H. Frenkel & F. Belanger), Chapter 1, pp. 1-55, Blackwell Publishing Ltd., Oxford, UK.

Thompson, C.C. (1987) Alcoholic Beverages. In: *Quality Control in the Food Industry* (ed. S.M. Herschdoefer), Ch. 2, Section Cider, p. 54. Academic Press, London.

Timberlake, C.F. & Bridle, P. (1976) Interactions between anthocyanins, phenolic compounds and acetaldehyde and their significance in red wines. *American Journal of Enology and Viticulture*, **27**, 97-105.

Timberlake, C.F. (1980) Anthocyanins - Ocurrence, extraction and chemistry. *Food Chemistry*, **5**, 69-80.

du Toit, W.J., Marais, J., Pretorius, I.S. & du Toit, M. (2006) Oxygen in must and wine: A review. *South African Journal of Enology and Viticulture*, **27**, 76-94.

Waterhouse, A.L. (2002) Wine phenolics. In: *Alcohol and Wine in Health and Disease*. (eds. D.K. Das & F. Ursini) pp. 21-36, New York Academy of Sciences, New York.

Wildenradt, H.L. & Singleton V.L. (1974) The production of aldehydes as a result of oxidation of polyphenolic compounds and its relation to wine ageing. *American Journal of Enology and Viticulture*, **25**, 119-126.

Woodman, J.S. (1985) Carboxylic acids. In: *Coffee Chemistry* (eds. R.J. Clarke & R. Macrae), Vol. 1, pp. 266-290. Elsevier Applied Science, London.

Chapter 4
Volatile Components

4.1 General

As already explained, the volatile organic compounds in wine are perhaps their most characterizing feature, both to the wine taster and certainly to the scientist. They are responsible for their so-called 'bouquet' on sniffing the head-space from a glass, and the odour/aroma component (palate-aroma) of the overall flavour perceived on drinking. Three sources of these compounds are recognized: (1) primary aromas, i.e. those persisting through from the grape; (2) secondary aromas, i.e. those arising from the vinification process, often a result of yeast or bacterial metabolism; and (3) tertiary aromas, i.e. those arising during subsequent storage of the finished wine as a result of chemical reactions and/or wood extraction especially during long-term storage in wooden barrels.

Recent years have seen a wealth of information regarding the formation of wine flavour compounds by yeasts during fermentation, summarized in a recent review (Ugliano & Henschke, 2009). The alcoholic fermentation involves the conversion of hexose sugars to ethanol and carbon dioxide, and in addition to ensuring yeast maintenance and growth, the glycolytic and associated pathways provide volatile and non-volatile metabolites that contribute to wine flavour. Factors influencing the formation of these compounds are nutrient content of the must, fermentation conditions and the species of yeasts involved in the fermentation. The role of yeasts has been shown to contribute significantly in the development of wine flavour, and the interactions between yeasts and grape compounds influence many aspects of wine quality, ranging from appearance, aroma and flavour of wine to its 'texture'. Both the fermentation bouquet (compounds typically resulting from fermentation and not necessarily related to the grape variety used) and the varietal character (compounds typical for the grape used for wine-making) are influenced by yeasts. When fermenting in wood, even wood extracted compounds can be modified by yeasts.

Wine Flavour Chemistry, Second Edition. Jokie Bakker and Ronald J. Clarke.
© 2012 Blackwell Publishing Ltd. Published 2012 by Blackwell Publishing Ltd.

The varietal character of a wine is an area of great interest to the wine maker, however, still difficult to define. It may not always be due to compounds directly derived from the grape, but has been reported to be due to yeast derived esters and alcohols (see Francis & Newton, 2005). In addition, the amino acid composition and concentration influences the volatiles produced during yeast fermentation and the varietal aspect of this is still under investigation. The current view is that for most cultivars, varietal aromas are due to quantitative differences in the aroma profiles.

Enhanced knowledge will increase the control wine makers may be able to exert over wine flavour production, by selecting yeasts and using different inoculation techniques. Currently, wine yeast strains are selected to be less susceptible to off-flavour formation and augment components for optimal wine flavour. Characterization of yeasts, including the genetic characterization of metabolic pathways and their regulation, will allow the wine maker more choice in optimizing wine flavour. Easy, rapid and reliable methods for measuring nutrients in grapes and must will increase the control. However, the balance of nutrients in grapes and must is thought to depend on viticultural conditions, so regions influencing this composition in any way may well contribute to the varietal character of a wine. Much more research is needed to fully explore and exploit the role of yeasts, nutrients and its influence on wine flavour and quality. More discussion on the formation of volatiles during fermentation is discussed in Chapter 7.

As a result of sophisticated analytical procedures using gas chromatography developed during recent decades, over 400 volatile compounds have now been detected and many of these also quantified in their amounts present in different wines. Similarly, these techniques enable the volatile composition of the original grapes to be determined, so that the primary aromas of the finished wines can be deduced (not otherwise possible). Some of these same aromas may, however, be produced additionally during the fermentation and storage stages.

Once these volatile compounds have been detected and, preferably quantified, it is possible to relate their presence to the perceived flavour of the wine and indeed to decide which compounds are determining the flavour of a particular wine; be it a Sancerre, Australian Chardonnay, a Bulgarian Cabernet Sauvignon, or whatever.

4.1.1 Sensory perception

The scientific understanding of the perception of volatiles has greatly increased over the last two decades, in particular the the award of the Nobel Prize for Physiology and Medicine in 2004 to Richard Buck and Linda Axel for their pioneering research on the genetics of the perception of odour (Buck & Axel, 1991) has given great impetuous to this research field. A short discussion of the main findings follows below, based on recent reviews of Swiegers et al. (2005); Ache & Young (2005); Mombaerts (2004); Meilgaard et al. (2007).

The perception of volatiles is of great importance to the wine drinker, since a great part of the enjoyment of wine depends on the smell or aroma. The wine industry is well aware that the sensory properties of wine are of great value, obviously the wine needs to be free of any off-flavour, but even more importantly a small difference in smell and taste of a wine can distinguish the wine from the average quality range and elevate it into the premium range, thus commanding a premium price! Nothing is more prestigious than winning a gold award in a well known wine competition.

Our sense of smell, or olfaction, is much more developed than our sense of taste. Our sense of taste seemed to be developed primarily to select energy and to avoid toxic compounds, which are mostly bitter, as discussed in Chapter 3. In contrast, we are able to detect thousands of different compounds and despite the great variation in chemical structure and molecular weight of these volatile compounds, we can differentiate between them and identify their individual smells. Volatile compounds travel from our food or drink via the nose, either nasally by sniffing or retronasally once the food or drink has been placed in the mouth volatile compounds move through passage in the throat, to the olfactory epithelium, which is the size of a postage stamp. Compounds are detected by bipolar olfactory receptor cells, which have cilia exposed where the air we breathe in carries the volatile compounds past the olfactory epithelium. The olfactory receptors to which the volatile components bind are located on the cilia. This binding leads to a number of transduction events thus generating a signal that is transmitted via the neuron of the olfactory cell to the olfactory bulb in the brain. The odour signals are sent to the olfactory cortex and to the higher cortical areas where perception takes place. It is thought that our sense of smell holds a central position in human physiology.

Buck & Axel (1991) discovered that a large family of genes encodes for the olfacory epithelium transmembrane proteins and were thereby the first to elucidate the molecular mechanism of odourant recognition. A wealth of information in this research area has been published since. Sequencing of the human genome revealed that the number of sequences for odour receptors would give only about 400 functional odour receptors, leaving the question: how we can detect and differentiate so many different volatile odour molecules? The answer seems to be that each odour receptor cell is specific for one odour and sends a specific signal to the brain. Despite this specific ability to recognize an aroma molecule, series of related molecules can be detected by one specific receptor cell, such as a series of aliphatic aldehyes. In addition, individual aroma molecules can react with several receptor cells. Hence it is thought that this explains why we can discriminate between so many volatile aromas. However, some volatile compounds can block our perception, although this is a temporary phenomenon. Questions still remain, for example why is there such a great difference in sensitivity for aroma compounds, ranging from $mg\,L^{-1}$ to $ng\,L^{-1}$ (see Section 4.1.2).

The perception of volatile components in mixtures does seem to hold additional difficulties. It appears that humans have difficulty in distinguishing more than three compounds in a mixture. In addition there are interactions in sensory perception and it seems so far not possible to predict how a blend of different compounds is perceived. Hence, despite our knowledge about smelling and volatile compound composition, the sensory assessment of a particular wine remains a key step in the production of wine!

4.1.2 Partition coefficients

Some of the information discussed in this section is discussed in more detail elsewhere (Pierotti, 1959; Perry, 1980; Clarke, 2001). In practice, many, if not all of these volatile compounds are present in exceedingly small quantity (expressed as $mg\,L^{-1}$, ppm; $\mu g\,L^{-1}$, ppb; or even $ng\,L^{-1}$, ppt). Their individual threshold flavour/odour values need also to be known. To reach the organo-leptic sensors of the nose, any particular compound must be present in the vapour phase, admixed in the air, which is carried to and passes through the front nostrils (as in 'nosing' bouquet), or additionally through the back (retro) passages of the throat (retronasal) or both (when actually drinking or 'slurping' the wine). In a wine, or an extremely dilute aqueous solution of a volatile substance, contained in a vessel, such as a half-filled wine glass, an approximate equilibrium condition will be established between the amount of the volatile compound 'j', present in the aqueous liquid phase, and that in the air/vapour space above, determined mathematically by the partition coeffi-cient, $K_{j,a\text{-}w}$, for this compound. This coefficient expresses the ratio of the weight of the component in unit volume in the air phase to that weight in the same unit volume in the water phase. The higher this value, the higher will be the amount of the compound in the vapour state in the air space for a given amount of the volatile compound in the aqueous phase. As these com-pounds are present in the aqueous liquid phase (e.g. as in wine) in very small quantities, so we have the concept of infinite dilution, which will be found to simplify enormously the understanding and calculation of this coefficient, including its mathematical derivatives.

$K_{j,a\text{-}w}$ can be determined by direct measurement by GC techniques; one sample is taken from the head-space above the aqueous liquid (e.g. the wine) and another from the aqueous liquid itself. Measurements are expressed per millilitre of each of the air and liquid, and the value of $K_{j,a\text{-}w}$ is merely and con-veniently the ratio of the GC peak areas in the two samples. The $K_{j,a\text{-}w}$ (or just K_j for simplicity, or $K_{j,a\text{-}w}^{\infty}$ more strictly) in a given circumstance, may also be predicted from other physical property measurements, as will be described. Several other methods of measurement are now available, as described in Section 4.5.

Volatile organic compounds with high boiling points, but low solubility in water (or wine) and other flavourful beverages, have, perhaps surprisingly, very high partition coefficients. In contrast, compounds with low boiling points and high water solubility have low values. They are, in fact, not much more volatile than water, at atmospheric pressure and normal drinking tempera-tures in which we are interested in assessing flavour. This phenomenon is a consequence of the inherent hydrophobicity of the volatile compound, reflected in the ratio of the number and size of non-polar to polar groups and their positioning in the particular molecule. It reflects the tendency of the molecules of such a compound to escape into the air space from their watery environment but is only true at very dilute solutions, close to infinite dilution. The concentration of their vapour in the air space will then be greater than above the same compound in the pure 100% liquid state (with a low vapour pressure at ambient temperature). This will of course, not be true at temperatures approaching boiling points but this situation is not relevant to the tasting of wines. As an example of an aroma/odoriferous volatile organic compound in wine, we can cite ethyl butanoate (n-butyrate), which shows a measured $K_{j,a\text{-}w}$ at 25°C of 1.87×10^{-2} (Pollien & Yeretzian, 2000). The antagonism to water of such a compound is shown in another useful mathematical derivative, known as its activity coefficient, $\gamma_{j,w}^{\infty}$ for the compound, 'j' in water 'w'. It is related to the partition coefficient by Equation 4.1, where P_j^s is the vapour pressure of the pure volatile compound at the same temperature.

$$K_{j,a\text{-}w} = \gamma_{j,a\text{-}w}^{\infty} P_j^s\, 0.97 \times 10^{-6} \tag{4.1}$$

The activity coefficient is a reciprocal measure of the saturation solubility of the compound in water, so that the lower the water solubility, the greater will be its activity coefficient. Activity coefficient is a dimensionless quality; the factor 0.97×10^{-6} ensures that the concentrations of 'j' are in the same weight/volume units in both the liquid and the air. From the measured $K_{j,a\text{-}w}$ of ethyl butanoate, and a vapour pressure (P_j^s) of 17 mm at 25°C, $\gamma_{a\text{-}w}^{\infty}$ calculates to 1134. Buttery et al. (1971b) described the derivation of this equation, and its application to the volatile compounds in wine is discussed in Section 4.5 and fully described in Appendix I.4.

Notably, therefore, in any homologous series, of say aliphatic esters, alcohols, ketones, etc., partition coefficients will be at their lowest for compounds at the bottom of the series, with the lowest molecular weights. Partition coefficients will rapidly increase with increasing molecular weight. However, as Buttery et al. (1971b) explained, this does not mean that, in a given homologous series, increasing molecular weight would indefinitely be associated with ever correspondingly increasing concentrations of the volatile compound in the air space. Eventually, such compounds would be found to be so insoluble in water that, at a certain level of concentration (say 30 ppm and less), the greater quantity of the compound present would be in the pure liquid state. They would then exert only a very low actual vapour pressure, with a low corresponding vapour concentration in the air space. It is only in the very dilute aqueous solution condition that high partition coefficients are possible. For example, the maximum aqueous solubility of this same ethyl butanoate is reported as 0.68% (Appendix II, Table II.1) at 25°C, or 6800 mg L^{-1}. Experimental data suggests a content in wine of 0.01–1.0 ppm (Table 4.11 which follows), figures clearly well below its maximum solubility. Similarly, ethyl hexanoate has a calculated maximum solubility (from Pierotti estimates, see Appendix I) of 0.032% at 20°C. A direct solubility measurement figure is not available; 320 mg L^{-1} is again higher than the quantity reported present in wine.

The concept and measurements of threshold flavour/odour levels for these same compounds in water and in wine is also important, and discussed in more detail in Section 4.1.2. Figures from various investigators suggest 0.1 ppm or less for the minimum detectable level of ethyl butanoate, when sniffed from an aqueous solution, though data is limited and absent for wine itself (see Table 4.11). Ethyl butanoate may, therefore, play some part in determining the perceived flavour of a wine, in which it is present, at higher amounts above the threshold concentration. Even more likely is ethyl hexanoate with a threshold of 0.036 ppm in water, and 0.850 ppm in wine (see Table 4.11).

Notably, volatile compounds of a high or relatively high molecular weight, and with a complex molecular structure, affect the nasal receptors much more efficiently than others, i.e. they have very low flavour threshold flavour levels. The best known of these is 2-methoxy-3-isobutylpyrazine, which is to be found in many vegetable plants, including some grape varieties, and is also present in green coffee beans. It is characterized as having a 'green-pepper' odour (green peppers themselves are an important source), with the incredibly low reported threshold figure of 0.002 ppb in water. Its measured partition coefficient ($K_{j,a-w}$) of 2×10^{-3} is similar to that of ethyl acetate (4×10^{-3}). However, the latter compound, which is a simpler molecule, has a much higher threshold, variously reported as 5–60 ppm (Table 4.11). This aspect of flavour impact will also be discussed later, together with the still controversial physiological mechanism of nasal (organo-leptic) perception, and indeed concerning why a given compound has its specific flavour-generating characteristic.

There are four factors still to be considered, however, in the use of threshold levels and partition coefficients for assessing volatile odour/flavour characteristics in a beverage: (1) the temperature of the aqueous solution containing the volatile compound; (2) the presence of any other constituents in substantive amount in the aqueous liquid, e.g. acids, sugars, and ethyl alcohol; (3) concentration of the volatile component in the aqueous solution; and (4) the presence of other volatile substances. The effect of increasing the temperature is usually to increase K_j for any given compound. Not unexpectedly, the concentration of the compound in the air space above the solution will usually be higher with a higher temperature, and K_j is governed by Equation 4.2, where T is the temperature in °C, and A and B are constants that depend upon the substance.

$$\text{Log } K_j = A \times T + B \tag{4.2}$$

Wine is normally 'nosed' and drunk at around 20°C, though, as is well known, white wines are recommended to be tasted and drunk chilled (e.g. 10–15°C) after storing in a refrigerator as necessary. The aroma impact actually would then be expected to be lower than at normal room temperature. Red wines are, however, recommended to be both tasted and drunk at room temperature (even up to 25°C). Of greater significance is the effect of the other factors mentioned above. Sucrose in solution increases the partition coefficients, as has been well described for a number of volatile compounds

(Chandrasekaran & King, 1972). Ethyl ethanoate (acetate) in pure water, for example, at 20°C, showed an activity coefficient of 65 in water, but 300 in a 70% aqueous sucrose solution, consequent upon its poorer solubility in the latter. The corresponding K_js were 4×10^{-3} and 2.6×10^{-2}. In finished wines, the sugar content is, however, very low (<0.3%), though in some sweet wines as high as $60 \, g \, L^{-1}$. Total soluble acid content (say at $10 \, g \, L^{-1}$) may also have some effect, but values will differ according to the particular aroma compound being assessed.

Ethyl alcohol in the aqueous solution, as in wine, at say 10–15% v/v, will have the opposite effect of lowering partition coefficients since, in general, volatile compounds are more soluble in alcohol–water mixtures than in water alone. The effect can be calculated, since (1) the activity coefficients of most of these compounds will be approximately one, in pure ethyl alcohol in which they are completely soluble, and (2) the logarithm of the activity coefficient (log γ) will generally be proportional to the mole-fraction of the constituents of the mixed solution. A graphical plot as in Figure 4.1 shows these relationships for ethyl hexanoate ($\gamma = 20\,900$ at 20°C, calculated value) and ethyl acetate ($\gamma = 78$ at 20°C) in 12% ethyl alcohol (0.04 mole-fraction), and in brandy (40% v/v ethyl alcohol, or 0.35 mole-fraction). An inspection of the plot indicates that the partition coefficient of ethyl hexanoate will be lowered by $(13\,200/20\,900) \times 100 = 63\%$; and similarly ethyl acetate to $56/70 \times 100 = 80\%$ (vapour pressures, P_j^s are unaltered). The threshold levels of these compounds are expected to increase correspondingly in the alcoholic solutions, as indicated in tests (Table 4.4 later in this chapter).

Another important aspect of this partition coefficient effect of ethyl alcohol present in wine is that the very high molecular weight substances may, however, now be brought into solution at higher amounts. Therefore a higher concentration of volatile compound will equilibrate in the air space and consequently contribute also to overall flavour.

There appears to be little satisfactory practical data available to determine whether the measured partition coefficient of a particular component is affected by the presence of other volatile components, of which there may be many in actual beverages. However, recently Pollien & Yeretzian (2001) demonstrated that the partition coefficient of 2-methyl-propanal and -butanal were little different whether determined in pure water or in a 0.5% instant coffee solution (added with its other volatiles present). However, actual threshold values can be markedly affected as shown in the next section.

4.1.3 Threshold flavour/odour levels

To examine properly changes in flavour contribution of different compounds due to the ethyl alcohol and other constituents of wine it is necessary first to examine the various methods of threshold determination that have been used. Guadagni et al. (1963b) carried out considerable experimental work on volatile aldehydes and some other compounds, with a trained panel, and came to the conclusion that the use of a plastic (Teflon) squeeze bottle was the

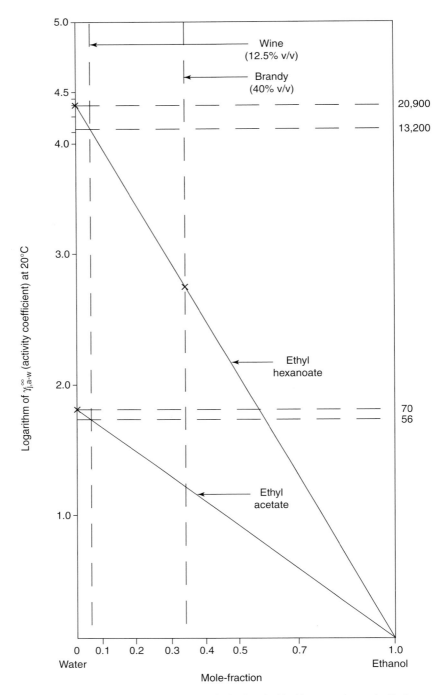

Figure 4.1 Lowering of activity coefficients of ethyl hexanoate and ethyl acetate in water–ethanol solutions. Pierotti-estimated values for activity coefficients in water; in ethanol, $\gamma = 1$. (Pierotti, 1959).

most satisfactory method. Vapour from the head-space over the dilute aqueous solution of the compound of interest, in a half-filled bottle, would be directed into the nostril(s), and its odour sensation compared with that from pure water at the same temperature. Threshold odour levels were thus obtained (with tolerance levels included), which for the homologous series of alkanals were found to be very low, i.e. in the ppb range (e.g. propanal at 9.5 parts $\times 10^{-9} \pm 1.0$ in 1 part water, or 9.5 ppb). The thresholds were rather higher in numerical value and much less reliable when obtained by typical sniffing techniques from the same solution held in a small glass vessel (e.g. propanal showed 12 ± 6 ppb). However, threshold *flavour* levels (perception or detection thresholds) determined from drinking experiments, from which the lowest concentration of volatile compound is detectable, are really more relevant. Some comparative results are available. Even more relevant are *flavour* threshold data for different compounds in water containing 10–15% v/v ethyl alcohol, and other non-volatile substances of wine or, better still, in actual wine from which all other volatile compounds have been stripped, leaving only the ethyl alcohol. This is not really feasible or even possible.

The concept of the Aroma Index or Unit has been developed, i.e. the ratio, amount of volatile compound present divided by the absolute threshold flavour level, when it may reasonably be expected that any compound showing an Aroma Index of substantially above 1.0 is contributing to the flavour of the wine. In practice, in wine, very few individual volatile substances appear to have a high aroma index; the content of substances with similar organoleptic properties, such as the 'fruity' esters, might well be added together for this purpose. The problems connected with the approach are discussed in Chapter 5.

Ribéreau-Gayon *et al.* (2000) describe the distinction between recognition and perception thresholds. When the former is employed, a descriptive term is used, rather than the mere appearance of a flavour/aroma. Numerous investigators have now compiled data on threshold values of all kinds, in water and other beverages, though generally they are those of perception, or not distinguished.

Units

Some confusion can arise over the usage of different units in expressing threshold levels, and indeed of partition coefficients (though these are usually in non-dimensional units). Threshold levels or values have been traditionally expressed in parts of the given compound in parts per thousand, per million, per billion (thousand million) or even per trillion, parts of pure water solution or other aqueous liquid. Preference is now given to expressing in grams (compound) per litre (aqueous liquid) units. Naturally, grams may be in the sub-units of mg, µg or ng. Equivalencies are illustrated in Table 4.1. There can be some advantage in the use of ppm, ppb, ppt units, as errors can arise with the use of 10^{-n}, where *n* may not be accurately read, or mis-transposed from the literature, as also with contents expressed in these units. Exponent values of 10 are now often quoted as $\times$ E (+ or − 1), etc. In citing actual values, it is

Table 4.1 Equivalencies of different threshold levels and content units.

Parts		Weight/Vol. or g/g Weight/Weight
ppm (parts per million parts of water, 10^6; or 10^{-6} parts in one part)	=	mg L^{-1}(kg) milligrams per litre (kg) i.e. 0.001g/ L^{-1} (kg) or 10^{-3} g/10^3 g $= 10^{-6}$ g/g
ppb (parts per billion of water, 10^9; or 10^{-9} parts in one part)	=	μg L^{-1} (μ – million), micrograms per litre (kg) 0.000001g L^{-1} 10^{-6} g/10^{-3} g $= 10^{-9}$
ppt (parts per trillion, 10^{12} or 10^{-12} parts in one part)	=	ng L^{-1} nanograms per litre 10^{-9} g/10^3 g $= 10^{-12}$
Percentages	**Parts**	**Weight/Vol**
0.1%	1000 ppm	1000 mg L^{-1}
0.01%	100 ppm	100 mg L^{-1}
0.001%	10 ppm	10 mg L^{-1}

desirable to quote them in the units originally given, not least on account of any possible differences between old British and US trillions. Threshold values are usually given as absolute values, and sometimes as difference values (where there is a known pre-existing content of the compound under examination).

Consistency of threshold odour levels

Reviewing the available data for a range of different compounds in wines, there appears to be a reasonable degree of consistency for figures given by different investigators for the same compound (Table 4.2). Factors of importance are, firstly, the relative skills or biases in the judgements of panels used (see Chapter 5) and, secondly, the purity of the test compound. As described by Meilgaard (1982), some compounds require elaborate technical procedures of purification, to achieve 100% purity. Contaminants can have a much lower threshold than the test compound and thus confuse assessment.

Threshold level difference between sniffing and tasting

It is important to determine whether a threshold is different, when detecting the presence of a compound only by the nose (through the nostrils alone) as with the work of Guadagni et al. (1963b), or actually measured by drinking the beverage or aqueous solution. Ahmed et al. (1978) obtained data in which threshold levels were determined by tasting (i.e. flavour level in water), and compared them with some other values determined by odour sniffing only. Table 4.3 gives Ahmed's tasting data for a number of different compounds, found in fruit juices, together with data derived purely from sniffing.

Table 4.2 Consistency of reported odour threshold levels data.

| Compound name | Odour threshold levels in pure water µg L⁻¹ (ppb), except where stated | | | | | |
	Ribéreau-Gayon (2000)	Ahmed[a] (1978)	Yokotsuka (1986) (Perception)	Rychlik[b] (1998)	Guadagni (1963b)	Other
Ethyl acetate (mg L⁻¹)	25	–	0.6	–	–	6.2[c], 5.0[d]
Propanal	–	–	9.5	10	9.5	10[e]
2-Methyl propanal	–	–	0.9	0.7	0.9	1.3[e]
3-Methyl butanal	–	–	–	0.4	0.15	0.35[e]
Hexanal	–	9.2	–	–	4.5	–
Methanyl thiol	–	–	–	0.2	0.02	2.1[f]
Furaneol (HDMF)	–	–	–	10	–	–
3-Isobutyl-2-methoxy-pyrazine	0.002 (2 ng L⁻¹)	–	–	0.005 (5ppt)	–	–
β-Damascenone	0.002	–	–	0.00075	–	–
Linalool	–	5.3	–	–	–	6[g]

[a]Also quoted by Shaw (1986); [b]quoted by Grosch (2001); [c]Mulders (1973); [d]Flath et al. (1967); [e]Grosch (1995); [f]Perrson & von Sydow (1973); [g]Buttery et al. (1969b).

Table 4.3 Comparison of threshold levels determined by taste/flavour and by odour (sniffing) alone in water.

| | By taste (flavour) (x)/odour (sniffing) (y), (x/y) | | | |
| | (Ahmed data) Units, ppb (µg L⁻¹) | | (Others) Units, ppb (µg L⁻¹) | |
Compound name	Mean	Range	Mean	Range
Acetaldehyde	22/17	-/4-21	-/-	-/15-120
Hexanal	3.7/9.2	5-30/1.4-30	-/4.3[a]	-/2
Ethyl 2-methyl butanoate	0.10/-		-/0.10[b]	-/-
Octanal	0.5/-	-/-	5[d]/0.7[a]	-/0.2
Nonanal	4.3/2.5	-/-	-/1[a]	-/0.2
Decanal	3.2/2	2.67-3.4/0.79-4.90	-/0.1[a]	-/0.04
Linalool	3.8/5.3	1.4-10/1.9-15	-/6[e]	-/-
Furaneol	-/-	-/-	160[g]/-, -/60[e], -/10[f]	

Ahmed data with confidence limits. [a]Guadagni *et al.* (1963b); [b]Flath *et al.* (1967); [d]Lea & Swoboda (1958); [e]Buttery *et al.* (1969b); [f]Grosch (2001); [g]Huber (1992).

Flament (2001) has provided a useful compilation of odour/flavour threshold data for most of the volatile compounds in green and roasted coffee. The differences are not large, with some indication that volatile compounds are more readily detected, at least in pure water, when the solution is taken by mouth, rather than in simple sniffing of the head-space gas. The difference has relevance to 'nosing' for 'bouquet' in fine wines. Similarly, Feneroli's compilation (Burdock, 2002) gives both values, where available.

Threshold levels in solutions of dissolved substances in water and in beverages

Ribèreau-Gayon *et al.* (1978 and 2006) in particular have described comparative tests for some compounds in water, or in a model solution containing substantial amounts of dissolved substances and which would be found in wine. For example, two models were used, (a) synthetic model solution (with 12.5% alcohol with added tartaric acid to bring the pH to 3.0), and (b) the same as (a) but also with 10% added sucrose; the results are shown in Table 4.4.

Even higher threshold levels are experienced when measurements are made in real beverages, which also will contain, in addition to non-volatile compounds, other volatile compounds than the particular one under examination. There is comparative data not only for wine but also for other beverages such as beer, orange juice and Japanese sake.

These determinations will, however, necessarily be carried out on the wine or beverage already containing a quantity of the same compound under examination, and so they do not give absolute values of threshold odour/flavour. Meilgaard (1982 and 1986) described the use of the Difference Threshold figure

Table 4.4 Threshold levels in alcoholic model solutions.

Compound name	Threshold in			Units
	Water	Solution (a)	Solution (b)	
Ethyl acetate	25	40	–	mg L^{-1}
Caproate (hexanoate)	36	37	56	µg L^{-1}
Amyl alcohols (mixture)	1.9	12.5	60	mg L^{-1}
β-Ionone	7	800	–	ng L^{-1}
β-Damascenone	2	45	–	ng L^{-1}

Data from Ribéreau-Gayon (1978 and 2006). Solution (a) water with 12.5% w.v alcohol, with tartaric acid, 10 pH3, (b) as (a) but with 10% with sugar.

and explained that the actual value obtained in a given solution depends upon the amount of that compound already present. Thus, quoting the work of Brown *et al.* (1978) on the threshold for butan-2,3-dione (diacetyl highly purified), determined in a beer already containing 30 µg L^{-1} (ppb), the perception threshold of added compound was reported to be 100 µg L^{-1}. The threshold was 300 µg L^{-1}, when the beer also contained 300 µg L^{-1}; similarly, the figure 1500 µg L^{-1} when the beer already contained 2000 µg L^{-1} of diacetyl. By linear extrapolation, the threshold back-calculates to 81 µg L^{-1}, when the initial diacetyl content in a beer is taken as zero (i.e. an absolute threshold figure). Other data gives a figure in water alone of 15 µg L^{-1} (Rychlik *et al.*, 1998). The difference between these two absolute values will be a result of the combination of the relatively small physico-chemical effect of the ethyl alcohol as already discussed, and a larger one due to the presence of the other volatile compounds on nasal perception. Meilgaard (1982) explained that the interactions were such that similar odour sensations were largely additive, whilst contrasting aroma substances could be antagonistic, increasing the threshold level of each, e.g. in solutions with octanoic acid and ethyl acetate. The effect of interactions is further considered in Chapter 5 (Section 5.7.1), along with psychological factors relating to other components present.

Comparative data for different beverages for different representative volatile compounds is given in Table 4.5. Different threshold figures are given according to a range of contents for which they are applicable; in the case of wine, the available data was not precisely related to content in the wine examined, except that it was 'a white wine'. We can clearly expect much higher thresholds for most components in actual wines, compared with those in plain water (a ratio of 50 or greater).

Data for ethyl acetate and hexanoate (caproate) in water alone and in alcoholic solutions shows the expected increases in threshold level that are observed, following a lowering of volatile compound concentration in the head-space air presented to the olfactory sensors in the nose. This observation is consistent with the figure already calculated for the lowering of partition coefficients (Fig. 4.1 and Section 4.1.2) by the presence of ethyl alcohol.

Table 4.5 Absolute and difference odour/flavour threshold levels for various compounds in different beverages.

Name of compound	Absolute Water[a]	Absolute Wine[b]	Difference[1] Beer[c]	Difference[1] Sake[d]	Difference[1] Orange Juice[e]	Units
Esters						
Ethyl acetate	25	160 (c.40)	20–40 (10–30)	15–30 (20–120)	–	ppm
Ethyl butanoate	0.001 / –	– / –	– / –	0.4 / 0.22	0.20 (0.16)	ppm
Ethyl hexanoate	0.036	0.850 (0.06–0.60)	0.15–0.25 (0.2–0.4)	(3–10)	–	ppm
Aldehydes						
Acetaldehyde	0.022 / 0.35	–	10–20 (2–10)	25–50 (50–120)	–	ppm / ppm
Benzaldehyde	–	3.0 (–)	(0–0.1)	–	–	
Ketones						
Diacetyl	0.015	2.0	0.07–0.15 (0.03–0.3)	–	–	ppm
Terpenes						
Geraniol	40	130 (50–500)	150 (0–60)	–	–	ppb
Linalool	3–6 / –	50 (6–450)	–	–	2000	ppb
α-Terpineol	–	400 (4–400)	–	–	(40)	ppb
Alkyl sulfides						
Dimethyl sulfide	0.33	5	25–50 (30–100)	– / –	7 / 7	ppb

[a]Various references (see under Tables for Compound Groups; esters, etc.). NB values for ethyl butanoate, very variably reported. [b]Ribéreau-Gayon *et al.* (2000). [c]Meilgaard & Peppard (1986). [d]Takashashi, K. & Akiyama, H. (1986). [e]Shaw (1986).
[1]Range of content in beverage in brackets, applicable to difference level given.

Relationship of threshold values to partition coefficients

However, the relationship between partition coefficients and threshold odour levels for a homologous series of compounds, even in water is not simple, as an inspection of the reliable data of Guadagni *et al.* (1963b and 1969) for both variables shows. Comparative data is given in Table 4.6 for aliphatic alkanals, which are, in effect, in infinitely dilute solutions.

Table 4.6 Partition coefficients and threshold odour levels for alkanals in aqueous solution (infinitely dilute).

Alkanal (number + name, 'j') (linear chain)	Partition coefficient $K_{j,a-w} \times 10^3$ at 25°C		Threshold level (T) 'j' in 'w' parts $\times 10^9$ ppb at 22°C		$T \times K_j \times 10^{12}$
	Mean	SD	Mean	Range[a]	Mean
C2 Ethanal	2.7	±0.5	10	–	27
C3 Propanal	3.0	±0.1	9.5	1	29
C4 Butanal	4.7	±0.3	9.0	2	42
C5 Pentanal	6.0	±0.1	12.0	2	72
C6 Hexanal	8.7	±0.6	4.5	1	39
C7 Heptanal	11.0	±1	3.0	0.1	33
C8 Octanal	21.0	±4	0.7	0.2	15
C9 Nonanal	30.0	±4	1.0	0.2	30
C10 Decanal	(34)*	(±4)*	0.1	0.04	–
C11 Undecanal	(45)*	(±5)*	5.0	2	–
C12 Dodecanal	(65)*	(±6)*	2.0	0.6	–
(C4 and C5 branched chain alkanals)[b]					
C4 2-Methyl-propanal	11.4	–	0.7	–	8
C5 2-Methyl-butanal	15.6	–	0.4	–	6
3-Methyl-butanal	16.7	–	1.9	–	31

[a]Data from Guadagni et al. (1963b), Buttery et al. (1971b) of 2–6 determinations, as quoted. [b]Data for 'C4 and C5 branched chain' from Pollien & Yeretzian (2001), and Rychlik et al., quoted by Grosch (2001). *Data in brackets, extrapolated.

The partition coefficients of these alkanals, C_3 to C_9 show an excellent logarithmic relationship with carbon number on a linear scale; the values used for C_{10}-C_{12} compounds have been extrapolated.

Compounds C_5-C_{10} also show a good logarithmic–linear relationship between threshold values and carbon number. Compounds C_2-C_4 exhibit a separate logarithmic relationship, indicating lower threshold levels than might have been expected, which could be due to the incomplete purity of compounds under test. Purity is much more difficult to achieve with inherently weakly odoriferous, low boiling point substances with high polarity, which are easily contaminated by higher molecular weight and more odoriferous homologues. Familiar substances often have an odour they do not really possess in the completely pure state, thus, acetamide has a 'mousy' smell probably due to some contaminant; similarly, aniline can have a powerful smell. Really pure acetaldehyde (ethanal) and propanal may well have higher threshold values, which is suggested by earlier data (Lea & Swoboda, 1958) in the literature. Similarly the data for the compounds C_{11} and C_{12} is apparently anomalous, but the most likely explanation has already been given, i.e. increasing insolubility in water to virtually like that of non-polar hydrocarbons.

Alkanals are not of particular relevance to wine flavour (except acetalde-hyde), but the data provides a good basis for determining any mathematical relationship that may exist between partition coefficients and threshold values, which could be of relevance to other homologous series of compounds. The threshold levels decrease much more markedly with molecule size, than do the head-space concentrations increase, e.g. the product $K_{j,a\text{-}w}$ threshold decreases with increasing molecular weight.

This implies that the organoleptic sensors in the olfactory organs in the nose are much more sensitive to the larger molecules than the smaller.

Also of relevance is that branched-chain alkanals have much lower threshold levels than their corresponding linear chain homologues, in accordance with their partition coefficients, values recently obtained by a new direct method (Pollien & Yeretzian, 2001). In contrast, unsaturated chain aldehydes tend to have higher threshold values, with lower partition coefficients; thus, the threshold for *trans*-2-hexenal is 17×10^{-9}, whilst that for hexanal is 4.5×10^{-9}, reflecting a slightly higher affinity for, and solubility in, water.

The relationship between partition coefficients and threshold odour values is shown in Figures 4.2 and 4.3. The data in Figure 4.2 accommodates the precision reported by these investigators, which varies according to the compound by usually around ±10% or less. At the moment, similar relationships cannot be established for other homologous series in other groups of compounds, such as the alkyl alkanoates, alkanols or acids. This is largely because of insufficient and reliable data on threshold odour values, though there is some available (Tables 4.11, 4.12, 4.13 and 4.14 later in this chapter). Buttery *et al.* (1969b) also showed that the logarithms of partition coefficients plotted against carbon number gave parallel lines for these homologous series on the same graph. Threshold odour levels might be expected to show similar parallel relationships for homologous series, giving scope for estimated values, when actual values are not available.

Volatile compound concentration in the vapour phase

The actual concentration of a given compound in the air conveyed by the nose or retro-nasal passage, to the olfactory receptors, is the figure directly determining its olfactory effect. It can, of course, be calculated, knowing the concentration of the volatile compound in the liquid phase (e.g. water or aqueous solution) and the partition coefficient ($K_{j,a\text{-}w}$) at the given temperature.

However, this figure can be determined directly during GC-olfactometry, when the eluted vapour is being sniffed at the exit port (see Chapter 5, Section 5.5). At the same time, the identity and purity of the test sample can be confirmed or otherwise, using pure reference components. Using aroma dilution experiments, the value can be obtained at a threshold odour point. Thus, butanal (Fors, 1983) shows threshold values varying from 0.013 to 0.042 mg m^{-3} of air, whilst odour thresholds measured from water are between 9.0 and 37.3 ppb (μg L^{-1}). This data

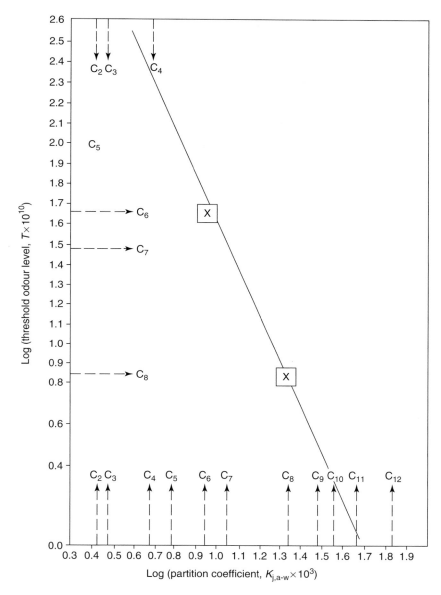

Figure 4.2 Relationship between partition coefficients and threshold flavour/odour levels for alkanals (data from Table 4.6).

is compatible with a measured partition coefficient of 4.7×10^{-3} (Buttery *et al.*, 1969b). Grosch and his colleagues (1995), quoted by Flament (2001), have provided air concentration data for a number of other compounds as shown in subsequent tables (Tables 4.9, 4.11, 4.12, 4.14, 4.15, 4.16, 4.17, 4.20, 4.21, 4.22, 4.23, 4.24, 4.25, 4.26, and 4.27).

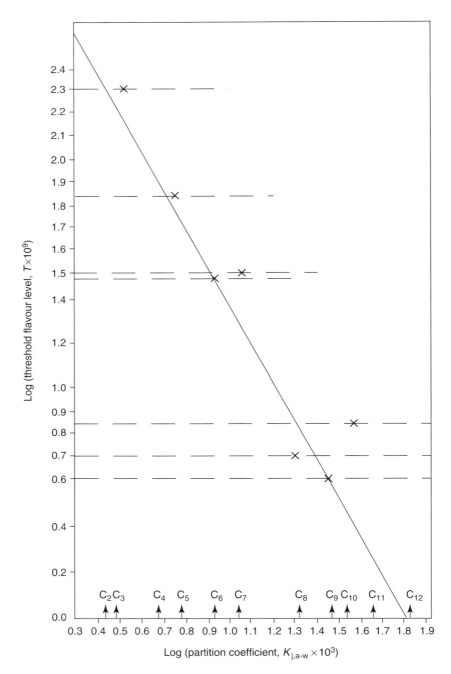

Figure 4.3 As in Figure 4.2 but showing only the flavour in water data from Table 4.14 (alkanals).

4.1.4 Flavour/odour descriptions

Use of word descriptions

Tables 4.11, 4.12, 4.14, 4.15, 4.16, 4.17, 4.20, 4.21, 4.22, 4.23, 4.24, 4.25, 4.26, and 4.27 give, for each of the various chemical compound groups, flavour/odour (i.e. olfactory only) descriptions of the individual volatile compounds found in wines, together with published threshold data (as available). Although flavour descriptions are, of course, somewhat subjective, nevertheless those given have a considerable degree of consensus amongst experts. Notably, however, many volatile compounds can present somewhat variable sensory impressions, depending upon their concentration in the air (or in aqueous solution) at which they are being assessed (Table 4.7). For detailed discussion and examples on this see Francis & Newton (2005).

Although the concentrations of volatile compounds are generally low, the method of presentation varies enormously. For example, compounds may be sniffed, when emerging from the discharge point of GC equipment when testing a particular beverage sample, or they can be sniffed from the static head-space in a squeezable plastic bottle (as described earlier). Of course, the volatile compounds are also perceived during actual drinking. In all these instances, the compounds will be present in very low concentrations, such as in wine and coffee. The odour of a substance, however, is often reported, whilst being assessed in the vapour above a large quantity of the pure liquid, when the head-space vapour concentration may be quite high, unlike for a small quantity of very dilute aqueous solution. Similarly, the perfumer, skilled in assessing and blending odoriferous substances (often called odorants), usually assesses odours of compounds dissolved in alcoholic solutions. Compilations and dictionaries of pharmaceutical compounds (e.g. *Merck Index*, 18th edn., 2000) and of perfume and flavouring chemicals (e.g. Feneroli, 2002) report the odour and colour of the pure liquids.

Arctander (1967) is particularly associated with a compilation of odour assessments of compounds generally, whilst Flament (2001) has provided a very detailed compilation of all available (and different types of) threshold data for volatile substances in coffee, many of which are also present in wines. Grosch and his colleagues, such as Blank & Semmelroch (1995), have special-ized in descriptions of odours eluting from sniffing ports during gas-chroma-tography of coffee brews. These descriptions may not be identical, for the same compounds, if a wine was being similarly tested. Arctander (1967), in particular, emphasizes the different odour description that can exist according to vapour concentration. This situation is especially evident with the alkanals, such as acetaldehyde to octanal described as having a 'nauseating', 'suffocat-ing', etc. odour in high concentration, but becoming much more pleasant (e.g. 'fruity') from very dilute solutions. The perfumer is well aware of the odoriferous natures of a substance such as skatole, which is 'putrid' at high concentration, but possesses useful and desirable odour notes at low concentrations.

Table 4.7 Typical examples of differing sensory (olfactory) impressions of volatile compounds with concentration in aqueous solution.

Name of compound	Flavour/odour	Concentration (ppb)	Method of assessment	Reference
trans-2-nonenal	Threshold	0.08	Odour from water	Guadagni *et al.* (1963b)
(*E*-non-2-enal)	Slightly plastic-like	0.2	Flavour in water at room temp.	Parliment *et al.* (1973)
	Woody	0.4–2.0	"	
	Fatty	3–16	"	
	Unpleasant oily	30–40	"	
	Strong cucumber	1000	"	
2-Methoxy-3-iso-butyl pyrazine	Threshold	0.002	Odour from water	Seifert *et al.* (1970)
	Green bell-pepper, peas, sl. earthy	–	Flavour in water	Chemisis (1995)
	Peasy	–	Odour in air	Czerny & Grosch (2000), quoted by Grosh (2001)
	Earthy	–	Odour in water	Grosch (2001)
	Red pepper	–	Odour in air	Belitz & Grosch (1999)
	Herbaceous	–	Odour/flavor	R.-G. *et al.* (2000)
Acetaldehyde	Threshold	(variable values)	Odour from water	–
	Pungent, ethereal, nauseating, but diminishing at high dilution – wine like		Odour from water	Arctander (1967) Motodo (1979)
	Pungent, fruity, malt	–	Odour at sniffing port	
	Green ethereal, fresh, fruity		Flavour in water	Chemisis (1998)

Descriptions are also available as the odour element of flavour on 'tasting' the aqueous solution. A specialized compilation of such data has been made by the Firmenich company in Switzerland, though the text Chemisis has not been published; Flament (2001) has quoted the available information for coffee volatile substances, many of which also occur in wines. *Feneroli's Handbook* (2002), for each flavour compound listed, provides information, as far is available, in three ways: (1) smell or odour description (generally from the pure substance); (2) aroma threshold level (detection) in aqueous solution, but source is not always referenced; and (3) taste characteristics at a particular concentration in solution, well beyond the detection threshold value. For example, under ethyl caproate (hexanoate) we have (1) 'powerful fruity odour with a pineapple-banana note'; (2) 0.3–5 ppb (cp data in Table 4.10, 1 and 36 ppb); and (3) 'fruity and waxy with a tropical nuance' at 10 ppm aqueous solution.

Intensity of flavour/odour

A further dimension to be considered in odour characterization is the perceived intensity of an odour at different concentrations, that is including and beyond threshold values. According to Steven's Law originally promulgated around 1957 (Stevens & Galanter, 1957), $R = k \times S^n$ where R is the perceived flavour intensity of a constituent, and S its concentration, whilst k and n are constants for a particular compound. The units of S are conventionally $g L^{-1}$, whilst R can be an intensity figure obtained by reference to a scale (typically, 0–9). The exponent, n, appears in practice to be between 0.12 and 1.7 (usually 0.4–0.7). This implies that doubling the concentration of a given compound does not mean that we will smell the compound twice as strongly; it could well be $2^{0.5} = 1.4$ times. However, for $n > 1$, we will smell the compound more than twice as strongly on doubling the concentration. Teghtsonian (1973), quoted by Meilgaard & Peppard (1986), suggested that n is a function of the range of concentrations under study; the value is higher the narrower the range, though this seems surprising.

These findings have particular implications in the use of Aroma Indices, such as the Flavour Unit concept of Meilgaard (1982), which are discussed in Chapter 5.

4.2   Volatile compounds detected in wines

Lists have been compiled of all the volatile compounds that have been detected and identified by instrumental analysis in wines by numerous different investigators over the last three decades, in a range of different types of wine. Information given at later dates, say after 1980, tends to be more reliable, due to the increasing capability and certainty of gas chromatographic techniques, which have been coupled with mass-spectral data (i.e. GC-MS), and have made use of capillary columns for better resolution. Such comprehensive lists of volatile compounds have been provided by Montedoro & Bertuccioli (1986), and by the T.N.O. Food Research Institute in Zeist in The Netherlands, with their publications (7th edn., 1996). Clearly, not every wine in commerce will contain all these listed compounds; some will derive from specialized wines, like botrytized wines, or are developed on ageing. Indeed, some compounds listed may unfortunately be mere artefacts of the GC technique used and the method of extraction from the sample. Assurance of correct identification needs a careful study of the original published papers. However, reliability is clearly increased by more than one investigator reporting the presence of the same given compound; three or more references has been the criterion for inclusion for the numerous esters reported, which are an important group of substances responsible for wine flavours. Since the methodology has been much improved (see Section 4.5), with most authors showing validation of their method and statistical data on the errors, therefore some compounds recently identified have been added, even when only one publication is available.

The definition of 'volatile' compound needs some explanation. Nearly all compounds can be said to be volatile from infinite dilution to some degree. Many relatively involatile compounds can be steam distillable, and therefore appear in an extracted sample in some GC techniques. The definition also describes compounds that are directly detectable by gas chromatographs, without use of derivatization (that is, not made volatile by some chemical combination process).

In the first instance, it is useful to provide information on the numbers of compounds in various groups, and sub-groups of chemical nomenclature. Tables 4.8 and 4.9 provide such information from the listing by Montedoro & Bertuccioli (1986). Table 4.8 details aliphatic compounds (by far the greatest number) and Table 4.9 shows compounds with a heterocyclic ring structure. No distinction is made between primary, secondary and tertiary aromas in the lists, which show the marked predominance of aliphatic esters, with much fewer heterocyclic compounds, except lactones/furanones. The volatile compounds described are those that frequently occur in numerous other beverages but the amounts and proportions in relation to each other can be very different.

Some of the main scientific investigators laying the foundation of the identification of volatile compounds of wine have been:

(1) Ribéreau-Gayon, and many colleagues and research students at the Faculty d'Oenology, University of Bordeaux, and other Universities and Institutes in France.
(2) Amerine, Ough, Webb and many others in the USA, mainly at the University of California, Davis.
(3) Drewart, Rapp, Schreier and others in Germany at the Institut für Rebenzüchtung Geilweilerhof, Siebeldinge and elsewhere.
(4) Bertuccioli and Montedoro, at the Istituto di Industrie Agrarie, University of Perugia and others in Italy.

Research institutes in Australia, South Africa and Spain have also been very active in wine technology and numerous aspects of wine science.

Several investigators have recently devoted and recorded their efforts in detail to specific wine types, such as the Bandol (southern France) wines based upon Mourvèdre grapes, and made valuable comparisons with wines based on Syrah grapes (Vernin *et al.*, 1993). Riesling and Sylvaner wines have been especially

Table 4.8 Volatile compounds found in wines: linear chain and homocyclic.

Group/subgroup	Aliphatic	Benzenoid	Alicyclic
Hydrocarbons (13)	4	5	–
Terpenes (19)			
Strict hydrocarbons	1	1	1
Others (alcohols, keto- etc.)	5	–	11
Alcohols (39)			
Mono-saturated	30	4	–
Mono-unsaturated	5	0	–
Aldehydes (25)			
Saturated	15	2	–
Unsaturated	2	1	–
Hydroxy	1	3	–
Alkoxy	–	1	–
Ketones (20)			
Mono-			
Saturated	10	1	see
Unsaturated	2	0	furanones
Hydroxy-	3	1	–
Hydroxy-methoxy-	0	2	–
Di-			
Saturated	1	–	–
Unsaturated	0	–	–
Acids (70)			
Saturated	27	5	–
Unsaturated	–	1	–
Hydroxy	10	15	2
Oxo-	3	–	–
Methoxy-	0	5	–
Keto-	1	1	–
Esters (163 total) (109, regarded as truly volatile and of potential aroma significance)	106	3	–
Acetals (dialkoxy) (12)	11	1	–
Nitrogen-containing (20)			
Amines			
Primary (mono and diamino)	14	2	–
Hydroxy	3	1	–
Sulfur-containing (11)			
Thiols	3	–	–
Thioethers (sulfides)	2	–	–
Esters	5	–	–
Acids	1	–	–
Phenols (14)	–	14	–
Total (333)			

Data from Montedoro & Bertuccioli (1986), all references taken.

studied by Rapp et al. (1985); Spanish white wines by Aldave et al. (1993); and wines based on Cabernet Sauvignon, by Ribéreau-Gayon (1990) and by Boison & Tomlinson (1990). A review by Ebeler & Thorngate (2009) showed the development of the flavour science, initially the emphasis was on determining the major

Table 4.9 Volatile compounds found in wines: heterocylic (oxygen and nitrogen containing).

Group/subgroup	Furans	Lactones (furanones)	Pyrones	Pyrazines
Substituted (alkyl-hydro)	–	19	–	–
Aldehydes	3	–	2	–
Alcohols	3	–	–	–
Acids	1	–	–	–
Esters	2	–	–	–
Ethers	2	–	–	2
Total	11	19	2	2

Data from Montedoro & Bertuccioli (1986).

volatile components, while current analytical techniques combining analytical and sensory information is more focused on identifying impact aroma compounds. Another focus is determining the effect on aroma of the matrix and other components in the wine giving so-called interaction effects.

The Spanish work of Aldave *et al.* (1993) is of particular interest, with its quantitative data on numerous samples over a whole range of different groups of volatiles. The French work of Vernin *et al.* on Mourvèdre grapes and wines is also of special interest, covering a wide range of volatiles, using valid identification techniques, though they are not fully quantitative. Numerous investigations have been carried out on individual compounds and specific groups of compounds like the terpenes, including quantitative data, often as exercizes in GC analyses by non-wine experts, such as Tressl of the University of Berlin, who has carried out similar determinations on various beverages. More investigators will be found in the study of the composition of grape brandies, Sherries and Port, which is relevant to that of the originating wines. A considerable amount of qualitative work has been reported by other investigators, such as in the IARC monograph (1988) on alcohol drinking, but the information is presented more for potential toxicity interest than flavour. This chapter leads on to the discussion in Chapter 5 of the wines of different grape varieties in relation to composition and individual flavour perception.

Both quantitative and qualitative data is now presented in the tables, grouped by the chemical nature of the compounds found, starting with the aliphatic/benzenoid esters, which are important compounds involved in any wine flavour assessment. Where possible, threshold odour/flavour data from different sources has also been tabulated, and in conjunction with actual content data a very approximate assessment can be made of the likely contribution of each compound to overall flavour, by comparing contents with threshold values.

4.2.1 Types of aroma in volatile compounds

The three main types of volatile compounds found in wines, according to origin, are included in the tables. These are (1) primary aromas (P), compounds already present in the grape and persisting through vinification; (2) secondary aromas (S), which are generated aromas primarily in the fermentation, by the action of the yeast (the main function of which is to produce ethyl alcohol) and bacteria, and make up, qualitatively and quantitatively, the largest amount of volatile compounds present in the wine; and (3) tertiary aromas, (T), which are generated during maturation or ageing processes, either in-cask (or vat/tanks) or in-bottle, subsequent to vinification. Secondary aromas may also include those really generated in pre-fermentation stages, such as in the process of crushing grapes, where notably, such compounds as the hexenals may be produced and largely persist through the remaining stages of vinification. The combination of secondary/tertiary aromas makes the wine markedly different in flavour to that of the grapes/must from which it originates.

Nevertheless, certain primary aromas characterize a wine. These are mainly the terpenes, the 'grapey' aromas present in relatively large amounts in the Muscat variety of grape, which has a number of sub-varieties with modified names. They may also be present to a lesser extent in other grape varieties, e.g. Riesling. There are some other primary aromas in grapes, which come through fermentation in the same way. These are often described as 'varietal' aromas, i.e. compounds in wine characteristic of a particular grape variety. Recently, the highly flavourful compound 2-methoxy-3-isobutylyrazine has been identified, at above recognition level, in wines made from Cabernet Sauvignon grapes, originating from the grape (mainly located in the skins). It imparts a variously described 'green' or 'bell-pepper' aroma with about 2 parts per trillion in water, though the threshold value in wine is higher. Nevertheless up to 50 ppt in wine has been detected by the most modern GC-MS techniques. This compound has been identified in a number of other grape varieties.

A 'blackcurranty' aroma is especially characteristic of Cabernet Sauvignon wines; its chemical origin is believed to differ from that of the blackcurrant fruit, and according to Marricon (1986) is based upon a complex terpene composition, which brings it in part to primary aroma status. Similarly thioketones and thioterpineols are now believed important. Jackson (2008) mentions also some other characterizing compounds of varieties, such as 4-vinyl guiaicol as 'spicy' in Gewürztraminer wines; but contrast Ribéreau-Gayon et al. (2006) with the description of odour 'smoky/clove-like'. The thioketone 4-methyl-4-mercaptopentan-2-one (guava-like) is particularly associated with Sauvignon Blanc wines and also Chenin Blanc and Colombard wines, with 2-phenyl-ethanol in Muscadine varieties and isoamyl acetate (fruity/bananas, pears) in Pinotage wines. Other specific varietal compounds have been identified recently, with accurate quantification data. The discovery of rotundone, giving the peppery aroma typical in Shiraz, was reported by Wood et al. (2008), and four ethyl esters contributing to sweet, fruity aromas in wines were identified by Campo et al. (2007). Recent reviews on wine flavour are Ugliano & Henschke (2009), Ebeler & Thorngate (2009) and Palaskova et al. (2008). It is, however, not always clear to what extent these substances are present in the original grapes, or are actually produced in the fermentation or by microbiological action from particular precursors in the grapes (see Chapter 7).

Varietal aromas can be present in other *Vitis* grape species, such as *Vitis labrusca*, where substances such as methyl anthranilate may be characterizing.

4.2.2 Stereochemical effects in aroma volatile compounds

The effect of slight changes in molecular structure of a given type of compound (e.g. with unsaturated/saturated carbon bonds) on flavour/aroma perception has already been discussed (Section 4.1.2). In recent years, the significance of chirality (i.e. the existence of molecular structures that are mirror-images, known as enantiomers, see Appendix I) on flavour/aroma has been examined (Pickenhagen, 1989; Lettingwell, 2002). In nature, however, biogenesis in fruits/vegetables usually generates only one of the enantomers possible, described as a specific chiral variant by the nomenclature of the letters *R* and *S* (see Appendix I), e.g. tartaric acid, with a characteristic optical rotation. Chemical or bacterial reaction may well produce both enantiomers or forms, e.g. lactic acid; as also in other compounds from fermentation processes cata-lysed by enzymes (e.g. amyl alcohols). Differences in flavour/aroma occur from one to another compound in a given pair.

The most marked differences in flavour/aroma occur with stereoisomers (see Appendix I), which arise when a compound has a 'cis' (Z) or 'trans' (E) molecular structure, in a molecule containing two carbon atoms joined by an unsaturated bond. The stereoisomers have the same atom groups attached to the two carbon atoms (the atom groups themselves, however, should not be identical). A typical example is 2-nonenal, also with perceived flavour differ-ences, (E), *trans*, odour threshold $0.5\,\mu g\,m^{-3}$ in air (0.5–1 ppb in water), and (Z) *cis*, 0.08–$0.23\,\mu g\,m^{-3}$ in air.

In lists of volatile compounds, the proper characterization of compounds in respect of possible chirality/stereoisomerism should therefore be included. The presence of asymmetrical carbon atoms is usually apparent from a structural formula, though they can be labelled by use of an asterisk.

4.3 Contents and sensory evaluation data

Tables 4.10, 4.11, 4.12, 4.13, 4.14, 4.15, 4.16, 4.17, 4.18, 4.19, 4.20, 4.21, 4.22, 4.23 and 4.24 record the available data for individual substances with their chemical struc-tures in groups of similar compounds. Many of these compounds, especially esters, have trivial and other names established by long usage, but chemical names according to IUPAC preferred recommendations are usually recorded first. Relevant physical properties of these same compounds are given in Appendix II.

4.3.1 Esters

As already mentioned, esters of all kinds are regarded as especially important to wine flavour, and are usually secondary aromas, arising from the fermenta-tion, and sometimes tertiary aromas arising from ageing, where alcohol–acid rearrangements can occur. Since there are many acids and alcohols in wine, there are many esters possibly formed in wine. Ethyl acetates are the most prevalent in wine, since there is a high concentration of ethanol and primary alcohols are quite reactive. Over 160 esters in wine have been identified, although not all are commonly present above their threshold values.

Structure

Esters result from the combination of organic alcohols R^1OH with organic acid, R^2COOH [e.g. $R^1-O-C(=O)-R^2$] with the elimination of water. R^1 and R^2 are usually alkyl or sometimes aryl radicals in mono-esters in wines, but some di-acid esters are also present.

Generally esters in wines can be divided in two groups. The first group consists of acetate esters, in which the acid group is derived from acetic acid and the alcohol group is ethanol or a complex alcohol. The second group is the ethyl esters, where the acid group is a medium chain fatty acid and the alcohol group is ethanol.

Presence in wines

Based on peak areas from GC traces, Vernin et al. (1993) found that about 38% of all the volatile compounds detected were esters, mainly aliphatic, in a red wine from Mourvèdre grapes grown in southern France, and similarly (35%) for a Syrah red wine; there were very few esters, in small amount, in the originating must.

Of the numerous esters (163) reported in the list of Montedoro & Bertuccioli (1986), some 109 could be regarded as properly volatile but only 59 of these are multi-referenced (i.e. three or more citations). At different times, 22 of these have been quantified, mainly by Bertrand (1975, unpublished data), as reported by Ribéreau-Gayon (1978) (Table 4.10 which follows). This data, covering a large number of wines, is of interest in that it shows on average a higher content of esters (lower C esters) in white wines compared with red, corresponding to the known effect of lower vinification temperatures used. The data of Vernin et al. (1993) is not fully quantitative, but the relevant amounts of some of the individual esters are of interest, as tabulated below (1984 vintage year, see Table 4.10). None of these compounds identified are new to the list in Tables 4.11 or 4.12 following.

Notably, individual ester content is generally in the ppm range, contrasting with most other volatile components, which are more usually at ppb levels.

However, four esters recently analysed and quantified in a range of wines, have very low thresholds at ppb levels (Campo et al., 2007). They suggested that ethyl 2-methylpentanoate, ethyl 3-methylpentanoate, ethyl 4-methylpentanoate and ethyl cyclohexanoate are formed by esterification reactions with ethanol and the corresponding acids formed by micro-organisms. Their concentrations vary, they are especially high in aged wines and can reach concentrations well above threshold levels.

Table 4.10 Percentages of some of the individual esters[a] for 1984 wines made from Mourvèdre grapes.

Diethyl succinate	12.90%
Ethyl lactate	5.56%
Ethyl-3-hydroxy butyrate	4.72%
Ethyl decanoate	4.68%
Ethyl octanoate	3.78%
Methyl-2-hydroxy 2-methylbutyrate	1.07%
Isoamyl lactate	0.92%
Ethyl phenyl acetate	0.58%
Ethyl hexanoate	0.55%
Others (20 in all)	3.5%
Total of 29	38.2%

[a] Tabulated using data from Vernin *et al.* (1993).

Francis & Newton (2005) surveyed the literature and compiled a list of compounds likely to contribute to wine aroma, based on their threshold values and the concentration ranges determined in wines. They listed the following esters all considered important in wine (Table 4.13).

It will depend on the grape variety, wine-making conditions and maturation which combination of esters will exert the greatest influence on the sensory properties of the wine. Some changes in content occur with malo-lactic fermentation, when practiced (e.g. increase of ethyl lactate).

Flavour characteristics

Tables 4.11 and 4.12 also give some reported odour/flavour characteristics of these esters. It is generally recognized that the lower aliphatic ethyl esters show fruity notes of different kinds (mostly tropical tree fruit, banana, pineapple, but also apple, pears, etc.) up to about ethyl heptanoate ($C_2 + C_7 = C_9$); whereas the higher homologues tend towards soapy, oily and candle wax characteristics. Other esters, such as ethyl benzoate, iso-amylacetate and hexyl acetate, will also show important fruity characteristics. However, none of these esters themselves appear to offer a number of other fruity characteristics found in many wines, such as blackcurrants, gooseberry or plums (see Chapter 5).

There is some quantitative threshold odour/flavour data for water around but little in wine at 10–15% alcohol content. It is apparent that some ester compounds will have high partition coefficients, and have very low threshold levels, even in alcoholic solutions, but others will not have so. It is difficult to establish a reasonably scientific basis of flavour assessment, based on Flavour Unit or other values as described in Chapter 5, Section 5.5. Whilst not many individual esters may show FUs or Aroma Indices greater than 1.0, in a given wine, a combination of the certain fruity esters may well give a combined high value, as recognized by wine tasters and, similarly, for the soapy esters from ethyl octanoate upwards to laurate. Ribéreau-Gayon *et al.* (2006) show how

Table 4.11 Main esters in wines: contents and olfactory characteristics (odour/flavour descriptions and thresholds). Origin, mainly secondary (S).

Compound — Esters: alcohol/acids	Wine type — White [content (mg L⁻¹) min, max]		Red [content (mg L⁻¹) min, max]		Odour/flavour characteristics	Threshold levels (ppm) — Odour from water	Flavour from water	Other
Methyl								
Methyl ethanoate (acetate)	(1) 0.00	0.11	0.08	0.15	Ethereal/ Fruity (at 60 ppm)	1.5–4.7^g	–	–
	(2) tr	2.22	–	–				
	(3) tr	5.45	–	–				
Ethyl								
Ethyl methanoate (formate)	(1) 0.02	0.84	0.03	0.20	Green/floral (rosy)/fruity	17^g	–	–
	(4) 0.01	2.5	–	–				
Ethyl ethanoate (acetate)	(1) 4.50	180	22	190	Ethereal/fruity/ [at low (100 ppm) concentrations]	60^d	3^b	125
						5^c	25^d	160^d
						6.2^a	6.6^h	(wine)
Ethyl propanoate (propionate)	(1) 0.0	7.50	0.07	0.25	Sweet/ethereal/fruity/ like rum [at 25 ppm, sharp, rummy, fermented]	9	–	+
	(4) 0.3	5.0	–	–		45		
	(6)		10.4	66.6				

						Threshold levels (ppb) — Odour from water	Flavour from water	Other
Ethyl n-butanoate (butyrate)	0.04		0.01	0.20	Ethereal/fruity (pineapple, banana)	1^c	450^b	–
	(6) 1.0		0.07	0.17		0.13^f	0.13^f	–
							15^h	
Ethyl 3-hydroxy-butanoate	(6)		0.05	0.52	Fruity/ winy, green			

(Continued)

Table 4.11 (Continued)

Compound — Esters: alcohol/acids	Wine type — White [content (mg L⁻¹) min, max]	Red [content (mg L⁻¹) min, max]	Odour/flavour characteristics	Threshold levels (ppm) — Odour from water	Flavour from water	Other
Ethyl 2-methyl propanoate (isobutyrate)	0.00 / 0.60 (6)	0.03, 0.03 / 0.08, 0.31	Fruity (apple, banana, pineapple) herbaceous	0.01-1[g]	–	–
Ethyl 2-methyl pentanoate	(5) 0.00 / 0.00 (6)	0, 12 ng L⁻¹ / 22 ng L⁻¹, 50 ng L⁻¹	Fruity, anise	0.003[i]		
Ethyl 3-methyl pentanoate	(5) 0.00 / 0.00 (6)	0 / 36 ng L⁻¹	Fruity	0.008[i]		
Ethyl 4-methyl pentanoate	(5) 14 ng L⁻¹ / 120 ng L⁻¹ (6)	175 ng L⁻¹, 60 ng L⁻¹ / 253 ng L⁻¹, 180 ng L⁻¹	Fruity, anise	0.01[i]		
Ethyl cyclohexanoate	(5) 0.00 / 0.00 (6)	0.00, 8 ng L⁻¹ / 5.5 ng L⁻¹, 12 ng L⁻¹	Licorice, anise	0.001[i]		
Ethyl n-hexanoate (caproate)	(1) 0.06 / 0.60 (6)	0.06, 0.03 / 0.13, 0.23	Fruity	36[d]	–	850[d] in wine
Ethyl-2-methyl-butanoate [NB enantiomers (S) and (R)]	(4) 0.1 / 2.0 ; 0.00 / 0.02 (6)	– ; 0.00, 0.00 / 0.08, 0.03	(pineapple/banana) ; Fruity/apple, banana, pineapple/herbaceous	1[e] ; 0.1[c]	–	–
Ethyl 3-methyl-butanoate (isovalerate)	0.00 / 0.04 (6)	0.00, 0.02 / 0.09, 0.03	(apple)/ethereal/vinous ; Fruity	0.1[e]-0.4[g]	0.01[b]	–
Ethyl -octanoate (caprylate)	(1) 1.10 / 5.10 ; (4) 0.4 / 2.5 (6)	1.00 / 6.00 ; – / –	Soapy/candlewax	8-12[g]	–	–
Ethyl nonanoate (pelargonate)	(4) tr / 0.3	0.02 / 010 ; – / –	Sl. fruity/oily/nutty	–	–	–

Compound					Odour			
Ethyl decanoate (caprate)	(1) 0.90 (4) 0.1 (6)	2.50 1.50	0.60 - 0.00	4.0 - 0.10	Oily/fruity/floral	8-12^g	-	-
Ethyl dodecanoate (laurate)	(1) 0.10 (2) tr	1.20 0.2	- -	- -	Oily/fruity/floral	-	-	-
Ethyl tetradecanoate (myristate)	(1) 0.10	1.20	-	-		-	-	-
Ethyl hexadecanoate (palmitate)	(1) 0.10	0.85	-	-		-	-	-
Ethyl mono-hydroxy propanoate (lactate) [NB R and S enantiomers]	3.80	15	9	17	Mild/ethereal/buttery/fruity	50-200^g	-	-
Ethyl cinnamate	(6)		0.00	1.22 µg L^{-1}	Honey, cinnamon		-	1.1^j
Ethyldihydrocinnamate	(6)		0.00	0.54 µg L^{-1}	Flower		-	1.6^j
Propyl								
n-propylethanoate (acetate)	(1) 0.00 (2) tr (3) tr	0.04 2.8 1.2	0.0 - -	0.08 - -	Fruity (pear) odour	2.7-11^g*	-	-
Diethyl								
diethyl-butan-dioate (diethyl succinate)	(1) 0.20 (4) 0.1	0.80 1.4	- -	- -	Faint, pleasant Odour	-	-	200000$^{(k)}$ (calculated in wine)
Isobutyl								
2-methyl-propyl (isobutyl) ethanoate	0.03	0.60	6.24	25.19	Fruity	-	-	-
	(6)		0.01 0.01	0.08 0.06			-	1600$^{(k)}$ (calculated)

(Continued)

Table 4.11 *(Continued)*

Compound	Wine type		Odour/flavour characteristics	Threshold levels (ppm)		
Esters: alcohol/acids	White [content (mg L⁻¹) min, max]	Red [content (mg L⁻¹) min, max]		Odour from water	Flavour from water	Other
Butyl						
n-butyl ethanoate (acetate)	0.04 / 6.10	0.01 / 0.02	Fruity, pears	50–500[g]	–	–
Isoamyl (pentyl)						
3-methyl-butyl (isoamyl or isopentyl) ethanoate	tr / 4.3 tr / 3.9 (6)	0.04 / 0.15 – – 0.12 / 0.22	Fruity (pears, bananas)	–	–	40[k] (in model wine)
3-methyl-butyl (isoamyl) lactate	(4) tr / 0.5	–		–	–	–
Hexyl						
Hexyl hexanoate	(2) tr / 1.3 (3) tr / 0.3	– –	Herbaceous odour/sweet/fruity (pears) flavour	6.4[g]	–	–
Phenyl-ethyl						
2-phenyl-ethyl ethanoate	(1) 0.20 / 5.10 (2) tr / 1.7 (3) tr / 2.1 (6)	– – – 0.02 / 0.04	Fruity (apricot)/honey-like odour. Butterscotch flavour.	650[g] (ppb)	–	250 (water with 10% ethanol)[k]

[a]Mulders (1973). [b]Keith & Powers (1968). [c]Flath et al. (1967). [d]Ribéreau-Gayon (1978). [e]Takeoka et al. (1995). [f]Ahmed (1978). [g]Feneroli (2002). [h]Sick et al. (1969). [i]Calculated from Campo et al. (2007). [j]Quoted by Francis & Newton (2005). [k]Quoted by Escudero et al. (2007). *Value not consistent with ethyl, butyl and hexyl acetates.

References: (1) Ribéreau-Gayon (1978) quoting Bertrand (1975). (2) Aldave (1993) from 33 samples among the Spanish varieties, Xarel-lo, Macabeo, Parallada. (3) Aldave (1993) from 21 samples among Airen, and Jaen Pardillo, and Viura (4) 54 samples of (2) and (3). (5) Campo et al. (2007). (6) Escudero et al. (2007).

Table 4.12 Other esters in wines identified (multi-referenced), but not quantified: odour/flavour characteristics and threshold flavour values.

Ester alcohol/acid	Odour/flavour characteristics	Threshold values (ppb)	
		Odour from water	Flavour in water
Methyl			
hexanoate (caproate)	Fruity (pineapple, apricot)	–	–
heptanoate (oenanthate)	Floral (orris) odour	10–87[g]	
octanoate (caprylate)	Currant-like flavour	4	
decanoate (caprate)	Fruity/orange like	200–870[g]	
Ethyl			
pentanoate (valerate)	Ethereal/fruity (apple)	5[c], 1.5–5[g]	94[b]
heptanoate (oenanthate)	Fruity (general) apricot	2[g]	–
dec-9-enoate	–	–	–
pyruvate (α-keto propaenoate)	Vegetal/caramel	–	–
4-hydroxy-butanoate	–	–	–
Benzoate	Floral/Fruity/warm odour. Flavour (at 30 ppm), sweet, medicinal, fruity, winter green	100[g]	–
3-phenyl-lactate	–	–	
3-methyl-butyl-succinate	–	–	
Butyl			
2-methyl propyl lactate	–	–	
2-methyl-propyl octanoate	Fruity		
Pentyl (amyl)		–	
pentyl (n-amyl) acetate (ethanoate)	Fruity (apple)	–	
3-methyl butyl (isoamyl) propanoate	Fruity (banana)	–	
3-methyl butyl butanoate	–	–	
3-methyl butyl hexanoate	Fruity (banana/apple)	–	
3-methyl butyl octanoate	–	–	
3-methyl butyl decanoate	–	–	
Hexyl			
Acetate	Fruity/(apple) beer flavour at 15 ppm	2–480[g]	
Diethyl			
Malonate	Faint, pleasant aromatic odour	N/a	
malate (hydroxy-butandioate)	Fruity odour with herbaceous undertones.	N/a	
Ethyl hydrogen Succinate (butan-dioate)		–	–

[b]Keith & Powers (1968). [c]Flath (1967). [g]Feneroli (2002). N/a not available.

Table 4.13   Esters likely to contribute to wine aroma, selected on wine concentration and sensory threshold information[a].

Ethyl esters	Acetate esters	Cinnamic esters
Ethyl isobutyrate	Isoamyl acetate	Ethyl dihydrocinnamate
Ethyl 2-methyl butyrate	Phenylethyl acetate	trans-Ethyl cinnamate
Ethyl isovalerate	Ethyl acetate	
Ethyl butyrate		
Ethyl hexanoate		
Ethyl octanoate		
Ethyl decanoate		

[a]Following Francis & Newton (2005).

the character of a white wine can be altered, by fermentation temperature and other factors, lower temperatures favouring the formation of 'fruity' esters, which are especially significant in young, white wines, contributing to their 'fruity' character. With red wines, higher fermentation temperature is used. These factors are further discussed in Chapter 7. The use of SO_2 in vinification (Aldave et al., 1993) and clarification procedures is also important.

From this data, it can be seen that ethyl acetate, on account of quantities often encountered (despite a relatively high odour/flavour threshold figure), could be an important flavour component of wines. However, though Ribéreau-Gayon (1978) regards 50–80 mg L^{-1} of ethyl acetate as being desirable in wines, with 125 mg L^{-1} a threshold flavour level, he considered that 160 mg L^{-1} contributes a 'hard' flavour, with an unpleasant pungent tang. Ethyl acetate is also listed as a potential off-flavour contributor (Chapter 5). Below threshold at 120 mg L^{-1} it can give a hot flavour in red wines, giving a bitter after taste. Acetic acid originates largely from undesirable spoilage (micro-organisms) by Acetobacter.

Ethyl lactate generally appears at relatively high amounts, especially after ageing, when lactic acid was formed in any malo-lactic acid fermentation, and/or from adventitious growth of lactic acid bacteria. It is not thought, however, that it contributes much to wine aroma, which is supported by a study of its physical property characteristics, as given in the Table II.1, in Appendix II along with those of all other esters present. The aroma effect of ethyl lactate in Tannat wines has been especially studied by Lioret & Versini (2002) in the Canary Islands. Similarly, diethyl succinate is another wine ester, again increasing in amount after ageing and, as already noted, representative of a high percentage of the wine distillate aroma (Vernin et al., 1993).

The recently described branched esters ethyl 2-, 3- and 4-methylpentanoate and one cyclic ester ethyl cyclohexanoate are thought to be important contributors to sweet-fruity aroma in wines (Campo et al., 2007). They are present in red and white wines, in particular in aged wines they reach concentrations well above their low thresholds (ppb levels) concentrations.

More studies focused on determining the contribution of the individual aroma compounds wine aroma, using analytical and sensory techniques as well as statistical analyses, for example Escudero *et al.* (2007) concluded that in the range of wine analysed, the berry fruit character could be related to the effect of nine fruity esters (ethyl butyrate, ethyl hexanoate, isoamyl acetate, ethyl 2-methylpropanoate, ethyl 2-methylbutyrate, ethyl 3-methylbutyrate, ethyl cyclohexanoate, ethyl 2-methylpentanoate, ethyl 4-methylpentanoate). It is likely that different groups of esters will contribute fruity flavours to different wines.

Table 4.12 gives some details of the ester compounds that have also been identified, but not quantified in the 1975 Bertrand data, which may have flavour significance in wine – many contributing 'fruity' characteristics. As most of these esters result from fermentations by yeasts (though not necessarily *S. cerevisiae*) they occur also in beer and such beverages as sake (rice wine). The composition of brandies (distilled product of grape wine) is also highly relevant.

4.3.2 Aldehydes

Presence in wine

Of the 25 aldehydes reported as being present by Montedoro & Bertuccioli (1986), only seven have three or more references to their detection. Ribéreau-Gayon *et al.* (2006) list 18 aldehydes (mostly alkyls) in wine, but state that, with the exception of acetaldehyde (ethanal) present at around $0.1\,g\,L^{-1}$, or $100\,mg\,L^{-1}$, they are only present in traces or in very small amounts. Whilst aldehydes can be present in grapes, under the conditions of vinification they will be largely oxidized to the corresponding alcohols. Acetaldehyde itself will be present as a fermentation product in differing amounts, dependent directly upon the quantity of sulfur dioxide also present, with which it combines. Acetaldehyde also combines with excess alcohol to form acetals. Only free acetaldehyde can be accorded flavour significance, e.g. as positively in Fino Sherries. Excess acetaldehyde confers 'flatness' in wines.

Hexanal, and the two hexenals [(*trans*) (*E*)-hex-2-enal and *cis* (*Z*)-hex-3-enal], the 'leaf aldehydes', are reported present in grapes and musts, together with 3-hexanol and *trans*-hex-2-enol by several investigators (Vernin *et al.*, 1993) in Mourvèdre wines. Their presence is due to the crushing of grapes, prior to vinification, when enzymatic oxidation of linoleic/linolenic acid can occur (Chapter 7). However, it is also stated that this wine flavour is a result of the use of unripe grapes (Ribéreau-Gayon *et al.*, 2006) and noted especially in the use of Grenache and Cabernet Sauvignon grapes. During fermentation, these aldehydes are transformed into the corresponding alcohols, which have a similar 'grassy' flavour at low concentration. As shown by Vernin, some amount of these aldehydes may remain unconverted. Octanal has been particularly mentioned in connection with Cabernet Sauvignon wines. Another aldehyde, (*Z*)-2-Nonenal, has recently been identified in a range of Spanish red wines, made from different grape varieties (Ferreira *et al.*, 2009), described as green and metallic.

Flavour characteristics

These higher aldehydes, if present, have strong flavours/odours, in particular nonen-2-al, which occurs in two stereoisomers, *E* (*trans*) and *Z* (*cis*) of which the former is the more important. It is of interest as being substantially present in green arabica coffee (Grosch, 2001), though negligible amounts remain in the subsequent roasted coffee. It has also been identified in malted barley (Meilgaard & Peppard, 1986), though again not in the subsequent brewed beer. In the latter case, it is converted into nonan-1-ol and nonan-l-yl acetate.

Nonanoic acid is of much less sensory interest. Nonanal has been reported present in wines but until recently, not the corresponding unsaturated aldehyde. However, Ferreira *et al.* (2009) identified (*Z*)-2-Nonenal in a range of Spanish wines. There was no information on formation, concentration or sensory contribution to the wines. It was detected in relatively high concentrations in all wine samples, giving a green, metallic note.

Except, therefore, for acetaldehyde the sensory significance of these alkyl aldehydes is regarded as low. Acetaldehyde, above its threshold level and in free form, is usually regarded as an off-odour ('flatness'); at high concentrations, it is characterized as pungent, even nauseating, but in dilute solution there are a number of more pleasant descriptions. A trace of acetaldehyde gives a recognisable apple like smell. It contributes significantly to the flavour of Sherry and of brandies.

Several aromatic aldehydes (or benzene derivatives) are of wine flavour importance, such as those developed during ageing in oak barrels. Thus, there are vanillin and cinnamic aldehyde, which are often recognized but little quantitative information appears available (further discussed in Section 4.4). Benzaldehyde (bitter almond) is a potential defect in wine but characteristic of certain grapes, such as Gamay. 2-Furfural and (5-hydroxy-methyl)-2-furfuraldehyde, resulting from carbohydrate oxidation, have also been reported, and increase in amount in in-bottle aged wines.

Only two aldehydes, acetaldehyde and phenylacetaldehyde, are listed as potential significant contributors to wine aroma by Francis & Newton (2005).

Detailed olfactory information is available for aldehydes (Table 4.14).

4.3.3 Ketones

Presence in wines

Of some 20 different ketones, only seven are multi-referenced in the listing by Montedoro & Bertuccioli (1986). β-Damascenone and α,β-ionones can be listed under terpenes, as also by Ribéreau-Gayon *et al.* (2006), but here (Table 4.15) they are listed simply as ketones.

Flavour characteristics

Diacetyl (buta-2,3-dione) may reach content levels to produce a sweet, buttery or butterscotch odour, in the range 1–4 mg L^{-1}, though it can be regarded in 'spoiled' wines as an off odour (at up to 7.5 mg L^{-1}). Low concentrations may

Table 4.14 Aldehydes in wines: content, general flavour/odour characteristics and threshold odour/flavour data.

Name of compound	Content[1]	General flavour/odour characteristics (at low concentration)[2]	Odour in air[2]	Threshold values		Other
				Odour from water (ppb)	Flavour/ water (ppb)	
Secondary aromas						
Ethanal (acetaldehyde)	Up to 0.1g L^{-1} 100 ppm in combined state	Pungent/suffocating at high concentration. Fruity at low (in free state) Freshly cut apples ('flatness')	0.005–0.12 mg m^{-3}	4–21[b] 120[a]	22[b]	25–30 ppm in sake[c] 10 ppm in beer[d] (3 ppm already present)
Propanal (propaldehyde)	tr	Fruity odour fresh green flavor	–	9.5[e] 10[f]	170[g]	
Butanal (butyraldehyde)	tr	Fruity, burnt, green odour, fatty, cocoa flavour	0.013– 0.042 mg m^{-3}	4–21[b]	5.3[b] 70[g]	–
2-methyl-propanal (iso butyraldehyde)	tr	Overripe, fruity, malty, more pleasant than butanal odour green, fermented flavour	0.015–0.014 mg m^{-3} 0.002– 0.004 mg m^{-3}	0.9[e] 0.9[h] 0.7[f]	–	–
Pentanal (valeraldehyde)	tr	Dry fruity/nut-like odour	–	12[e]	70[i]	–
3-methyl-butanal (isovaleraldehyde)	tr	Fruity odour peach-like flavour, also cheesy, malt	2–4 μg m^{-3}	0.15–0.2[e]	170[j]	0.66 ppm in beer[d]
Hexanal (caproaldehyde)	–	Cut grass, herbaceous, unripe fruit odour, leafy, green fruit	65–98[n] μg m^{-3}	4.5[e] 9.2[b]	3.7[b] 30[g] 16[i]	
Heptanal (oenanthic aldehyde)	tr	Fatty rancid: fermented, fruit-like odour	–	3[e]	31[c]	–
(*E*)-hex-2-enal (in grapes only)	?	Green, fatty, fruity flavour	125 μg m^{-3}	17[b] 24[b]	49[b]	–

(*Continued*)

Table 4.14 (Continued)

Name of compound	Content[1]	General flavour/odour characteristics (at low concentration)[2]	Odour in air[2]	Threshold values		
				Odour from water (ppb)	Flavour/water (ppb)	Other
Tertiary aromas						
Octanal (capryl aldehyde)	?	Green, fatty, orange, juicy flavour	5.8^n–$13.6\,\mu g\,m^{-3}$	0.7^e 0.7^e	0.52^b 5^g	
Nonanal (pelargonaldehyde)	?	Soap-like metallic odour	5.2^n–$12.1\,\mu g\,m^{-3}$	1^e 2.5^b	4.25^b	–
(E)-non-2-en-al (trans 2-nonenal)	?	Paper, planks Fatty, cardboard	$0.5\,\mu g\,m^{-3}$	0.08^e	–	1.11 ppb in beer[d]
(Z)-non-2-en-al[n] (cis 2-nonenal)	?	green/sl rancid	0.08–$0.23\,\mu g\,m^{-3}$	–		
Decanal	?	Soapy, citrus-like	–	0.1^e 2^b	7^g 3^b	–
Dodecanal	–	Soapy	–	$2^{k,e}$	1^b	40 ppb in beer[d]
Vanillin (4-hydroxy-3-methoxy-benzaldehyde)	tr	Vanilla	0.6–$1.2\,\mu g\,m^{-3}$	25^b	–	–
Benzaldehyde (benzoic aldehyde)	–	Bitter/almond odour cherry, pistachio flavour	–	350^k	1500^j	1000 ppb in beer[d]
Syringaldehyde (1-hydroxy, 2,6-dimethoxy-aldehyde)	–	–	–	–	–	–
Coniferaldehyde	–	–	–	–	–	–
Cinnamic aldehyde (phenyl propenal)	–	–	–	–	–	–
Phenyl-ethyl aldehyde (phenyl-acetaldehyde)	1.5–$9.9\,\mu g\,L^{-1m}$	Floral/rose/honey odour	–	4^k	–	–

References: (1) Data from Ribéreau-Gayon et al. (2000). (2) Data from Flament (2001). [a]Mulders (1973); [b]Ahmed (1978) also quoted by Shaw (1986); [c]Takahashi & Ayiyama (1993); [d]Meilgaard (1982); [e]Guadagni et al. (1963b); [f]Grosch (1995); [g]Lea & Swoboda (1958); [h]Yakotsuka (1986); [i]Sick et al. (1971) ; [j]Keith & Powers (1968); [k]Buttery et al. (1971a); [l]Parliment (1981); [m]Escudero et al. (2007); [n]Ferreira et al. (2009).

Table 4.15 Ketones in wines: contents, general olfactory characteristics and threshold data.

Name of compound	Range of content (mg L⁻¹, ppm)	General flavour/odour characteristics[i]	Threshold value data			
			Odour in air[i] µg m⁻³	Odour from water (ppb, µg L⁻¹)	Flavour in water	Other
Propan-2-one (acetone)	–	Light ethanol	–	300 ppm[a] 500 ppm[b]	–	–
Butane-2,3-dione (diacetyl)	0.05–3.14 white wines[c] 0.02–5.4[c] 0.2–2.5 mg L⁻¹ᵏ red wines traces[i]	Buttery (a negative above threshold value)	10–20[d]	6.5[a] 15[d]	5.4[g] ppb	2 mg L⁻¹ in wine[f]
3-Hydroxy-butan-2-one (acetoin)	0.002–0.3 French wines 0.01g L⁻ⁱ 11–55 mg L⁻ⁱᵏ	Creamy, fatty odour	–	–	–	–
Pentane-2,3-dione (acetyl acetone)	0.007–0.4 white wines 0.01–0.88 red wines	Buttery odour	10–20	30[d]	–	–
Hexan-2-one (methyl butyl ketone)	–	Fruity, blue cheese flavour	–	–	–	–
Heptan-2-one	–	Fruity odour	–	140[e]	–	–
Octan-2-one	–	Unripe appley	–	–	1.6 ppm[f]	–
Nonan-2-one	–	Flowery/green odour	–	–	0.15 ppm[g]	–
4-Hydroxy-3-methoxy-acetophenone (phenyl methyketone)	–	Green herbal flavour	–	–	–	–

(Continued)

Table 4.15 (Continued)

Name of compound	Range of content (mg L⁻¹, ppm)	General flavour/odour characteristics[1]	Threshold value data			
			Odour in air[1] µg m⁻³	Odour from water (ppb, µg L⁻¹)	Flavour in water	Other
α, (β), ionones, 4-(2',6',6'-tri-methyl-cydohex-1-2'-en-1'-yl)-buten-3-en-2-one	ng L⁻¹ Av. 13, SD 19- Dry white wines Av. 381, SD 396 Red wines	Violets (β; soft, strawberries) Cedarwood	–	7 ng L⁻¹ (0.007 ppb) (7 ppt)	–	1800 ng L⁻¹ in model alcoholic solution 1500 ng L⁻¹ in wine (1.5 µg L⁻¹) (recognition) 2.6 µg L⁻¹ in beer[i]
α ionone	0.017–0.54 µg L⁻¹[j] 0–0.67 µg L⁻¹[k]					
β ionone	0.032–30 µg L⁻¹[j] 0.09–0.23 µg L⁻¹[k]					4.5 µg L⁻¹[j] 0.09 µg L⁻¹[k] in model wine
β-Damascenone,	0.23–3.5 µg L⁻¹[k] in red wine Av. 709, SD	Fruity, honey odour	0.002–	9 ppt 0.75 ppt[h]	–	45 ng L⁻¹ in
E-isomer l-(2',6',6'-trimethyl-cyclohex-1',3'-diene-1'-yl)-but-2-en-l-one	561 Dry white wines Av. 2160, SD 1561 Red wines	Fruity, juicy red-fruit, woody, sweet flavour	0.004 µg m⁻³	2 ng L⁻¹ (2 ppt)		model solution 5000 ng L⁻¹ in red wine (recognition)

Av. = average value. SD standard deviation of range. *References:* [1]Data from Flament (2001). [a]Mulders (1973); [b]Perssons *et al.* (1973); [c]Nykanen & Soumaileinan (1983); [d]Blank *et al.* (1992), quoted by Grosch (2001); [e]Buttery *et al.* (1969a); [f]Lea & Swoboda (1958); [g]Siek *et al.* (1969), quoted by Flament (2001); [h]Data of Rychlik *et al.* (1998), quoted by Grosch (2001); [i]Ribéreau-Gayon *et al.* (2000). [j]Quoted by Swiegers & Pretorius (2005). [k]Escudero *et al.*, 2007, for five red wines. All other data for ionones/damascenone from Chattonnet & Dubordieu (1997), quoted by Ribéreau-Gayon *et al.* (2006).

impart yeasty, nutty, toasted aromas (see a review by Bartowsky & Henschke, 2004). They quoted that taste threshold levels for diacetyl are wine dependent, ranging from 0.2 to 2.8 mg L^{-1}, thought to be due to other compounds present in wines. Control of its formation is another factor (see Chapter 7).

Acetoin (3-hydroxybutan-2-one; or acetyl methyl carbinol in some literature) has a similar slightly milky odour, and may be perceptibly present in wines. The other simple aliphatic ketones, though present and formed during fermentation, are not considered to have much flavour significance in wines.

The complex ketones, β-damascenone and α,β-ionones (so-called isoprenoids derived by oxidative degradation from carotenoids) are present, partly as a result of crushing the grapes. β-Damascenone has a variably described odour, sometimes rose-like, and is believed to contribute to the aroma of wines from grape varieties such as Chardonnay, but it is probably present in all wines. A study on the aroma impact of β-damascenone in red wines by Pineau *et al.* (2007) places doubt on how large its contribution on wine aroma is. They studied 23 French wines from six different regions and comprising eight different grape varieties and determined a concentration range from 545–2307 ng L^{-1}, whilst they quote a range based on literature between 1–1.5 μg L^{-1}. They also accurately analysed the sensory threshold value of β-damascenone. The threshold in water/ethanol solution was 50 ng L^{-1}, but in wines these values were significantly higher (model white wine 140 ng L^{-1}, model red wine 850–2100 ng L^{-1} and red wine 7000 ng L^{-1}). Hence the perception threshold appears to depend significantly on the matrix used and in red wine it ranges probably between 2–7 μg L^{-1}, within the range of the value quoted in red wine in Table 4.15. Hence they concluded that there may not be a direct impact of β-damascenone on red wine aroma, although they suggested that it may enhance the fruity notes in particular of ethyl cinnamate and caproate, which need to be further investigated. Escudero *et al.* (2007) also suggested a flavour enhancing role for β-damascenone. In contrast, the detection threshold of β-damascenone in gas chromatography olfactometry is very low, which may have led to the conclusion in other research papers that β-damascenone does contribute to red wine aroma.

Also, it has been shown by Grosch (2001) to be present in green coffee, and an important contributor to the flavour of coffee brews from roasted coffee, since its threshold level is very low (in water). Similarly, α- and β-ionones occur in Riesling grapes, again, having a somewhat variably described odour, but notably, of violets. Considerable data on these ketones in wine has been obtained by Chatonnet & Dubourdieu (1997), and reported by Ribéreau-Gayon *et al.* (2006). The latter tabulate content and threshold information for 12 white wines and 64 red wines, showing that the content of both these ketones is much higher in red wines than in white (especially β-ionone). Both these compounds are present in considerable amount in brandies (4.I and 4.II).

β-Ionone

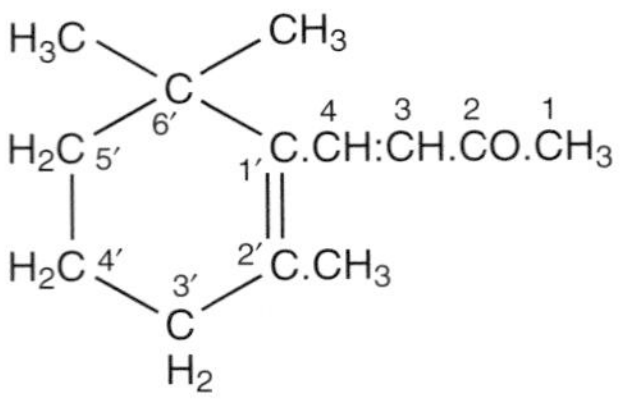

4-(2′,6′,6′-trimethyl-cyclohex-1′-en-1′-yl)but-3-en-2-one
α-ionone is, -2′-en-1-yl, otherwise identical with β

4.I

β-Damascenone (*E* isomer)

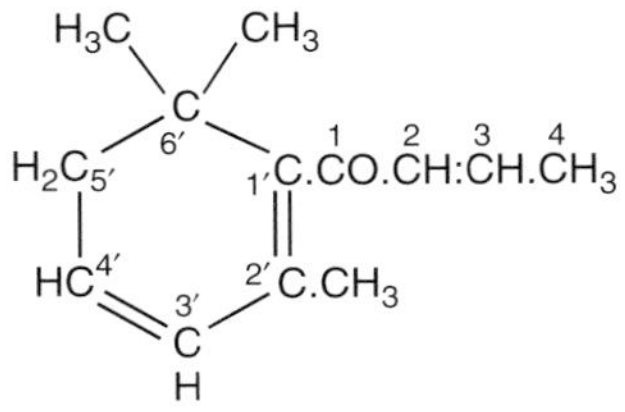

1-(2′,6′,6′-trimethyl-cyclohex-1′,3′-dien-1-yl)but-2-en-1-one

4.II

Certain ketonic substances are associated with the ageing of wines, such as the oak-lactones, but these are described under lactones/furanones (Section 4.3.6).

Based on a literature survey, Francis & Newton (2005) listed 2,3-butanedione, acetoin, β-damascenone and β-ionone as potental significant aroma ketones in wines.

4.3.4 Acetals

Acetal (1,1-diethoxyethane) is formed by the reaction of acetaldehyde with ethanol and regarded as flavour significant. Similarly, other acetals are formed from other alcohols and aldehydes; though only 12 acetals are mentioned by Montedoro & Bertuccioli (1986) up to 20 are suggested as having been detected in wines by Jackson (2008) and Ribéreau-Gayon *et al.* (2006). They have an herbaceous-like character but are regarded to have little significance in wine flavour (except Vin Jaune, of the Jura) and they are more important in Sherry and aged Ports (Chapter 6), in which conditions for their formation are more favourable.

4.3.5 Alcohols

Some 39 alcohols (plus ethyl alcohol) have been identified and listed by Montedoro & Bertuccioli (1986) but only 16 are multi-referenced. Ribéreau-Gayon *et al*. (2006) list 28 different alcohols (mainly alkyl). Contents of various alcohols in wines have been determined, and are shown in Table 4.16, together with some sensory description information, and threshold values.

Presence in wines

There is some detailed quantitative information (Aldave *et al*., 1993) on young Spanish white wines (Table 4.17).

The data of Vernin *et al*. (1993) on Mourvèdre grape wine shows the proportions of the different alcohols present in a distillate, following GC analysis, as shown in Table 4.18.

Methanol is normally present at low concentrations, 0.1–0.2 g L^{-1} quoted by Jackson (2008), between 30 and 35 mg L^{-1} quoted by Ribéreau-Gayon *et al*. (2006) and does not contribute to the sensory properties of wine. These concentrations are too low to pose a toxicity risk. Even its contribution to esters formed in wine is small. Methanol is formed due to pectin degradation and as the pectin levels in grapes are low, the amount of methanol formed is low. The amount formed depends on the maceration used, hence red wines have higher concentrations than rosé or white wines. Ethanol is of course the most abundant alcohol in wine and discussed in Section 3.2.

Most higher alcohols, having more that two carbon atoms, are formed during yeast fermentation, and often constitute about 50% of the volatile aroma compounds in wine, excluding ethanol. Typical concentrations in wine are 150–550 mg L^{-1} (quoted by Ribéreau-Gayon *et al*., 2006). Spoilage by yeasts and bacteria can increase the concentration of higher alcohols.

The alcohols (C$_4$ upwards), mainly the branched chain compounds, such as 2-methylpropan-1-ol, and the 2- and 3-methylbutanols, together with the normal straight chain compounds, such as butan-1-ol, as well as the benzenoid alcohol phenylethanol are known as fusel oils, and are usually present in fair quantity. The most important of these are the amyl alcohols, C$_5$H$_{11}$OH, of which there are three main isomers, pentan-1-ol (normal amyl), 3-methyl butan-1-ol (isoamyl) and 2-methyl-butan-1-ol (optically active amyl alcohol). The word 'amyl' comes from the Latin 'amylum' and means starch, so these alcohols are known as fermentation alcohols. Other pentanols such as the -2, and -3-ols are believed present, but only in traces.

The quantity of fusel oils present has been related (like esters) to the vinification conditions by Ribéreau-Gayon (1978) (quoting the works of Soufleros, 1978; Doctoral Thesis); that is, lower fermentation temperatures (e.g. 20°C and pH 3.4) produce higher amounts of total alcohols (i.e. 201 mg L^{-1}) than higher temperatures. Similarly, higher amounts are also produced by not clarifying juice by racking before fermentation. See also Chapter 7.

Table 4.16 Higher alcohols in wines: contents, general olfactory characteristics and threshold value data

Name of compound	Range of content [mg L⁻¹ (ppm)]	General olfactory characteristics	Threshold value data			
			Odour in air[1]	Odour/ water (ppm)	Flavour/ water (ppm)	Other (ppm)
Propan-1-ol (propyl alcohol)	(1) 11–68 (2) 30 (3) 9–18 (4) 11–52	Fruity/alcoholic odour	9[a]	–	–	500 (wine solution)[h]
Butan-1-ol (butyl alcohol)	(3) 1.4–8.5 (4) 2.1–2.3 (2) traces (5) 1.9–2.5	Alcoholic	–	40[b] 0.5[b]	–	150 (calculated in wine)[j]
2-Methyl-propan-1-ol (isobutyl alcohol)	(1) 6–174 (2) 100 (3) 28–170 (4) 45–100 (5) 28.7–89.6	Ethereal/Fruity	–	3.2[b]	–	40 (in model wine)[i]
Pentan-1-ol (pentyl or amyl alcohol)	(1) <0.4 (2) traces	Fusel	–	0.5[a]	–	–
2-Methyl-butan-1-ol (active amyl, (–) rotation)	(1) 19–96 (2) 50 (3) 17–82 (4) 48–150	Pungent but pleasant Earthy-musty (R) Ethereal – fruity (S)	–	–	–	–

Compound	Concentration (references)		Odour/flavour	Threshold	Threshold
3-Methyl-butan-1-ol (isoamyl alcohol)	(1) 83–400 (2) 200 (3) 70–320 (4) 117–490 (5) 112–277	–	Fruity-winey	0.25[f] 0.77[b]	300 (wine solution)[h] 30 (in 10% ethanol)[i]
Hexan-1-ol	(1) 0.5–12 (2) 10 (3) 3–10 (4) 3–10 (5) 0.7–1.5	–	Fatty/fruity odour Green/fatty flavour	0.5[a]	4 (in wine)[i]
Hex-3-en-l-ol (Z) (leaf alcohol) (cis)	(2) (5) 0.04–0.23	–	Herbaceous	–	0.4 (in model wine)[i]
Hex-3-en-l-ol (E) (trans)	–	–	Herbaceous	–	–
Hex-2-en-l-ol (Z) (cis) 1	–	–	Herbaceous	–	–
Octen-1-en-3-ol (enantiomers S and R)	–	–	Mushroom odour/flavour (R)	2.3 –5.3 ppb[g]	–
Octan-1-ol	(1) 0.2–1.5	54 ppb[e]	Green, fatty coconut flavour	190 ppb[e] (0.19 ppm)	–
2-Phenyl-ethan-1-ol (β-phenyl ethyl alcohol)	(2) 50	–	Floral/woody (rose-honey-like)	–	10 (in 10% ethanol)[i]
3-Thio-propen-l-ol	(5) 46–96 (1) 0.05–2	–	–	–	–
See also under sulfur compounds					

References: (1) Amerine & Ough (1983); (2) Ribéreau-Gayon *et al.* (2000); (3) Bertrand (1975) – red wine quoted by Ribéreau-Gayon *et al.* (1978). (5) Escudero *et al.* (2007), range of five5 red wines.

[a] Flath (1967); [b] Mulder (1973); [c] Yakotsuka (1986); [d] Ribéreau-Gayon (1978); [e] Ahmed (1978); [f] Buttery *et al.* (1971a); [g] Grosch and Würzenberger (1985), quoted by Flament (2001). [h] Quoted by Swiegers *et al.* (2005). [i] Quoted by Escudero *et al.* (2007).

Table 4.17 Higher alcohols found in Spanish white wines (additional data of Aldave et al., 1993).

Alcohol	Average amount (mg L⁻¹)		SD value[a]		
	A	B	A	B	
Propan-1-ol	21.7	21.7	5.7	8.7	
2-Methyl-propan-1-ol (isobutyl alcohol)	36.9	47.3	15.6	13.3	
3-Methyl-1 butanol (isoamyl/pentyl alcohol)	134.6	206.2	45.9	54.6	
2-Methyl-1 butanol (active amyl alcohol)	30.2	53.9	12.8	17.0	
Phenyl-2-ethanol	32.6	66.2	14.2	27.6	
Hexan-l-ol	0.3 min.		8.5 max.		for 54 samples
cis-3-Hexen-1-ol	0.1		2.5		for 54 samples
Octanol-1-ol	Tr		0.1		for 28 samples

[a] SD Standard deviation of mean value. Group A: 33 samples from Xarelelo, Meccbeo and Paralleda grapes used for sparkling wines. Group B: 21 samples, Airen, Jean, Pardillo and Viura grapes.

Table 4.18 Percentages of alcohols in a distallate following GC analysis of wine made from Mourvèdre grapes in 1984[a].

2- and 3-Methyl butanols	31.60
1-Butanol	3.32
β-Phenylethanol	2.77
2-Methylpropanol	0.84
1-Hexanol	0.47
Benzyl alcohol	0.27
cis-3-Hexen-1-ol	0.20
trans-3-Hexen-1-ol	0.17
trans-2-Hexen-1-ol	0.13
2-Pentanol	0.30
1-Penten-ol	0.05
Total (11)	39.95% of all volatile compounds present

[a] Taken from data in Vernin et al. (1993).

Table 4.19 Numbering of atoms in linear and cyclic lactones.

No in linear chain	No in ring
1	2
2	3
3	4
4	5

Flavour characteristics

Higher alcohols generally are thought to contribute to the complex aroma of wines, however, they are not considered to be desirable at high concentrations and may mask the fruitiness of the wine. These fusel alcohols have a characteristic pungent odour. At a high concentration ($>300\,mg\,L^{-1}$) these are negative quality factors but at lower levels they add to the desirable aspects of wine flavour. Their esters, in particular isoamylacetate, may contribute a fruity banana note to young wines.

Hexanol is known as 'leaf' alcohol, and has a 'grassy' flavour that is reflected in the flavour of wines in which it is present in a sufficient amount, together with *cis*- and *trans*-3-hexen-1-ol and *trans*-2-hexen-1-ol. It is often originally present in the grape must, due to enzyme action on linoleic acid, resulting from grape crushing techniques (see also Section 4.3.2). Vernin *et al.* (1993) have compared content in Mourvèdre must with wine and generally showed a decrease in the amount of hexanol and the enols, as a result of fermentation. Oct-1-en-3-ol is reported to be present in wines, due to *Botrytis* action (Chapter 7) and has a mushroom aroma.

4.3.6 Lactones and furanones

Lactones/furanones are widely distributed in nature, particularly in the fruit of plants. The organic substances known as lactones are particularly associated with Sherry but some lactones originate from ageing in oak barrels.

Molecular structures

Chemically, lactones are inner anhydrides of hydroxy acids, resulting from the splitting off of one molecule of water. The so-called γ- and δ-lactones in older literature are the most easily formed compounds. They are generated from aliphatic hydroxy-acids with the hydroxy group in the 4 or 5 position (4.III).

(a) 4-hydroxy-butanoic acid
(or γ-hydroxy-)

(b) γ-butyrolactone

4.III

NB Numbering of the atoms has changed as in Table 4.19 below. (Other groups can be attached on linear carbon = 4-ring carbon-5.)

γ-Lactones have also been known as '-olides'; thus γ-butyrolactone was also known as 'butanolide' but they are now more correctly described as furan derivatives, as can be expected by the basic furan ring present. Thus, γ-butyrolactone is now ascribed as dihydro-3(*H*)-furan-2-one.

The term '3(*H*)' means that a hydrogen atom has been moved from the 2- to the 3-position in the ring (as the result of carbonyl formation) and that there

Table 4.20 Lactones/furanones in wines: contents and olfactory characteristics (descriptions and threshold data).

Name (furanone)	Content	Odour/flavour characteristics	Origin	Threshold data
-Furan-2-one				
Dihydro-3(*H*)-(γ-butyrolactone, a saturated γ-lactone; tetra hydro-fur-2-one)	A few mg L^{-1} 17–34 mg L^{-1}	Aromatic, faint, sweet character.	S	High level
Dihydro-5-methyl-3(*H*)-(γ-isovalero lactone)	–	Sl. buttery, acrid buttery Warm, sweet, herbaceous odour fatty flavour	S	–
Dihydro-5-acetyl-3(*H*)-(solarone)	Below threshold	–	S	–
Dihydro-3-hydroxy-4,5-dimethyl-3(*H*)-(2,4-dihydroxy-3,5-dimethyl γ-butyrolactone) (Pantalone)	?	–	S	–
Dihydro-5-butyl-4-methyl-3(*H*)-(β-methyl-γ-octalactone) (Oak-lactones)	*cis* (−) can be a few mg L^{-1} *cis* (+) 0–589 µg L^{-1} *trans* (+) *trans* (−)	Herbaceous coconut Sweet coconut Spicy coconut, walnut Strong coconut, leather, woody	T	*Cis* compds with lower threshold than *trans*. In white wines *cis* (−) 90 µg L^{-1}, *trans* 490 µg L^{-1}
Dihydro-5-pentyl-3(*H*)-(γ-nonalone)	3.3–71 µg L^{-1}	*R*. Coconut fatty-milky *S*. Less intense odour *R*. Herbal, coconut *S*. Dairy, coconut flavour	S	30 ppb[b] in wine
Dihydro-5-hexyl-3(*H*)-(γ-decalone)	3.1–15 µg L^{-1} 0–2.9 µg L^{-1}	*R* and *S*	S	90 ppb in water

Compound	Occurrence / concentration	Flavour / odour		Odour threshold
7-Hydroxy-6-methoxy-2(H)-pyron-2-benzo-furanone. 7-hydroxy-6-methoxy-coumarin [hydroxy cinnamic acid – δ-lactone] (Scopoletin)	2.2–73 µg L^{-1}[l]	Sweet, coconut odour, fruity, milky, peach flavour	T	–
Dihydro-5-vinyl-5-methyl-3(H)-Dihydro-	In Muscat/Riesling wines	–	P	–
3-vinyl 5-methyl-3(H)-(2-ethenyl pentanoic acid γ lactone). The raisin lactone	–	–	P	–
3-Hydroxy-4,5-dimethyl-5(H)-(tautomer with 4,5-dimethyl furan-2,3-dione) (Sotolon) [NB unsaturated γ-lactone]	+ in botrytized wines and Vin Jaune (Jura) in Port wine 5–958µg L^{-1}[j] in Madeira 100–1000µg L^{-1}[k] in young white wine 5.4 µg L^{-1}[j]	Spicy toasty walnut odour nutty, sweet flavour	S	Odour in air 0.01–0.20 µg m^{-3} odour in water 20 ppb flavour in water 1–5 ppb
-Furan-3-one 4-hydroxy-2,5-dimethyl-2(H)-(Furanol, HDMF) [Unsaturated, but not a lactone]	2.1–36µg L^{-1}[i]	Fruity, caramel odour (–) (+) isomer, faint sweet, 'roasty'[h]	P	Odour 10 ppb[d], 60 ppb[f], flavour in water, 160 ppb[g]
4-Methoxy,2,5-dimethyl-2(H)-	Varietal to *Vitis lambrusca*	Fruity/strawberries	P	–

References: [a]Ribéreau-Gayon *et al.* (2000). [b]Nakamura *et al.* (1988). [c]Keith & Powers (1968). [d]Semmelroch & Grosch (1996). [e]Rychlik *et al.* (1998). [f]Buttery (1999). [g]Huher (1992). [h]Bruche (1995). [i]quoted by Francis & Newton (2005). [j]Ferreira *et al.* (2003). [k]Camara *et al.* (2006). [l]Escudero *et al.* (2007) determined in red wines.

are two additional hydrogen atoms present, at positions 4 and 5, in addition to those in a simple furan ring (4.IV).

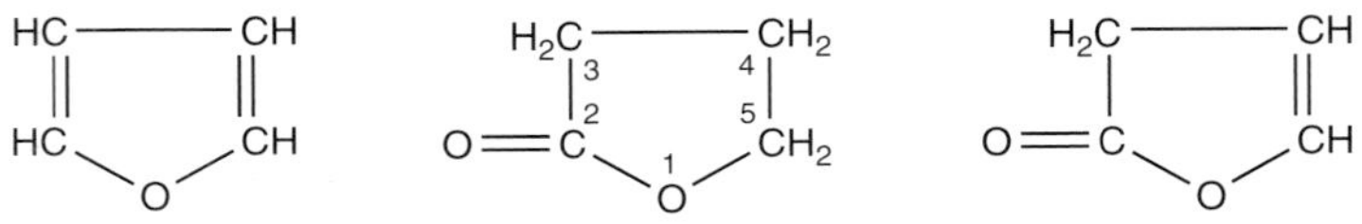

(a) furan   (b) dihydro-3(H)-furan-2-one   (c) furan-2-one

4.IV

Dihydro compounds are so-called saturated γ-lactones and are also referred to as tetrahydro-furanones.

The symmetry of the furan ring also means that furan-2-one is identical with furan-5-one and similarly, for substitutes at positions 3 and 4.

A compound such as 4-hydroxyisovaleric acid (4.V) forms a furanone (4.VI).

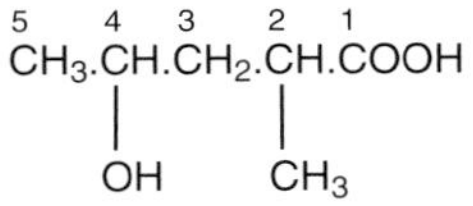

$$\overset{5}{C}H_3.\overset{4}{C}H.\overset{3}{C}H_2.\overset{2}{C}H.\overset{1}{C}OOH$$

OH   CH₃

2-methyl-4-hydroxypentanoic acid
(4-hydroxy-isovaleric acid)

4.V

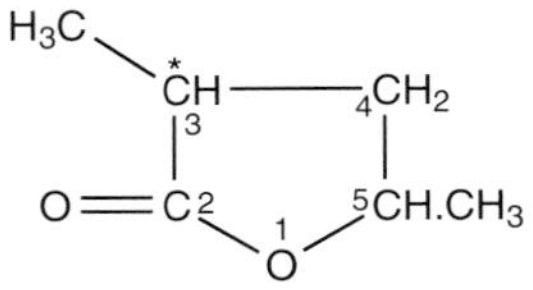

3,5-dimethyl-dihydro-(3H)furan-2-one

4.VI

Some lactones are derived from unsaturated hydroxy-acids.

Several other chemical reactions lead to the formation of furanones, and where the beta position is at C3 (or C4) they are thus known as furan-3-ones, which are not, strictly speaking, lactones.

Stereoisomers of these lactones are possible with *cis* and *trans* geometric forms, and optical variants with flavour implications (see the next section on *Flavour characteristics*).

When linkage takes place with a hydroxyl- group in the 5 (linear) position, the resulting lactone is the δ-lactone type, now forming a 6-membered ring, i.e. a pyran ring. Thus, the '5' (linear) position may be in a benzene ring (4.VII and 4.VIII).

These substances may still be referred to as either γ or δ-lactones, but their origins can be determined from their furanone/pyranone structure.

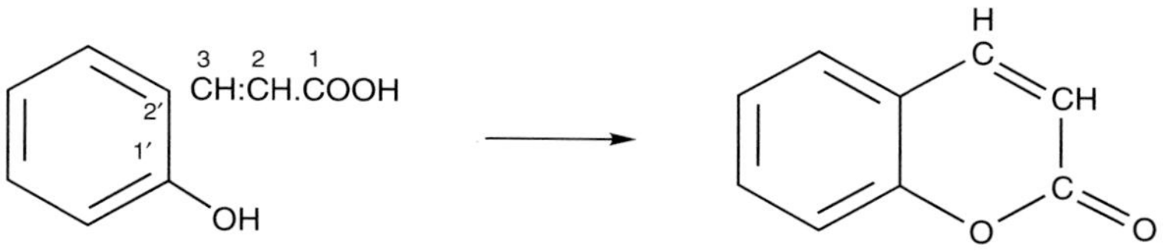

1-hydroxycinnamic acid
3-(1′-hydroxy-2′-phenyl)
propenoic acid

4.VII

Benzo-pyran-2-one
(coumarin)

4.VIII

Presence in wines

Some 20 lactones have been identified in wines and are listed by Montedoro & Bertuccioli (1986) and several others have been recently identified, both in grapes and extracted from oak barrels during ageing. Most lactones/furanones appear to be formed during fermentation. γ-Butyrolactone is present at about $1\,mg\,L^{-1}$ (Ribéreau-Gayon et al., 2006), but it is not considered an important flavour component in wines. Certain lactones are believed to be present in certain kinds of grapes, thus:

(1) Riesling and Muscat grapes; 2-vinyl-dihydrofuran-2-one.
(2) HDMF or furanol is believed present particularly in Merlot wines, but also in others, including *Vitis lambrusco* wines.
(3) Sotolon is particularly associated with *Botrytis* infected grapes and wines (Noble Rot) occurring at above the sensory threshold of 7.5 ppb (Mauake et al., 1992), also identified in the Vin Jaune from the Jura, France (Ribéreau-Gayon et al., 2000).
(4) Sotolon has also been found in fortified Port wines (Fereira et al., 2003). In Madeira wines sotolon levels identified were up to $2\,mg\,L^{-1}$ in wines aged 25 years (Camara et al., 2004).

However, Vernin et al. (1993) report few lactones/furanones in the Mourvèdre and Syrah grape must and wines they examined. There appears to be relatively little quantitative information and it is thought that, unlike in Sherries, their flavour importance in wines is low. Lactones tend to be high-boiling compounds and their solubility in water is relatively high with consequent low volatility. Based on a literature review, Francis & Newton (2005) concluded that the following lactones may contribute to the aroma of some wines since their concentration was near or above threshold concentration: *cis*-oak lactone, γ-nonalactone, γ-decalactone, γ-dodecalactone, 4-hydroxy-2,5-dimethyl-3(2H)-furanone and sotolon.

Nevertheless, the so-called oak-lactones, sometimes referred to as whisky lactones, present in aged wines have aroused considerable interest, with concentrations at several $mg\,L^{-1}$. The main oak-lactone is 5-butyl-4-methyl-dihydro-3(H)-furan-2-one. There are two geometrical isomers (*cis* and *trans*) and two optical isomers for each, (+) and (−), with differing sensory characteristics.

Table 4.21 Volatile acids in wine: contents, general olfactory characteristics and threshold values data.

Name of compound (acid)	Content (mg L⁻¹)			General olfactory characteristics	Odour/water (ppm)	Flavour/water (mg L⁻¹)	Other
	(1)	(2)	(3)				
Methanoic (formic)	<60	–	–	Stinging above 50 ppm	–	–	–
Ethanoic (acetic)	Tr <60	–	69–489	Vinegary	–	54[a] 22[b]	200 in 10% ethanol[e]
Propanoic (propionic)	–	2.7–8.4	4.1–113	Animal (goaty)	–	–	8.1 in 9.5% ethanol[d]
Butanoic (butyric)	<0.5	0.4–1.3	0.4–4.4	Rancid	0.24	6.8[a] 6.2[b]	0.173 in 10% ethanol[e]
2-Methyl propanoic acid (iso butanic)	–	0.6–4.2	0.4–7.6	Fruity pleasant odour, buttery – cheesy	–	–	2.3 in 10% ethanol[e]
Pentanoic (n-valeric)	–	Tr	–	Animal especially when dilute, fruity when very dilute	–	–	–
3-Methyl butanoic (isopentanoic)	–	1.6–2.1	–	Cheesy, herbaceous (dependent upon concentration)	1.6[c]	–	–
2-Methyl butanoic (R and S)	–	0.09–0.26	–	Cheesy odour (R) Fruity (R) Fruity-sour flavour (<10 ppm)	–	–	–
Hexanoic (caproic)	1–73	1.0–3.3	0.8–9.2	Sweet like odour	–	5.4[a]	8 in wine[d] 0.4 in 10% ethanol[e]
Heptanoic (oenanthic)	–	Tr	–	Fatty odour when pure fatty, fruity flavour	–	15[b]	–
Octanoic (caprylic)	2–717	0.5–2.1	0.5–13	Fatty odour cheesy-sour flavour	–	5.8[a]	10 in wine[d] 0.5 in 10% ethanol[e]
Nonanoic (pelargonic)	–	–	–	Nutty odour	–	–	–
Decanoic (capric)	0.5–7	0.09–3.0	0.06–2.1	Rancid like	–	3.5[a]	6 in wine[d] 1 in 10% ethanol[e]
Phenylacetic acid	–	–	0.02–0.11				
Benzoic acid	–	–	0–0.07				

References: (1) Amerine & Roessler (1983). (2) Escudero *et al.* (2007) determined in red wines. (3) Quoted by Francis & Newton (2005). [a]Patton (1964); [b]Sick *et al.* (1971); [c]Keith & Powers (1968). [d]Swiegers & Pretorius (2005). [e]Quoted by Francis & Newton (2005).

Ribéreau-Gayon *et al.* (2006) report in detail the findings of Chatonnet (University of Bordeaux) and others on these characteristics, and furthermore on their extraction from oak barrels including the effect of barrel treatment (fresh, steamed and toasted).

Some of these lactones/furanones have also been found in green coffee, roasted coffee and coffee brews (Grosch, 2001).

Flavour characteristics

Lactones/furanones show a range of flavour characteristics. Though of similar chemical structure, sotolon (an unsaturated 3-hydroxy-2-one) and HDMF or furanol (also unsaturated, 4-hydroxy-3-one) have very different odour/flavour characteristics (see Table 4.20). Thus, the latter isomers have a fruity/caramel odour impression, whilst the former has some spicy notes. The 2-ones tend to be fatty and herbaceous, especially the lower molecular weight compounds, the higher molecular weight compounds include coconut notes. The *trans* oak-lactone is distinctive in showing spicy notes.

Many flavour investigators have described enantiomers of each compound as having often quite different odour/flavour characteristics. Unless they were well separated when assessing their flavour, there may be some anomalies in their descriptions.

Recently, HDMF has been shown to be an important flavour contributor to roasted coffee and its brews, with sotolon and other furanones important in both green and roasted coffee (Grosch, 2001).

A listing of these compounds is provided in Table 4.20 together with olfactory information, as available.

4.3.7 Acids

The main concentration of acids are non-volatile and are discussed in Chapter 3. Montedoro & Bertuccioli (1986) list some 116 acids as having been identified in wine, but of these only some 14 are volatile liquid substances, while the remainder like the hydroxy benzoic acid are generally crystallizable solids with higher melting points. Though some acids may have recognizable odours, these may be due to lower boiling point contaminants. In addition, these acids will be rather soluble in water. Of the 14 separately identified nearly all are multi-referenced (three or more). Some quantitative data is included in Table 4.21.

Presence in wines

Ribéreau-Gayon *et al.* (1982, quoted by Ribéreau-Gayon *et al.*, 2006) analysed all fatty acids in the table to be present in trace amounts. In a Mourvèdre wine, Vernin *et al.* (1993) identified all these acids up to octanoic acid but in this aroma complex only three acids, butanoic, 3-methyl-butanoic and hexanoic, were included in a total GC peak area assessment, consisting

Table 4.22 Amines found in wine: contents together with general odour character and threshold values.

Amine	Concentration range ($\mu g\ L^{-1}$)		General odour character	Threshold value
	Min	max		
Methylamine	13.5	491	Ammoniacal	–
Dimethylamine	5	110	Ammoniacal	–
Ethylamine	560	8600	Ammoniacal	–
Diethylamine	Tr	<1	Ammoniacal	–
3-Butylamine	<1	360	–	–
2-Phenethylamine	1	189	–	–

Content data from Ough *et al.* (1981), quoted in Montedoro & Bertuccioli (1986).

of 0.78% of the total. In their examination of 13–16 young Spanish white wines Aldave *et al.* (1993) only reported quantitative information on octanoic acid, averaging $1.3\,mg\,L^{-1}$ in wines made where SO_2 had not been used and $2.6\,mg\,L^{-1}$ in wines made with SO_2. Patton (1964) has described some work on the odour thresholds of fatty acids. Escudero *et al.* (2007) analysed the acids in five different Spanish red wines and Francis & Newton (2005) published a list based on a literature survey, including odour threshold values (see Table 4.21).

Flavour characterisitcs

At concentrations near their threshold value, fatty acids contribute to the complexity of wine but at higher concentrations they have a negative impact. High acetic acid concentrations indicate spoilage but yeasts will also form some, so acetic acid will constitute the volatile acidity in wine. Propanoic and butanoic acid can be associated with bacterial spoilage. Hexanoic, octanoic and decanoic acids are formed by yeasts and seem to have generally negative aroma descriptors in wines.

4.3.8 Nitrogeneous compounds

Montedoro& Bertuccioli (1986) identified some 31 different amines and amides in wines, though many are non-volatile and mainly single referenced. In addition, Ough (1981) has quantified their presence in wine, as shown in Table 4.22, with odour data. Note that the concentrations of these listed components are all very low, i.e. $\mu g\ L^{-1}$.

Their concentration decreases during fermentation, since yeasts can metabolize amines. They can also be formed by yeasts. Their contribution to wine flavour has not been established but they can contribute a harsh flavour to beer (see Jackson, 2008).

4.3.9 Phenols

As noted in Chapter 3, phenolic and polyphenolic compounds abound in grapes, musts and wines but our interest here is only in volatile phenolic compounds.

Presence in wines

Some 17 phenols were listed by Montedoro & Bertuccioli (1986). Tressl (1976) listed 13, together with their content, which were present in the order of µg per litre, rather than the expected mg per litre. Later information on certain of these phenols has been obtained by Chatonet (1992, 1993) (quoted by Ribéreau-Gayon et al., 2006) (Table 4.23). A detailed method for accurate determination of ethyl phenols in eight wines ranged from 394–1292 µg L^{-1} for 4-ethylphenol, 42–107 µg L^{-1} for 4-ethylguiacol and 0–99 µg L^{-1} for 4-ethylcatechol (Carillo & Tena, 2007).

Flavour characteristics

Of these phenols, vinyl-4-phenol, vinyl-4-guaiacol, ethyl-4-phenol and ethyl-4-guaiacol are regarded by Ribéreau-Gayon et al. (2006) as being especially important in an olfactory defect known as 'phenol' character, which occurs, unfortunately, relatively frequently. Changes in wine-making, in particular maturation, contribute to their occurrence. Recent research regarding their formation, sensory characteristics and occurrence in wines is reviewed by Suarez et al. (2007) and discussed in Chapter 7. Vinyl compounds may be formed during fermentation, whereas ethyl compounds are much more likely to develop during barrel ageing (further mentioned in Chapter 7).

In addition to the phenolic medicinal character, several such compounds have also been described as having a 'smoky' or 'tarry' or even 'bacon-like character', though not many appear to be present in sufficient quantity to affect flavour. Their origin is described in Chapter 7.

Detailed sensory information is included in Table 4.23.

4.3.10 Terpenes

Terpenes and their derivatives (generally called terpenoids) are widely distributed in nature, and occur in different amounts in grapes; they persist through vinification and can be thus described as essentially primary aromas, or even varietal. It is not possible here to describe all the different types of terpenoids, including bicyclic forms (not present in grapes).

Chemical structure

Terpenes consist of a wide range of substances, which have been regarded as deriving from a basic structure of a linear chain of five carbon atoms, as in isoprene (C_5H_8): $CH_2=C(CH_3)CH=CH_2$.

Table 4.23 Volatile phenols in wines: content, together with sensory information.

Name of phenol	Concentration (µg L⁻¹) (ppb) Av./range			Origin	General odour[e] character	Threshold values [µg L⁻¹ (ppb)]
	(1)[a]	(2)[b]	(3)[b]			
Hydroxy-benzene (phenol)	Tr–30	–	–	S	Phenolic, medicinal, smoky	Odour from water[e], 5.9 ppm 47 ppb in air
2-Methyl- (o-cresol)	Tr–20	–	–	T	Balsamic medicinal	0.25 ppm in flavour; Odour in water, 65–250 ppb
3-Methyl- (m-cresol)	5–10	1	1	T	Tarry leathery	Odour in water, 68–200 ppb[f] 68 ppb[e]
4-Methyl- (p-cresol)	1–10	–	–	T	Tarry smoky	2–550 ppb[f]
4-Vinyl- (4-ethenyl-)	1–20	1150–73	0–111	S	Clove like guache paint (unpleasant)	–
2-Ethyl-	1–50	–	–	T	Smoky odour phenolic flavour	–
4-Ethyl-	350	0–28	1–6047	T	Sweaty/saddles horse manure in wine (unpleasant)	Odour/water 660 ppb[f] Flavour/ water 100 ppb[f]
2-Methoxy- (guaiacol)	5	–	–	T	Odour in water phenolic flavour smoky/woody	In water 25 ppb[c] 3 ppb[g]
2-Methoxy-4-vinyl- (4 vinyl guaiacol)	10–50	15–496	0–57	T S	Carnations smoky/clove-like powerful	In water 20[c] ppb 3 ppb[h]
2-Methoxy-4-ethyl- (ethyl guaiacol)	–	0 7	0 1561	T	Carnations smoky/clove-like powerful	In water, 50 ppb[c]
2-Methoxy-4- (prop-1-enyl) (2-iso-eugenol)	–	–	–	T	Warm/spicy clove-bud oil	In water 6 ppb[g]
2,6-Dimethoxy- (syringol)	10–80	–	–	T	Smoky, phenolic	Odour[i] 1.85 ppm[i] Flavour[i] 1.65 ppm
2,6-Dimethoxy-4-ethyl-	40	–	–	T	–	–
2,6-Dimethoxy-4-isopropyl-	–	–	–	T	–	–
Trihydroxy-3,5,4-stilbene (resveratrol)[d]	–	1–3 mg L⁻¹	–	P	–	–

[a]Data from Tressl et al. (1976). [b]Data from Chatonnel et al. (1992 and 1993) for (2) white wines and (3) red wine, quoted by Ribéreau-Gayon et al. (2000). [c]Data from Rychlik (2001) quoted by Grosch (2001). [d]Data from Ribéreau-Gayon (1978). [e]Maga (1978). [f]Kim Ha & Lindsey (1991). [g]Buttery et al. (1971a). [h]Buttery et al. (1976). [i]Wasserman (1966).

Table 4.24 Main types of terpenes in wines: contents according to grape variety and location ($\mu g\ L^{-1}$).

Grape variety	Linalool	α-Terpeneol	Citronellol	Nerol	Geraniol	Ho-trienol
Scheurebe	(1) 70–370	160–260		Tr–40	Tr–30	100–240
Ruländer	(1) Tr–60	Tr–180	–	Tr	Tr	Tr–40
Gewürztraminer	(1) 6–190	30–35	–	Tr–20	20–70	Tr–40
	(2) 6	3	12	43	218	ND
Morio-Muscat	(1) 160–280	240–400	–	Tr–30	Tr–10	80–140
Muscat of Alexandria	(2) 435	78	ND	94	506	ND
Muscat of Frontignan	(2) 473	87	ND	135	327	ND
Riesling	(1) 60–140	100–280	–	Tr	Tr–10	50–130
	(2) 40	25	4	23	35	25
SauvignonBlanc	(2) 17	9	2	5	5	ND
Müller-Thurgau	(1) 100–190	100–210	Tr	Tr	Tr–0.01	40–80
Chardonnay	(3) 100	500	–	–	–	–
Airén	(3) 500	250	–	–	–	–
Viura	(3) 100	1000	–	–	–	–
Muscadelle	(2) 50	12	3	4	16	ND
Albariño	(2) 80	37	ND	97	58	127

Data: (1) Schreier *et al.* (1977). (2) Ribéreau-Gayon *et al.* (2006) quoting others. (3) Aldave *et al.* (1993).

There are many terpenes, which all have the basic formula of $C_{10}H_{16}$ with two double bonds but there are also compounds with a cyclic structure (monocyclic terpenes). The basic ring structure is shown by 4.IX.

Terpenes

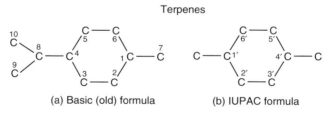

(a) Basic (old) formula   (b) IUPAC formula

4.IX

The main compounds of wine flavour interest are the hydrocarbons themselves, but more particularly their alcohols, both with linear and monocyclic structures.

The structure (4.IXa) conforms generally with an older numbering system for the carbon atoms, while 4.IXb has the IUPAC system of numbering the carbon atoms of the ring separately, e.g. those for which attached radicals have separate numbers, as will be seen in the compounds to be described (e.g. α-terpineol).

These structural formulae also indicate a relationship to the hydrocarbon p- (or 1,4) cymene, which is 4-isopropyl-1-methylbenzene and to hexa-hydrocymene (menthane). Some chemical names of the unsaturated compounds may show a menthene connection. Straight chain compounds are then regarded as having a molecular structure opened up at the C1 and C6 carbon atoms, with a slightly different numbering carbon system. Structurally, many may contain asymmetric carbon atoms (e.g. at C8), so that optical isomers exist, though in nature, a particular form is usually prevalent (+), or (–). Geometrical isomers (Z and E), due to presence of double bonds, may also occur. Terpenes therefore have complex molecular structures.

The correct chemical names and structures of the important terpene compounds found in wines, are as follows:

- Straight chain monoterpenols.
- Linalool, 3,7-dimethyl-1,6-octadien-3-ol (4.X).

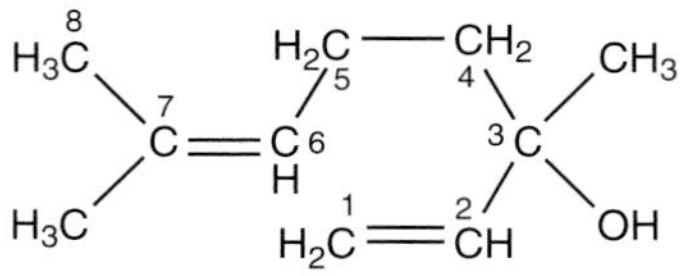

$$H_3C-C(CH_3)=CH(6)-CH_2(5)-CH_2(4)-C(3)(CH_3)(OH)-CH(2)=CH_2(1)$$

4.X

The structure as illustrated to show relationship with a corresponding cyclic compound otherwise, in a linear form (4.XI).

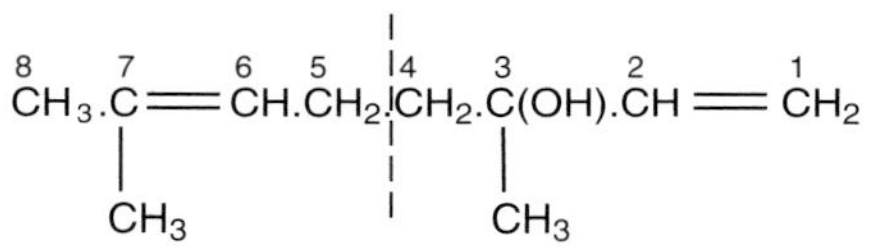

$$\overset{8}{C}H_3.\overset{7}{C}=\overset{6}{C}H.\overset{5}{C}H_2.\overset{4}{C}H_2.\overset{3}{C}(OH).\overset{2}{C}H=\overset{1}{C}H_2$$
with CH_3 groups at C7 and C3.

4.XI

Enantiomers result from the C3 position, whilst geometrical isomers result from the C6 (and C2) position. Isomerism can occur at the double bond C1 position to C2:

- Geraniol, 3,7-dimethyl-octo-2-6-dien-1-ol (4.XII).

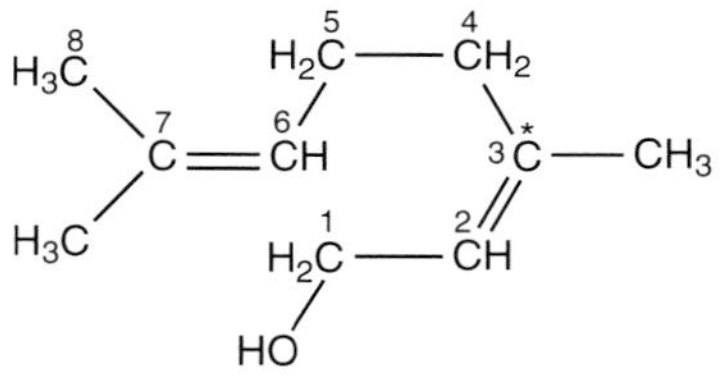

$$H_3C-C(CH_3)=CH(6)-CH_2(5)-CH_2(4)-\overset{*}{C}(3)(CH_3)=CH(2)-CH_2(1)-OH$$

4.XII

Nerol is a structural isomer (-8-ol).

- Citronellol, 3,7-dimethyl-octa-6-en-1-ol (4.XIII); (+) and (–) isomers known.

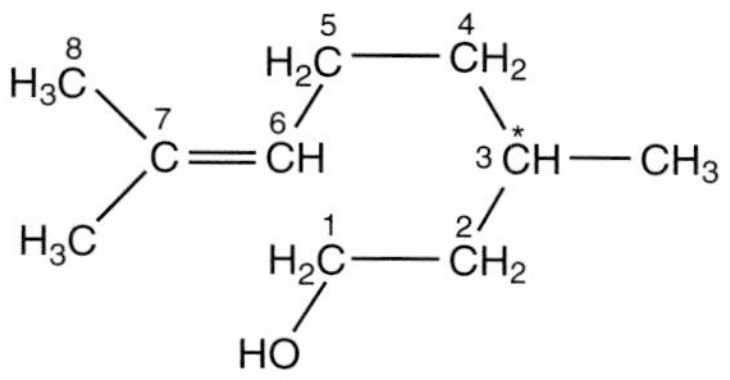

$$H_3C \quad {}^8$$
$$\overset{8}{H_3C} \diagdown {}^7 \quad {}^5_{H_2C} - {}^4_{CH_2}$$
$$\quad\quad C = CH \quad {}^3_{*}CH - CH_3$$
$$\overset{}{H_3C} \diagup \quad {}^1_{H_2C} - {}^2_{CH_2}$$
$$\quad\quad HO$$

4.XIII

- Ho-trienol, 3,7-dimethylocta-1,5,7-trien-3-ol (4.XIV).

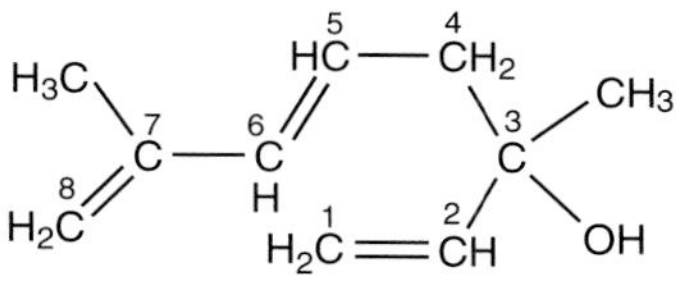

$$H_3C \diagdown {}^7 \quad {}^5_{HC} - {}^4_{CH_2} \quad CH_3$$
$$\quad\quad C - {}^6C \diagup {}^3C$$
$$\overset{8}{H_2C} \diagup \quad H \quad {}^1_{H_2C} = {}^2_{CH} \quad OH$$

4.XIV

- Cyclic terpenols: α-terpineol (terpineol-8), 2-(4′-methyl-3′-cyclohexene-l′-yl) propan-2-ol or p-menth-1-en-8-ol (4.XV); both (+) and (−) isomers are known.

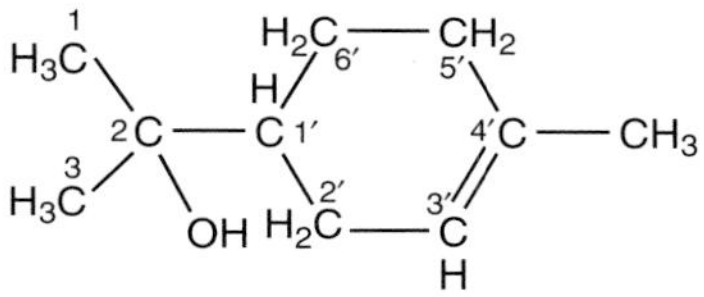

$$\overset{1}{H_3C} \diagdown \quad {}^{6'}_{H_2C} - {}^{5'}_{CH_2}$$
$$\quad H/$$
$$\quad {}^2C - C\,1' \quad {}^{4'}C - CH_3$$
$$\overset{3}{H_3C} \diagup \quad OH \; H_2C - \overset{3'}{C}$$
$$\quad\quad\quad\quad {}^{2'} \quad H$$

IUPAC

4.XV

Presence in grapes/wines

In the Bertuccioli–Montadoro (1986) list, 19 different terpenoids have been identified in wine by different investigators, many, however, with only one reference. Ten had been well investigated and their content quantified. According to Ribéreau-Gayon et al. (2006), about 40 different terpenes have now been identified in grapes but only six compounds (linalol, α-terpeneol, citronellol, nerol, geraniol and ho-trienol) are regarded as having flavour/aroma significance in wines. Ebeler & Thorngate (2009) in their review report that over 50 monoterpenes having been identified in grapes and wines. Rotundone, a complex terpene recently identified, contributes significantly to Shiraz, Mourvèdre, Durif and possibly some Cabernet Sauvignon wines (Wood et al., 2008).

They exist in grapes in three forms: volatile forms are free monoterpene alcohols or oxide, a variable proportion is non-volatile and complexed to glycosides, or may occur as di- or tri-ols, in which forms they are nonvolatile and will not contribute to aroma. They are largely present in the skins of grapes. There is little detailed information on the content of any separate stereoisomers.

Table 4.25 Terpenes in wines: olfactory characteristics (odour/flavour descriptions and threshold values).

Terpene	Odour/flavour description	Threshold values [µg L^{-1} (ppb)]	
		Water[b] (abs)	Wine[a] (diff)
Linalool	Floral/rose[a] odour	6[c] (odour)	50[c]
	Floral/woody	5.3[d] (odour)	
	Faintly/citrusy	3.8[d] (flavour)	
	Floral/green flavour[e]		
Geraniol	Floral/rose	40	130
Nerol	Floral/rose	–	400
Citronellol	Citronella	–	18
Hotrienol	Linden	–	110
α-Terpineol	Lily of the Valley[a] Lilac[e]	350[f]	400
		280[d] (odour)	
		300[d] (flavour)	
Linalool oxide	Musty/pine-like	–	–
cis		–	7,000
trans			65,000
Rose oxide	–	–	–
Rotundone[g]	Peppery	8 ng L^{-1g}	16 ng L^{-1g}

[a]Ribéreau-Gayon et al. (2006); [b]Shaw (1986); [c]Buttery et al. (1969a); [d]Ahmed (1978); [e]Chemisis (1989); [f]Buttery et al. (1971a); [g]Wood et al. (2008).

Their content in grapes varies with different varieties/cultivars of *Vitis vinifera* and is highest in the Muscat variety but they are also present significantly in some 'noble' varieties, e.g. Riesling. Monoterpenes contribute to the varietal character of Viognier, Albariño and Muscadelle. Though other 'neutral' grapes may have zero or low content such as the white grape Chardonnay as already noted, grapes for red wines usually have a low terpene content, except Black Muscat. During ageing, the types and proportion found in wine change; Table 4.24 gives some available data on contents from different sets of investigations, and is discussed further in Section 4.4.

The marked content of terpenes in a Muscat wine is notable, though other varieties of grape have high contents of linalool. Ribéreau-Gayon (1978) has also provided some data on terpene contents in Muscat grapes themselves, e.g. average 0.4 mg L^{-1} of linalool and 0.8 of geraniol, as the main contributors, whilst α-terpineol is at a low level (0.06). Changes in terpene content can take place on ageing (see Section 4.4).

Data on young white Spanish wines (i.e. not aged) of their terpene content from Aldave et al. (1993) for some different grape varieties grown in Spain are included in this table, though the reported contents seem rather high by comparison. The method of determination is not given.

Other components identified by two or more investigators are limonene (a hydrocarbon) and the linalool oxides. β-Damascenone and the α- and β-ionones are included in lists of ketones.

Table 4.26 Methoxypyrazines in wines: contents and sensory characteristics (flavour description and threshold data).

Name of pyrazine	Concentration range found	General flavour characteristics[g,f]	Origin	Threshold value
2-Methoxy-3-isobutyl-(2-methyl propyl)-	0.5–50 ng L^{-1a} 1.9–15 ng L^{-1i}	Green; bell-pepper: peas; slightly earthy flavour in water. Peasy odour earthy	P	Odour in water 2 ppt[a,c,d] 5 ppt[b]/ odour in air[e] 2–4 μg m^{-3} odour/flavour in wine, 15 ppt[a]
2-Methoxy-3-isopropyl-(methyl ethyl)-	Lower than above[a]	Green bell-pepper; but between 0.1–10 ppb, green peas, vegetable (potatoes) earthy, nutty, green, potato flavour	P	Odour in water 2 ppt[a,c]/ odour in air[e] 1–2 μg m^{-3}
2-Methoxy-3-sec-butyl-(1-methyl propyl)-	Lower than above	Green bell-pepper, peasy, earthy odour	P	Odour in water 1 ppt[h]/ odour in air[e] 24 μg m^{-3}
2-Methoxy-3-ethyl-	–	Green pepper, earthy odour	P	Odour in water 400 ppt

Data: [a]Ribéreau-Gayon et al. (2000); [b]Rychlik et al. (1998); [c]Seifert et al. (1970); [d]Parliment (1981); [e]Wagner et al. (1999); [f]Chemisis (1995); [g]Flament (2001); [h]Murray & Whitfield (1970); [i]Cullere et al. (2009).

- Limonene, 4-isopropenyl-l-methyl-l-cyclohexene, with R(+) and S(−) enantiomers present.

Flavour characteristics

Terpenes are complex substances, which tend to have floral aromas. Threshold values are probably high in wines, but the cumulative or even synergistic effect of a number of different terpene compounds may well determine their actual flavour contribution. Some available information is shown in Table 4.25. The effect of the presence of stereoisomers on flavour/threshold levels is not fully known, or much reported for wines.

Linalool has quite a low threshold level even in grape must but any transformation into, or presence of, its oxides will show terpene compounds with a much higher threshold level, especially the trans-isomer and so is sensorially less important.

A bicyclic sesquiterpene, rotundone, has been identified to be a potent aroma compound, responsible for the typical pepper aroma in Shiraz grapes and wines (Wood et al., 2008). Its correct name is (-)-rotundone, the S stereoisomer, so the correct full chemical name is (3S)-3,4,5,6,7,8-Hexahydro-3,8α-dimethyl-5α-(1-methylethenyl)azulen-1(2H)-one (personal communication, 2010, Alan Pollnitz). Interestingly, these authors also identified this compound for the first time in peppercorns, to which it gives the pepper aroma. Detection threshold levels are very low, in red wine 16 ng L^{-1} and in water 8 ng L^{-1}. Grape samples analysed for

rotundone ranged from 10–60 ng kg⁻¹. In wines the levels were from 29 ng L⁻¹ in a
sensory non-peppery year to 145 ng L⁻¹ from wine in a peppery year, well above
the reported threshold concentration. Indeed, the authors reported an excellent
correlation between sensory panel ratings of peppery intensity and the rotundone concentration, further support of its aroma contribution. Rotundone was
also found above threshold levels in wines made from other grapes, Mourvèdre
(134 ng L⁻¹), Durif (128 ng L⁻¹) and lower levels in some Cabernet Sauvignon wines.

4.3.11   Pyrazines

Methoxy-alkyl-pyrazines are well known substances found in plants with important flavour/odour characteristics, of high intensity and low threshold values.

Chemical structure

They are known constituents, together with other pyrazines (4.XVI) of simpler
molecular structure, in roasted coffee, generated largely through roasting,
though some are present in the original raw coffee beans (Grosch, 2001).

4.XVI

Presence in grapes/wines

According to Ribéreau-Gayon *et al.* (2006), 2-methoxy-3-isobutyl-pyrazine
(strictly speaking, 2-methoxy-3-(2-methylprop-1-yl)pyrazine) was first identified in grapes (Cabernet Sauvignon) by Bayonave *et al.* in France around 1975,
and also in wine. Subsequently it has been identified in a number of wines, but
is only of particular interest in wines from Sauvignon Blanc and Cabernet
Franc grapes. Ribéreau-Gayon *et al.* (2006), however, consider that the concentrations of these pyrazines are significantly above their recognition
threshold levels but that their presence is more apparent when the grapes are
under-ripe. Therefore, they conclude that pyrazines are not welcome or
appreciated in 'red Bordeaux wines' and that at higher concentrations the
herbaceous character spoils the the wine aroma. The other methoxy-pyra-
zines are much less influential for flavour as can be seen from the data assem-
bled in Table 4.26. Relevant physical data can be found in Appendix II.

Flavour characteristics

Quantitative content data is limited. Part of the reason is the very low concentrations in wine having an odour impact. Ebeler & Thorngate (2009) quote
concentration ranges in wines from 4 to 30 ng L⁻¹, depending on growing

Table 4.27 Sulfur compounds in wines: contents, general flavour/odour characteristics and threshold value data.

Name of compound	Content range [μg] L⁻¹ (ppb)]	General flavour/ odour character	Origin	Threshold value data			Other in wine [ppb (μg L⁻¹)]
				Odour in air (ppb)	Odour in water (ppb)	Flavour/ water	
Hydrogen sulfide (1)	Up to 0.3 for 'clean' wine	Rotten eggs	S	4.7	5^a	–	–
Alkyl thiols							
Methane thiol (methyl mercaptan) (1)	0.7	Stagnant water, rotten cabbage	S	–	0.3^b 0.02^c	–	0.3^e
Ethane thiol (ethyl mercaptan) (1)	0	Onion	S	–	2.1^d 0.19–1.0^d	–	0.10
2-Mercapto ethanol (1)	72	Burnt rubber	S	–	–	–	130
2-Methyl thio ethanol (1)	56	Cauliflower	S	–	–	–	250
3-Methyl thio-1-propanol	0.500–10.8 ng L^{-1i}	Sweet, potato	S				1 ng L^{-1a} in model wine
4-Methyl thio butanol (1)	36	Earthy	S	–	–	–	80
3-Mercapto-3-methyl butan-1-ol (3MMB) (1)	Up to 128 ng L^{-1} (0.128 ppb) 20–150 ng L^{-1a}	Leeks soup-like 'cabbagy'	P,S	–	2.6 ppb	8–10	1500 ng L^{-1} (1.5 ppb)
4-Mercapto-4-methyl-pentan-2-ol (4MMPOH) (1)	Up to 111 ng L^{-1} 15–150 ng L^{-1a}	Citrus-zest	P,S	–	20 ng L^{-1f}	–	55 ng L^{-1}
3-Mercapto hexan-1-ol (3MH) (1)	Up to 1178 ng L^{-1} (12 ppb) 150–3500 ng L^{-1a}	Grapefruit	P,S	–	1 ng L^{-1f}	–	60 ng L^{-1}

(Continued)

Table 4.27 (Continued)

Name of compound	Content range [μg] L⁻¹ (ppb)]	General flavour/ odour character	Origin	Threshold value data			
				Odour in air (ppb)	Odour in water (ppb)	Flavour/ water	Other in wine [ppb (µg L⁻¹)]
8-Terpen-thiol p-menthene-8-thiol	–	Warm lactone/ spicy, tropical		–	–	–	5 × 10⁻⁴ ppb
Benzenemethanethiol (1)	5–20 ng L⁻¹ [a] 10–40 µg L⁻¹ [i]	Gunflint, smoke		–	–	–	0.3 ng L⁻¹ [i]
2-Methyl-3-furanthiol	27–95 ng L⁻¹ [i]	Cooked meat					5 ng L⁻¹ [i] in model wine
2-Furanemethanethiol	Up to 100 ng L⁻¹ [g]	Roast coffee					0.4 ng L⁻¹ [h]
Alkyl thio ketones							
4-Mercapto-4-methyl pentan-2-one (4MMP) (1)	Up to 40 ng L⁻¹ in Sauvignon Blanc 1–120 ng L⁻¹ [a]	Boxwood – broom	P,S	–	–	–	0.8 ng L⁻¹ 3.3 ng L⁻¹ [f] in model soln.
Alkyl thio esters							
3-Mercapto hexyl ethanoate (A3MH) (1)	Up to 451 ng L⁻¹ 0–500 ng L⁻¹ [a]	Boxwood/ Passion fruit		–	4.2 ng L⁻¹ [f]	–	4 ng L⁻¹
Alkyl thio ethers							
Dimethyl sulfide dimethyl (sulfane) (1)	0.0 in clean wine	Quince, asparagus	P.S	–	0.33 ppb[c] l ppb[d]	–	2.5 mg L⁻¹ in wine
Others							
Methyl-2-tetra thiophenone (1)	68 in clean wine	'Gas'				–	90
Ethyl methionate (1)	1.6 in clean wine	Metallic				–	300
Methionyl acetate (1)	1.5 in clean wine	Mushrooms				–	90
Methionol,2-3-(methylthio)-propan-l-ol	838 in clean wine	Cooked cabbage		–	–	–	1200 µg L⁻¹ in wine

References: (1) Ribéreau-Gayon et al. (2006); [a]Ribéreau-Gayon et al. (2006); [b]Shankaranarayana et al. (1982); [c]Guadagni et al. (1963a); [d]Persson & Sydow (1973); [e]Semme Ravel et al. (1995); [f] Quoted by Swiegers & Pretoruius (2005); [g]Tominaga & Dubourdieu (2006); [h] Tominaga et al. (2000); [i] Quoted by Francis & Newton (2005).

conditions, maturity and grape variety. Concentrations of 2-*sec*-isobutyl-3-methoxypyrazine and and 3-isopropyl-2-methoxypyrazine were reported at least eight times lower. In a recent study using multidimentional gas chromatography-mass spectrometry authors were able to achieve detection limits in wines between 0.09 and 0.15 ng L^{-1} (Cullere *et al.*, 2009). They reported that the 36 wines and 17 must samples analysed contained between 1.9 and 15 ng L^{-1} 2-methoxy-3-isobutyl-pyrazine.

Whilst the 2-methoxy-3-isobutyl compound is described as having a 'green pepper' aroma, in fact the more characterizing feature, as with the others, is 'green', 'earthy' and 'peasy'. Good threshold data is available.

4.3.12 Sulfur compounds

Various sulfur compounds have been found in wines, ranging from simple alkyl thiols or mercaptans to the more complex to thiolactones and terpenthiols. The latter have been identified in recent years and found to confer strong flavours with very low threshold levels, especially when compared with their oxygen-containing oxy counterparts.

Chemical structure

These compounds may be divided into simple alkyl thiols, where the OH (hydroxyl) group in an alcohol is replaced by the –SH (mercapto-) group, which applies also to certain terpen-thiols (4.XVII). Additionally, mercapto-groups may also be present in alkyl alcohols, ketones, aldehydes and esters. Other compounds may have –S– groups (thio-) as in simple thioesters or sulfides), or also in alcohols, aldehydes, ketones and esters. Further complexity can come in the form of disulfides (–S–S–) or even trisulfides (–S–S–S–).

| 2-Mercapto ethanol | Ethyl 3-mercapto-propanoate | 4-Mercapto-4-methyl-pentan-2-one | Methionol |

4.XVII

Presence in wines

The more complex sulfur compounds, which have the more desirable flavour notes, have not only been detected but also quantified. Some mercapto compounds (described as 4MMP, A3MM, 3MM, 4MMPOH and 3HMB) by

Ribéreau-Gayon et al. (2006) have varietal character. They are particularly associated with Sauvignon Blanc and their contents are shown in Table 4.27 together with their full chemical names. They also describe their presence in differing amounts in Alsace wines, based upon Gewürztraminer, and Riesling and Muscat grapes.

Some more sulfur containing compounds have been determined in wines, contributing positively to wine character. Tominaga et al. (2003) identified benzenemethanethiol in red and white wines, in particular in Sauvignon Blanc, Semillon and Chardonnay. Two other highly odoriferous sulfur compounds, 2-methyl-3-furanthiol and 2-furanmethanethiol were isolated and quantified by a method specially adapted for these compounds (Tominaga et al., 2000; Tominaga & Dubourdieu, 2006). The concentrations of benzenemethanethiol, 2-furanmethanethiol and ethyl 3-mercaptopropionate increase during bottle ageing in Champagne wines and contribute their typical flavour quality (Tominaga et al., 2003a). A number of sulfur compounds have also been associated with wines made from grapes affected by Noble Rot (see Chapter 7). The sulfur thiols, including hydrogen sulfide, are often present, as also shown in Table 4.27 together with other thio compounds contributing to wine flavour.

Flavour characteristics

Sulfur compounds of the thiol and thio-type are generally associated with flavour defects in wine (Table 4.27, under general odour/flavour characteristics). These are divisible into 'light' sulfur compounds (low boiling points) such as dimethyl sulfide and 'heavy' compounds (higher boiling points), such as methionol, both responsible for reduction odours (formed during bottle-ageing), as discussed by Ribéreau-Gayon et al. (2006), when concentrations present can exceed threshold levels. Sulfur compounds have low or even very low threshold levels.

Some new information suggests that some sulfur compounds at suprathreshold concentrations can be perceived as pleasant in some wines (see the review by Ugliano & Henschke, 2009). For example dimethylsulfide can contribute to pleasant corn, molasses and asparagus in some wines, and quince, truffles and metallic in other wines. In some red wines dimethylsulfide enhanced the berry fruit aromas. There are also suggestions that some sulfur compounds at near threshold concentrations can contribute to the varietal character of some wines, for example sulfides, disulfides, benzothiazole and thioalcohols were determined at higher concentrations in some Merlot wines. More data are needed to confirm this positive contribution of sulfur compounds at very low concentrations.

Of particular interest are those mercapto-compounds previously described in this Section, characterizing the Sauvignon Blanc aroma. The individual odour/flavour characteristics together with threshold values are given also in Table 4.27. Some of these compounds in particular, 4-mercapto-4-methyl-pentan-2-one (4MMP) can reach levels well in excess of its threshold to give the 'boxwood', 'broom' aromas often described in Sauvignon Blanc wines. Interestingly, 4MMP has been described as giving a 'catty' flavour in beer (Meilgaard & Peppard, 1986).

Another sulfur compound, 3-mercaptohexan-1-yl acetate, is associated with some tropical and citrus fruit flavours, such as passion fruit and grape fruit, and indeed with blackcurrant flavour.

p-Menthene-8-thiol (by a substitution of –SH for OH in α-terpineol) has a very intense flavour, with an extremely low threshold figure, with a marked complex smell of a 'spicy/ tropical note with a warm lactone background', over the floral notes of the terpenol compound.

Tominaga *et al.* (2003) identified benzenemethanethiol, giving a smokey character to wines. Its sensory threshold is very low and its concentration in wine was determined 30 to 100 times higher than the threshold. In particular Chardonnay wines contained 30–40 ng L^{-1}.

Furfurylthiol has been identified in red Bordeaux wines, white Petite Manseng as well as in toasted barrel staves (Tominaga *et al.*, 2000). Further isolation and quantification gave very low threshold values, and a contribution of 2-methyl-3-furanthiol to a cooked meat note in red wines, whilst 2-furanmethanethiol contributed to coffee notes in certain wines Tominaga *et al.*, 2000; Tominaga & Dubourdieu, 2006).

Although the concentrations in wine are low, the contribution of many sulfur compounds can be significant in wines due to their very low threshold values. Francis & Newton (2005) listed seven compounds that could be significant in wine (3-mercapto hexyl ethanoate, 4-mercapto-4-methyl pentan-2-one, 3-mercapto hexan-1-ol, 2-Methyl-3-furanthiol, 3-methylthio-1-propanol, benzenemethanethiol, dimethylsulfide).

4.4   Changes during maturation

Changes in composition of the non-volatile compounds, but including the volatile acids (like acetic acid) responsible in part for acidity, have been discussed in Chapter 3, Sections 3.8.1 and 3.8.2. Wines can be fermented as well as stored in oak vats, and although storage will give the largest changes, some effects of fermentation in vats have been documented. During maturation slow chemical reactions in wine occur. Some do not require any other input than time, others require oxygen (oxidative changes) or oak (oak wood compound extraction). In-bottle maturation only allows the anaerobic changes that typify the 'in-bottle' aged bouquet, whilst 'in-barrel' maturation allows also oxidative and extractive changes. Consequently, information on changes in the volatile components composition needs to be sub-divided into sections on 'in-barrel' (also 'in-vat' and 'in-tank') and 'in-bottle' storage.

4.4.1   Fermentation and storage of wines 'in-vat (tank)' and 'in-barrel (cask)'

Fermentation

During fermentation in oak, some compounds extracted from the wood can be changed by enzymic action of the fermenting yeast, for example the reduction of vanillin to vanillic alcohol (see review Ugliano & Henschke, 2009), thus giving a less woody character to a wine fermented in an oak vat. Yeasts can metabolize ferulic and p-coumaric acids to aromatic phenols, contributing to off-flavours (see Chapter 7).

Fermentation in vats is also an established method of conditioning new barrels, although time consuming and costly. Phenols and ellagitannins are readily extracted, but tend to combine with wine compounds and precipitate and hence are lost with the removal of lees. Desirable aromatic compounds such as oak lactones dissolve more slowly. Other methods are available; barrel preparation, care and hygiene are pertinent for quality wine maturation, discussed in depth by Jackson (2008) and Ribéreau Gayon et al. (2006).

Storage

The additional aromas that develop after fermentation are conveniently referred to as tertiary aromas and are derived from two sources. Firstly, there is a slight oxidation of some existing components, either primary or secondary, by the dissolved oxygen present and absorbed by wine during storage. Ribéreau-Gayon (1978) has used the term 'oxidative bouquet' arising in this type of ageing. Certain non-oxidative changes may also take place. Secondly, we have the chemical/physical extraction from the wood of the barrels when used, usually oak, producing so-called oak aroma. The characteristics of different types of oak barrels have been described in Chapter 3. Thirdly, oak absorbs compounds. These aspects are briefly discussed next. However, given the number of variables, it appears not an easy matter to mature in oak in order to enhance the aroma quality of a wine. A quote from Ribéreau-Gayon et al. (2006) says it all: 'Wood is capable of enhancing a wine's intrinsic qualities and may hide certain faults. Used unwisely, barrel ageing may produce disastrous results.'

Oxidation is characterized by increases in aldehydic compound content (including acetaldehyde itself, by oxidation of ethyl alcohol) to give potential rise to quince, apples, dried nuts, buttery, rancid and Madeira flavour notes. These flavours are especially associated with brandy and fortified wines. Such changes are likely, however, to be influenced by the amount of sulfur dioxide present at any time (a reducing agent), irreversibly reducing oxygen content and other factors. Sulfur dioxide should not, though, be present in sufficient quantity to generate thiol compounds (e.g. methyl/ethyl mercaptan), which have unpleasant odours. As described above (Section 4.3), acetaldehyde (ethanal) is not regarded as a desirable component of ordinary table wines – if present in free form at great excess over its threshold it produces so-called 'flatness'.

The general nature of the oxidative changes occurring during in-vat or cask storage has been discussed in Chapter 3 in relation to non-volatile compounds. This discussion, including the concepts of oxidation–reduction potential, as reviewed by Ribéreau-Gayon et al. (2006), is relevant to the volatile compounds, in particular acetaldehyde.

However, the precise mechanism whereby particular volatile substances are generated such as ethanal, from ethanol and acetoin from butan-2-diol, is not clear, i.e. whether by the direct addition of hydrogen through dehydrogenase catalysis, or through hydrogen peroxide intermediate action. For further information, see Appendix I.

During storage in oak barrels, a range of flavour/aromas can develop as a result of extractives from the wood, dependent upon the grape variety/vinification conditions, the type of oak (e.g. American or French oak, type of drying, degree of 'toasting') and the age/condition of the barrels. Quantitative data is somewhat scarce for content in actual wines, but aldehydes, ketones, furanones (oak-lactones) and phenols are clearly identified, which have been included in the corresponding tables of Section 4.3. Regarding flavour, a wine is often referred to as 'oaky', but separate flavour distinctions are made with the threshold level data available, as also shown in the tables. Particular aged flavours are associated with certain grape varieties (more fully described in Chapter 5), thus there are 'vanilla/coconut' flavours of Chardonnay wines, 'cedar' for Cabernet Sauvignon and 'toasted/spicy' for Rioja red wines. The exact origin of these compounds from the oak wood is not always clear, involving some chemical changes in the actual extractives – thus, degradation of lignin extracted from wood to produce aldehydes such as vanillin.

Jackson (2008) has recently overviewed oak and its use in wine flavour. An 'oaky' flavour can also be introduced by fermentation in an oak barrel, or fermentation in the presence of oak chips. Pérez-Coello et al. (2000) have studied the chemical and sensory changes in white wines (American and European). The wines gained a distinctive oak aroma and lost their fresh and green apple aromas. The quantity of oak-lactones (cis and trans) was found to increase with increasing concentration of wood chips. More cis oak-lactone was extracted from American oak than from two of the three European ones tested.

However, Ribéreau-Gayon et al. (2006) comment unfavourably on the separate use of oak chips and of other wood extractives, for artificial ageing. They reported a high degree of extraction of cis β-methyl-γ-octalactone and vanillin when using chips. Robinson (1995) also comments on their use, saying 'that they are not necessarily a bad thing'; 'providing consumers with the sort of flavours for a fraction of the cost'; but not 'able to provide the physical properties of barrel fermentation and maturation'.

An overview on the use of oak chips listed numerous factors relevant when using oak chips to flavour wine (Campbell et al., 2006), such as size of fragments, chip heating time and temperature, timing of oak addition and the time with wine, pretreatment of chips with sulfur dioxide and shelf life of chips. Chip size is an obvious parameter, with smaller chips giving faster extraction due to the larger surface area. However, the chips are heated before use and the combination heat and extraction gives both qualitative and quantitative differences in extraction. The order of treatment can have a significant effect on the extraction of compounds in wine, for example oak shaved and then heated gave a much larger extraction of vanillin than wood heated and then shaved.

Time after oak maturation may also influence the oaky character of a wine. Wilkinson *et al.* (2004), who showed that the odourless and open-ring precursor of *cis*-oak lactone, 3-methyl-4-hydroxyoctanoic acid, takes time to close. Hence the full oak aroma of *cis*-oak lactone extracted from shavings may not become apparent until some time after removal of the shavings. The development of the oak lactone flavour in vats will also take time to establish.

Adsorption of compounds by oak is particularly relevant for water and ethanol (see also Chapter 3), but Ramirez-Ramirez *et al.* (2004) reported that significant amounts of fruity esters can be adsorbed by wood from model wines. However, more studies are needed to assess the effect of selective adsorption of oak on the aroma of wine.

Further insight (R. de Heide, 1986) comes from the study of grape brandy storage, where storage in oak barrels is arranged over many years, where even sugars (such as xylose, arabinose) can be obtained by acid hydrolysis of the hemicelluloses and then to furfural derivatives over many years. Vanillin and syringaldehyde increase markedly with time and similarly, so do γ-lactones, phenolic acids, ethyl esters of higher aliphatic acids. Oak-lactones are of particular interest in brandy (distilled wine), since a large amount was noticed in brandies aged in American oak barrels compared with those in French.

Maga (1978) has published an extensive review of the contribution of wood to the flavour of alcoholic beverages. Whisky is a particular spirit, the flavour of which is strongly determined by oak barrel storage.

4.4.2 'In-bottle' ageing

Compositional changes relate almost entirely to the volatile compound content, especially noticeable in the so-called 'bouquet' of the wine. Here, Ribéreau-Gayon (1978) uses the term 'reductive bouquet', since during bottle-ageing chemical changes take place in the absence of oxygen (though there will be residual SO_2); of course, there is no possibility of extractives in glass bottles. Ribéreau-Gayon *et al.* (2006) associate reductive changes when the oxidation–reduction potential has fallen below 200 mV, that is, after any initial oxygen has disappeared (see also Chapter 3). A wine's bouquet is formed by complex reactions corresponding to the formation of reducing substances and harmonization of aromas.

Reactions during bottle ageing depend on the storage temperature, at 12°C wine develops slowly and at 18°C the process is faster. A constant storage temperature helps to keep the bottles well sealed and it is pertinent that the corks do not dry out and keep the wine airtight. Higher temperatures (up to 25°C) may accelerate ageing, although it is thought that the characteristic bouquet does not develop. Wine should be kept in the dark.

Rapp & Marais (1993) have provided a useful report of the chemical changes that they found occurring in a Riesling and other white wines in storage by comparing vintages of different years. Comparison was made also with a wine stored in a freezer (−30°C), and a cellar-stored wine from the same fermentation in 1976, and then examined in 1983. The main changes found are listed below.

Changes in ester content

These consist of:

- a decrease in alkyl acetates, e.g. ethyl acetate, fruity esters such as isoamyl and isobutylacetates;
- an increase in mono- and dicarboxylic ethyl esters of other acids, e.g. diethyl succinate (noted by others), and ethyl hexanoate (also fruity).

These ester composition changes are a re-establishment of chemical equilibrium with time, according to the percentage contents of the acid, alcohol and esters present immediately after fermentation. In particular, young white wines tend to contain an excess of fruity esters. During ageing the content of higher alcohols hardly changed. These changes will be greater for white wines, rather than red, since lower fermentation temperatures during white wine vinification produces higher contents of these fruity esters, as already noted.

Substances produced by carbohydrate degradation

An increase in 2-furfural content was noted (dehydration in acid aqueous solution) over a number of years; also 2-formyl-pyrrole as a Maillard reaction product, though higher temperatures would be required.

Sulfur compounds

Dimethyl sulfide (DMS) is an important volatile compound in beer, formed during fermentation, but in wines it forms only during bottle ageing. These investigators examined its formation in wines made from Chenin Blanc, Riesling and Colombard grapes. The concentration increased with both time (up to 16 weeks) and temperature (examined up to 30°C), and was found to correlate well with taste panel findings on so-called 'maturation' bouquet. There were considerable differences in the effect of temperature and storage on the different cultivars. They believed that the presence and content of DMS is an important consequence of ageing in relation to the bouquet formed.

The concentrations of benzenemethanethiol, 2-furanmethanethiol and ethyl 3-mercaptopropionate increase during bottle ageing in Champagne wines, and contribute their typical flavour quality (Tominaga *et al.*, 2003a).

Changes in terpenoids

During ageing, terpenes become involved in a very complex set of reactions:

(1) Decreasing concentration of the monoterpene alcohols (linalool, geraniol and citronellol).
(2) Increasing concentration of the isomeric linalool oxides (four in number), nerol oxides, trimenol, α-terpineol, hydroxylinalool and hydroxycitronellol.

Linalool is one of the main terpenes in Riesling, which gradually decreases in amount whereas the oxide content increases. Linalool has a floral rose odour, whilst α-terpineol has a floral lilac odour and the linalool oxides have a musty pine-like aspect. For example, a Riesling with a stated linalool content 400 μg L^{-1} (or 0.4 mg L^{-1}) contains only 50 μg L^{-1} after three years, which is below its odour threshold of 100 μg L^{-1}. This decrease of concentration is, reportedly, noticeable in the bouquet. There is an increase of linalool oxide content, which has, however, a much higher threshold perception level than linalool itself. Odour perception is also influenced by an increase in α-terpineol. Quantitative changes of some eleven terpenes in a Riesling wine during storage are also reported.

Formation of substances from carotene breakdown

Riesling wines during ageing develop a strong, petrolly kerosene-like aroma, which is detrimental to wine quality, if present at too high a level. It is more apparent in wines from hot wine-producing countries (e.g. South Africa) than from cooler European countries. The substance responsible is believed to be 1,1,6-trimethyl-1,2-dihydronaphthalene (TDN) (4.XVIII), which has a threshold of 20 ppb (20 μg L^{-1}) in wine. Postulated possible formation mechanisms are, essentially, from carotenoids, by enzymatic or acid catalysed reactions. Other volatile compounds are generated from carotene, i.e. C$_9$, C$_{10}$, C$_{11}$ and C$_{13}$ isoprenoids, such as β-damascenone, α- and β-ionone, vitispiranes (see also Section 4.3 on ketones). However, the carotene content is not the determining factor, since Chenin Blanc grapes have a high carotenoid but low terpene content and a low TDN level on ageing. Ribéreau-Gayon et al. (2006) claim that the concentration of this substance can reach 200 μg L^{-1}.

4.XVIII

Aldave et al. (1993) in reviewing the literature on the shelf-life of young white wines, with particular relationship to Spanish wines, covers similar information to that given by Rapp & Marais (1993).

Ribéreau-Gayon et al. (2006) have also reported changes occurring during in-bottle ageing, especially of red wines. They also mention that the important esters, ethyl esters of higher alcohols with their 'fruity' flavours such as hexyl acetate, tend to decrease on ageing, thus being responsible for 'fading' effects.

A study on accelerated ageing of model wines with added aroma precursors showed continuous increases in vanillin derivatives, furan linalol oxides, β-ionone, and some compounds with a possible negative impact 4-ethylphenol and guiacol (Loscos *et al.*, 2010). Numerous other compounds increased and then decreased in concentration. Varietal differences could be observed and remained present even after accelerated ageing. Once more information regarding aroma changes during ageing is available, possibly the time required for maturation can be predicted.

4.5   Aroma detection and quantification

The main method for the identification of the volatile aroma volatiles is based on gas chromatography, used to separate the compounds in a sample, followed by a detection method aiding the identification of the compounds. Prior to analysis, there is the sample preparation, a very important part of the procedure. Great progress over the last decade has been made in both the preparation of reliable samples, as well as the separation and detection techniques. Techniques for the identification of compounds, at low concentrations, have also improved.

4.5.1   Gas chromatography

Gas chromatography (GC) allows the analysis of volatile compounds, or those made volatile by derivatization. The equipment basically consists of an inlet for the sample, a column on which the sample is separated and a means of detecting the volatile compounds. A controlled gas flow through the column and a gradually increasing temperature are used to separate the volatile compounds over the column. Great advances have been made in GC equipment, allowing the identification and quantification of very low concentrations of compounds. The rapid development of computer hard- and software, used to program GC equipment, and handle enormous data sets, has contributed to the wealth of information now available on aroma compounds in wines. Some aspects will be discussed next. The sensitivity of the equipment requires careful sample preparation, avoiding the introduction of artefacts and maximizing the accuracy of the quantification.

Volatile aroma compounds in wines (or grapes) are separated, detected and quantified by gas-chromatography techniques. A mass spectrometry detector can be added and allows the accurate identification of compounds corresponding to the peaks shown in a GC trace. GC coupled to a mass spectrometer (MS) allows the tentative identification of eluting volatile compounds, by matching the fragmentation pattern and molecular ion obtained by MS with specialist reference libraries of such information by using a computer. Peak area determination, by comparison with a peak from a known amount of a pure compound, enables quantification of the compound in the original sample and so, therefore, for all the compounds detected from peaks. Internal standards are routinely used to minimize errors in quantification.

These techniques are now well described in the literature, including analyses of particular beverages, with improvements taking place continuously. An in-depth overview of the development and contributions mass spectrometry has made to advances in knowledge of grape and wine constituents is given by Hayasaka *et al*. (2005), including discussions on sample inlet and preparation techniques. In its infancy mass spectrometry was suitable to analyse volatile compounds, hence mostly suited to analyse aroma compounds and taints, provided samples of sufficient quality (purity) and quantity could be prepared. However, a combination of more sophisticated inlet techniques and increased sensitivity and selectivity has allowed the analysis of ever smaller samples. In addition the development of new techniques in mass spectrometry has also allowed the analysis of charged compounds, such as anthocyanins, and non-volatile compounds, such as proteins, phenols. Significant contributions in the analysis of pyranoanthocyanins were also made using mass spectrometry.

Nuclear magnetic resonance (NMR) techniques are also useful in identification but see Flament (2001). Although the sensitivity of GC and GC-MS techniques has greatly improved, sample preparation remains an important step.

4.5.2 Sample preparation

It is pertinent to mention here that the means of preparation of the sample for testing and the equipment used for testing, should be separately studied. The sample presented to the GC equipment is often a so-called head-space sample, which is a volume of gas/vapour drawn off from above the head-space of the liquid beverage (with which it is in equilibrium). In the purge-and-trap method this gas is trapped in a solid absorbent and then released (by controlled heating) to the GC equipment for analysis of its content. These methods are clearly directly relatable to the vapour/gas that passes to the human nose (olfactory region) and therefore to the perceived aroma/flavour. However, there can be a disadvantage to this method, since it may not be possible to process sufficient vapour to ensure that even extremely small quantities of particular compounds are detected. These compounds may have extremely low threshold flavour levels and, therefore, be potentially important contributors to flavour.

Despite possible drawbacks with sensitivity and selectivity, the sample preparation methods have been much improved both in quality of the sample and ease of preparation, in particular the use of solid-phase extraction techniques has increased. It avoids the use of solvents and allows samples without solvent induced artifacts. Solid-phase extraction is widely used to extract, clean and concentrate samples, has proven to be robust and can be automated. For example a paper validating a solid-phase micro-extraction method of Pinot Gris and Chardonnay wine aroma compounds using headspace analysis followed by gas chromatography and mass spectrometry of with automated sample handling (Howard *et al.*, 2005) showed good sensitivity and reproducibility for esters, alcohols and terpenes. Detection limits were in the µg L^{-1} range, covering aroma compounds and concentrations typical in wines. This progress in sample preparation technique no doubt has contributed to publications with detailed qualitative as well as quantitative data, often accompanied by data showing reproducibility and errors. However, further advances in the technique can still be made if the selectivity of the method can be improved, in order to achieve more selective samples of greater amounts.

The alternative method, which has been widely used, is to extract the volatile compounds by a solvent, followed by distilling and condensing. A liquid concentrate is thus obtained for presentation to the GC equipment. In the Likens-Nicholson combined extraction-distillation apparatus, pentane-methylene chloride is a popular solvent (Vernin *et al.*, 1993). Other solvents, such as the freons have been described for use with wine volatiles (Rapp & Marais, 1993; Fisher *et al.*, 2000), along with operation under vacuum to lower the extracting temperature in extraction-distillation (Grosch, 2001). The main advantage of distillation is the large quantities of volatile compounds that can be handled, but care is needed in assessment since some of the compounds subsequently detected may merely be artefacts of the extraction-distillation operation. Again, it is important to assess how the composition of the distillate vapour corresponds to what is presented to the olfactory organs when drinking the wine to assess its flavour. For example, much work on coffee has been conducted by exhaustive extraction of the roasted coffee, rather than of actual brews, which have been prepared by simple aqueous extraction, and have lower and different yields of volatile compounds.

Numerous variants of design in the GC equipment have been produced by different commercial suppliers. The important aspect is the choice of 'column' - capillary columns are now regarded as essential and appropriate for effective separations of components.

The measurement of partition coefficients has a different aim from identifying aroma compounds in wine, for the latter the experimenter introduces a concentration step to ensure sufficient concentration of the volatile is present to be analysed and detected. For partition coefficient measurement the actual concentration is relevant and headspace analyses have been used, often based on a form of static air sampling method. The accurate determination of such measurements is time consuming. Martin *et al.* (2009) proposed a quick new method with which they demonstrated the measurement of congeners in wine. Their method involved was based on distillation in a type of equilibrium cell, and data obtained were comparable with other methods. The availability of more accurate partition coefficients would help in the theoretical prediction of flavour in wines and other foods.

4.5.3 Olfactometry

Recent reviews by Plutowska & Wardencki (2007 and 2008), Francis & Newton (2005) Palaskova *et al.* (2008) describe the chromatography methods using olfactometry, a measuring method of volatile aroma compounds using the nose, currently used to analyse the aroma compounds in alcoholic beverages, and assess their sensory properties. This method allows the separation and identification of volatile compounds in wines and combined with olfactometry it can reveal which components give typical wine aroma descriptors. The human detection methods linked to such analyses gives direct information regarding the actual smell of the samples, and various methods are available to quantify the aroma thus determined. Although the detection methods used for gas chromatographic analyses are very sensitive, the article emphasized that sometimes compounds detected with the nose are not detectable with instrumental detectors, illustrating the enormous sensitivity of the nose.

These techniques have been applied to a wide range of investigations, such as giving so-called odour profiles of wines of a particular region, early detection of compounds giving possible off-flavours, monitoring terpene concentration in wine, role of fermenting yeast on wine aroma and monitoring the undesirable effect of oxidation of white wines. One technique reviewed was developed to determine the aftertaste of a wine, involving sampling the oral cavity of persons just having evaluated the sensory properties of wine. It is anticipated that the evolvement of ever more sophisticated and sensitive methods, including the use of the nose, will lead to a better understanding of the aroma volatiles in wines contributing to their sensory properties.

Another method of detecting aroma compounds has been developed by different commercial companies, often in conjunction with university engineering departments and is referred to as electronic nose or artificial nose. Detection is based upon changes in the electric resistances of odour sensors incorporated in a sensor chamber exposed to the vapour. A typical sensor chamber may contain up to six different exchangeable metal oxide surfaces (with 18 surfaces in total) of semi-conducting character. Computer software is needed to analyse the signals received but the information obtained does not yet compare in detail with that from GC methods.

In contrast to the use of gas chromatography, where individual aroma compounds are analysed, sensor based instruments analyse the aroma as a whole, without dividing it into individual aromatic compounds (Wardencki *et al.*, 2009). Instrumentation is relatively cheap, and analysis can be done very fast. This method is more like the human sensory analysis, still used to analyse drinks, despite drawbacks, such as low repeatability, reproducibility and compounds cannot be quantified or qualified. Sensors are being designed to be analogues of the protein receptors in the olfactory epithelium. The electronic system which transfers the electric signals from the sensors is comparable with human sensory neurons, using artificial neural networks. Hence it is thought that electronic and human noses function similarly, despite the many differences in the processes involved. Considering its sensitivity, speed of analysis and reproducibility, future developments of this method will iron out the current drawbacks and in time electronic noses may play a significant role in assuring wine aroma quality.

4.6 Chemical structure and physical properties

The tables in Appendix II set out the chemical structure (molecular weight and molecular formula), together with reported physical properties, that are or may be relevant to their study in wine, focusing on apparently important volatile compounds already described in Sections 4.2 and 4.3.

The significance of the partition coefficient of a particular aroma substance in very dilute aqueous solution ($K_{j,a-w}$), together with the related relative volatility ($\alpha^{\infty}_{j,a-w}$), has been discussed in Section 4.1. Methods to determine these quantities are described by Clarke (1990), Buttery *et al.* (1969b, 1971b) and detailed descriptions of gas chromatographic procedures are given by Buttery (1969b), and Gretsche *et al.* (1995), together with some determined values. Unfortunately, published data by direct determination of K_j at given temperatures for different compounds are very few compared with the number of compounds reportedly present in wine. Recently, a new direct method has been published (Pollien & Yeretzian, 2001), and results presented for some 15 different volatile compounds, along with comparison of values made by other methods.

However, estimates can be made as already discussed. Methods of estimation are fully described in Appendix I.7.

Bibliography

General

American Chemical Society Symposium Series. Flavour subjects No. 388 (1989): 490 (1992); 596 (1995); 714 (1998); 794 (2001), American Chemical Society, Washington.

Burdock, G.A. (ed.) *Feneroli's Handbook of Food Flavour Ingredients* (2002), 4th Edn. CRS Press, Orlando, FL.

Jackson, R.S. (1994) and 2nd Edn. (2000) *Wine Science, Principles, Practice, Perception*. Academic Press, San Diego.

Ribéreau-Gayon, P., Glories, Y., Maujean, A. & Dubourdieu, D. (2006) *Handbook of Enology, Volume I, Wine and Wine Making*. John Wiley & Sons, Ltd, Chichester.

Ugliano, M. & Henschke, P.A. (2009) Yeasts and Wine Flavour. In: *Wine Chemistry and Biochemistry* (eds. M.V. Moreno-Arribas & M.C. Polo), pp. 313–392. Springer ScienceBusiness Media, LLC.

Sensory perception

General

Ache, B.W. & Young, J.M. (2005) Olfaction: Diverse species, conserved principles. *Neuron*, **48,** 417–430.

Buck, L. & Axel, R. (1991) A novel multigene family may encode odorant receptors: A molecular basis for odor recognition. *Cell*, **65**, 175–187.

Meilgaard, M., Civille, G.V. & Carr, B.T. (2007) *Sensory Evaluation Techniques*. 4th Edn. CRC Press, Florida.

Mombaerts, P. (2004) Genes and ligands for odorant, vomeronasal and taste receptors. *Nature Reviews Neuroscience*, **5**, 263–278.

Swiegers, J.H., Chambers, P.J. & Pretorius, I.S. (2005) The genetics of olfaction and taste. *Australian Journal of Grape and Wine Research*, **11**, 109–113.

Identification and contents

General

Amerine, M.A. & Roessler, E.B. (1983) *Wines: Their Sensory Evaluation*, loc. cit. Chapter 1. WH Freeman & Co., San Francisco.

Nijssen, L.M. (Ed.) (1996) *Volatile Compounds in Food. Qualitative and Quantitative Data*. 7th Edn., TNO Nutrition and Food Research Institute, Zeist, The Netherlands.

Vernin, G., Pascal-Mousselard, H., Metzger, J. & Párkányi, C. (1993) Aromas of Mourvèdre wines. In: *Shelf-Life Studies of Foods and Beverages* (ed. G. Charalambous), Developments in Food Science No. 33, pp. 945–974. Elsevier, Amsterdam.

References

Aldave, L. (1993) The shelf-life of young white wines. In: *Shelf-Life Studies of Foods and Beverages* (ed. G. Charalambous), Developments in Food Science, 33, pp. 923–944, Elsevier, Amsterdam.

Bartowsky, E.J. & Henschke, P.A. (2004) The 'buttery' attribute of wine -diacetyl-desirability, spoilage and beyond. *International Journal of Food Microbiology*, **96**, 235–252.

Buttery, R.G., Seifert, R.M., Guadagni, D.G. & Ling, L.C. (1971a) Characterization of additional volatile components of tomato. *Journal of Agricultural and Food Chemistry*, **19**, 524–9.

Camara, J.S., Marques. J.C., Alves, M.A. & Ferreira, A.C.S. (2004) 3-Hydroxy-4,5-dimethyl-2(5H)-furanone levels in fortified Madeira wines: relationship to sugar content. *Journal of Agricultural Food Chemistry*, **52**, 6765–6769.

Camara, J.S., Alves, M.A. & Marques, J.C. (2006) Changes in volatile composition of Madeira wines during their oxidative ageing. *Analytica Chimica Acta*, **563**, 188–197.

Campbell, J.I., Pollnitz, A.P., Sefton, M.A., Herderich, M.J. & Pretorius, I.S. (2006) Factors affecting the influence of oak chips on wine flavour. *Wine Industry Journal*, **21**, 38–42.

Carillo, J.D. & Tena, M.T. (2007) Determination of ethylphenols in wine by in situ derivatisation and headspace solid-phase micro extraction-gas chromatography- mass chromatography. *Analytical Bioanalytical Chemistry*, **387**, 2547-2558.

Cullere, L., Escudero, A., Campo, E., Cacho, J. & Ferreira, V. (2009) Multidimensional gas chromatography-mass spectrometry determination of 3-alkyl-2-methoxypyrazines in wine and must. A comparison of solid phase extraction and headspace solid phase extraction methods. *Journal of Chromatography A*, **1216**, 4040-4-45.

Ebeler, S.E. & Thorngate, J.H. (2009) Wine chemistry and flavour: Looking into the crystal glass. *Journal of Agricultural and Food Chemistry*, **57**, 8090-8108.

Escudero, A., Campo, E., Farina, L., Cacho, J. & Ferreira, V. (2007) Analytical characterization of the aroma of five premium red wines. Insights in the role of odour families and the concept of fruitiness of wines. *Journal of Agricultural and Food Chemistry*, **55**, 4501-4510.

Ferereira, A.C.S., Barbe, J.C. & Bertand, A.T. (2003) 3-Hydroxy-4,5-dimethyl-2(5H)-furanone: A key odorant of the typical aroma of oxidative aged Port wine. *Journal of Agricultural and Food Chemistry*, **51**, 4356-4363.

Francis, I.L. & Newton, J.L. (2005) Determining wine aroma from compositional data. *Australian Journal of Grape and Wine Research*, **11**, 114-126.

Loscos, N., Hernandez-Orte, P., Cacho, J. & Ferreira, V. (2010) Evolution of the aroma composition of wines supplemented with grape flavour precursors from different varietals during accelerated wine ageing. *Food Chemistry*, **120**, 205-216.

Loiret, A. & Versini, G. (2002) Aroma variation in Tannet wines: Effect of malo-lactic fermentation on ethyl lactate level. *Italian Journal of Food Science*, **XIV**, **2**, 175-180.

Montedoro, G. & Bertuccioli, M. (1986) The flavours of wines, vermouth and fortified wines. In: *Food Flavours, Part B* (eds. I.D. Morton & A.J. Macleod), Elsevier Science Publishers BV, Amsterdam.

Nykanan, L. & Soumaileinan, M. (1983) Composition of wines, quoted by the IARC Monograph, *Alcohol Drinking* (1988) Vol. 4, pp. 79-86. International Agency for Research into Cancer, Lyon.

Palaskova, P., Herszage, J. & Ebeler, S.E. (2008) Wine flavor: chemistry in a glass. *Chemical Society Reviews*, **37**, 2478-2489.

Ramirez-Ramirez, G., Chassagne, D., Feuillat, M., Voilley, A. & Charpentier, C. (2004) Effect of wine constituents on aroma compounds sorption by oak wood in a model system. *American Journal of Enology and Viticulture*, **55**, 22-26.

Rapp, A. & Guntert, M. (1985) *Vitis*, **24**(2), 87-98.

Rapp, A. & Marais, J. (1993) The Shelf-life of Wine: Changes in aroma substance during storage and ageing of white wines. In: *Shelf-Life Studies of Foods and Beverages*, (ed. G. Charalambous), Developments in Food Science, 33, pp. 891-921, loc. cit. Elsevier, Amsterdam.

Ribéreau-Gayon, P. (1978) Wine flavour. In: *Flavour of Foods and Beverages* (eds. G. Charalambous & G.E. Inglett), pp. 355-380. Academic Press, New York.

Ribéreau-Gayon, P. (1990) *Rev. Fr. Oenol.*, **123**, 25-33.

Ribéreau-Gayon, P., Glories, Y., Maujean, A. & Dubourdieu, D. (2006) *Handbook of Enology, Volume I, Wine and Wine Making*, Ch. 1. John Wiley & Sons, Ltd, Chichester.

Schreier, P., Drawert, F. & Junker, A. (1977) Gaschromatographische bestimmung der inhaltsstoffe von gärungagetranke. Part X. *Chem. Mikro. Technol. Lebensm.* **5**, 45-52, quoted by Montedoro & Bertuccioli (1986).

Suarez, R., Suarez-Lepe, J.A., Morata, A. & Calderon, F. (2007) The production of ethylphenols in wine by yeasts of the genera *Brettanomyces* and *Dekkera*: A review. *Food Chemistry*, **102**, 10–21.

Swiegers, J.H. & Pretorius, I.S. (2005) Yeast modulation of wine flavour. *Advances in Applied Microbiology*, **37**, 131–175.

Tominaga, T., Blanchard, L., Darriet, P. & Dubourdieau, D. (2000) A powerful aromatic volatile thiol, 2-furanmethanethiol, exhibiting roast coffee aroma in wines made from several *Vitis vinifera* grape vaireties. *Journal of Agricultural and Food Chemistry*, **48**, 1700–1802.

Tominaga, T., Guimbertau, G. & Dubourdieau, D. (2003a) Role of certain volatile thiols in the bouquet of aged Champagne wines. *Journal of Agricultural and Food Chemistry*, **51**, 1016–1020.

Tominaga, T., Blanchard, L., Darriet, P. & Dubourdieau. D. (2003) Contribution of Benzenemethanethiol to smokey aroma of certain *Vitis vinifera* grape varieties. *Journal of Agricultural and Food Chemistry*, **51**, 1373–1376.

Tominaga, T. & Dubourdieau, D. (2006) A Novel method for quantification Of 2-methal-3-furanthiol and 2-furanemathanethiol in wines made from *Vitis vinifera* grape varieties. *Journal of Agricultural and Food Chemistry*, **54**, 29–33.

Tressl, R., Remer, R. & Apetz, M. (1976) Volatile phenolic components in beer, smoked beer and Sherry. *Z. Lebensm – Unters – Forsh*, **162**, 115–122, quoted by Montedoro & Bertuccioli (1986).

Wilkinson, K.L., Elsey, G.M., Prager, E.P. & Sefton, M.S. (2004) Rates of formation of *cis*- and *trans*- oak lactone from 3-methyl-4-hydroxyoctanois acid. *Journal of Agricultural and Food Chemistry*, **52**, 4213–4218.

Flavour descriptions and odour values

General texts

Arctander, S. (1967) Perfume and flavour chemicals, quoted by I. Flament (2001), *Coffee Flavour Chemistry*. John Wiley & Sons, Ltd, Chichester.

Belitz, H.D. & Grosch, W. (1999) *Food Chemistry*. pp. 319–377. Springer-Verlag, Berlin.

References

Chemisis Data base on organoleptic properties of perfume and flavour chemicals, Firmenich, S.A., Geneva, unpublished, quoted by I. Flament (2001), loc. cit.

Fors, S. (1983) Sensory aspects of volatile Maillard reaction products and related compounds. *American Chemical Society Symposium Series*, Series 215, pp. 185–286. American Chemical Society, Washington.

Grosch, W. (2001) Volatile compounds. In: *Coffee: Recent Developments* (eds. R.J. Clarke & O.H. Vitzthum), pp. 68–89. Blackwell Publishing Ltd., Oxford.

Lettingwell, J.C. (2002) Chiralty and flavour perception, http://www.lettingwell. com/chiralty/chiralty.htm (last accessed: 1 May 2010).

Meilgaard, M. (1982) Prediction of flavour differences between beers from their chemical compositions. *Journal of Agricultural and Food Chemistry*, **30**, 1009–1017.

Meilgaard, M.C. & Peppard, T.L. (1986) The flavour of beer. In: *Food Flavours, Part B* (eds. I.D. Morton & A.J. Macleod), pp. 99–170, loc. cit.

Motodo, S. (1979) Formation of aldehydes from amino acids by polyphenoloxi-
dase. *Journal of Fermentation Technology*, **57**, 395–399, quoted by Flament
(2001).

Pickenhagen, W. (1989) Enantioselectivity in odour perception. In: *Flavour
Chemistry, American Chemical Society Symposium Series*, No. 714, pp. 151–157.
American Chemical Society, Washington.

Shaw, P.E. (1986) The flavour of non-alcoholic fruit beverages. In: *Food Flavours*,
pp. 337–368, loc. cit.

Odour/flavour thresholds

General texts

Buttery, R.G. (1999) Flavour chemistry and odour thresholds. In: *Flavour Chemistry:
Thirty years of progress* (eds. R. Teranishi, E.L. Wick & J. Hornstein),
pp. 353–365. Kluwer Academic/Plenum, New York.

Campo, E., Cacho, J. & Ferreira, V. (2007) Solid phase extraction, multidimensional
gas chromatography determination of four novel aroma powerful ethyl esters.
Assessment of their occurrence and importance in wine and other alcoholic
beverages. *Journal of Chromatography* A, **1140**, 180–188.

Flath, R.A., Black, D.R., Guadagni, D.G., McFadden, W.H. & Schultz, T.H. (1967)
Identification and organoleptic evaluation of compounds in *Delicious* apple
essence. *Journal of Agricultural and Food Chemistry*, **15**, 29–35.

Huber, U.A. (1992) Homofuranol. *Perfume and Flavour* **17**(7/8), 15–19.

Keith, E.S. & Powers, J.J. (1968) Determination of flavour threshold levels and
sub-threshold, additive and concentration effects. *Journal of Food Science*,
33, 213–218.

Kim Ha, J. & Lindsay, R.C. (1991) Volatile alkyl phenols and thiophenol in species-
related characterizing flavours of red meats. *Journal of Food Science*, **56**,
1197–1202.

Mulders, E.J. (1973) Odour of white bread. Part IV. *Z. Lebensm. Unters. Forsch.*, **151**,
310–317.

Murray, K.E. & Whitfield, F.B. (1970) 2-Methoxypyrazines and the flavour of green
peas. *Chemistry and Industry (London)*, 973–974.

Nakamura, S., Crowell, E.A., Ough, C.S. & Totsuka, A. (1988) Quantitative analysis of
7-nonalactone in wines and its threshold determination. *Journal of Food
Science*, **53**, 1243–1244.

Parliment, T.H., Clinton, W. & Scarpellino, R. (1973) Trans-2-Nonenal. Coffee
compound with novel organoleptic properties. *Journal of Agricultural and
Food Chemistry*, **21**, 485–487.

Parliment, T.H. (1981) The chemistry of aroma. *Chemtech*, **25**(8), 38–47.

Perrson, T. & von Sydow, E. (1973) Aroma of canned beef. *Journal of Food Science*,
38, 377–385.

Seifert, R.M., Buttery, R.G., Guadagni, D.G., Black, D.R. & Harris, J.G. (1970) Synthesis
of some 2-methoxy-3-alkyl pyrazines with strong bell-pepper-like odours.
Journal of Agricultural and Food Chemistry, **18**, 246–249.

Semmelroch, P. & Grosch, W. (1996) Studies in character impact compounds in
coffee beans. *Journal of Agricultural and Food Chemistry*, **44**, 537–543.

Sick, T.J., Albin, J.A., Satter, L.A. & Lindsey, R.C. (1971) Comparison of flavour
thresholds of aliphatic lactones, with those of fatty acids, esters, aldehydes,
alcohol and ketones. *Journal of Dairy Science*, **34**, 344–346.

Swiegers, J.H., Bartowsky, E.J., Henschke, P.A. & Pretorius, I.S. (2005) Yeast and bacterial modulation of wine aroma and flavour. *Australian Journal of Grape and Wine Research*, **11**, 139–173.

Takeoka, G.R., Buttery, R.G., Turbaugh, J.G. & Benson, M. (1995) Odour thresholds of various branched esters. *Lebensmittel Wissenschaft und Technologie B*, 153–156; also (1998) **31**, 443–448.

Wasserman, A.E. (1966) Organoleptic evaluation of three phenols present in wood smoke. *Journal of Food Science*, **31**, 1005–1010.

References

Ahmed, E.M., Dennison, R.A., Dougherty, R.H. & Shaw, P.E. (1978) Flavour and odour thresholds in water of selected orange juice components. *Journal of Agricultural and Food Chemistry*, **26**, 187–191.

Buttery, R.G., Seifert, R.M., Guadagni, D.G. & Ling, L.C. (1969a) Characterization of some volatile constituents of bell-peppers. *Journal of Agricultural and Food Chemistry*, **17**, 1322–1327.

Flament, I. (2001) *Coffee Flavour Chemistry*. John Wiley & Sons, Ltd, Chichester.

Grosch, W. (1995) Instrumental and sensory analysis of coffee volatiles. *Proceedings of 16th ASIC Colloquium (Kyoto)*, ASIC, Paris, 147–156.

Guadagni, D.C., Buttery, R.G., Okano, S. & Bun, H.K. (1963a) Additive effect of sub-threshold concentrations of some organic compounds associated with food aromas. *Nature*, **200**, 1288–1289.

Guadagni, D.C., Buttery, R.G. & Okano, S. (1963b) Odour threshold of some organic compounds associated with food flavours. *Journal Science Food and Agriculture*, **14**, 761–765.

Lea, C.H. & Swoboda, P.A.T. (1958) The flavour of aliphatic aldehydes. *Chemistry and Industry (London)*, 1289–1290.

Maga, J.A. (1978) Simple phenol and phenolic compounds in food flavour. *CRC Critical Reviews Food Technology*, **7**(2), 147–192.

Patton, S. (1964) Flavour thresholds of fatty acids. *Journal of Food Science*, **29**, 629–680.

Pineau, B., Barbe, J.C., van Leeuwen, C. & Dubourdieu, D. (2007) Which impact for β-damascenone on red wines aroma? *Journal of Agricultural and Food Chemistry*, **55**, 4103–4108.

Rychlik, M., Schieberle, P. & Grosch, W. (1998) Compilation of odour thresholds, odour qualities and retention indices of key odorants. Quoted by W. Grosch, (2001), loc. cit.

Shankaranarayana, M.L., Raghavan, B., Abraham, K.O. & Natarajan, C.P. (1982) Sulphur compounds in flavours. In: *Food Flavours* (eds. I.D. Morton & A.J. McCleod). Elsevier Science, Amsterdam.

Stevens, S.S. & Galanter, M. (1957) Ratio scales and category scales for a dozen perceptual conditions. *Journal of Exp. Psychology*, **84**, 377–411.

Takahashi, K. & Ayiyama, H. (1993) Shelf-life of Sake. In: *Shelf-Life Studies of Foods and Beverages* (ed. G. Charalambous), loc. cit.

Wagner, R., Czerny, M., Bielochrodsley, J. & Grosch, W. (1999) Structure-odour activity relationships of alkyl pyrazines. *European Food Research and Technology*, **208**, 308–16, quoted by Flament (2001).

Wood, C., Siebert, T.E., Parker, M., Capone, D.L., Elsey, G.M., Pollnitz, A.P., Eggers, M., Meier, M., Vossing, T., Widder, S., Krammer, G., Sefton, M.A. & Herderich, M.J. (2008) From wine to pepper: Rotundone, an obscure sesquiterpene, is a potent spicy aroma compound. *Journal of Agricultural and Food Chemistry*, **56**, 3738–3744.

Yokotsuka, T. (1986) Chemical and microbiological stability of Shoyu (fermented Soya sauce). In: *Handbook of Food and Beverage Stability* (ed. G. Charalambous), pp. 518-621. Academic Press Inc. Orlando; quoting Sasaki, M. & Nunomura, N. (1981) *Journal of the Chemical Society Japan* (no. 5), 736-745.

Physical properties

General texts

Clarke, R.J. (2001) Instant Coffee – Physical properties. In: *Coffee: Recent Developments* (eds. R.J. Clarke & O.G. Vitzthum), pp. 133-137. Blackwell Publishing Ltd., Oxford.

Perry, J.H. (1980 and later editions; ed. with D.W. Green, 1997, 7th Edn.) *Chemical Engineers' Handbook*. McGraw-Hill, New York.

Pierotti, G.J. (1959) Activity coefficients and molecular structure. *Industry and Engineering Chemistry*, **51**, 95-102.

References

Buttery, R.G., Guadagni, D.E. & Ling, L.C. (1969b) Volatilities of aldehydes, ketones and esters in dilute solution. *Journal of Agricultural and Food Chemistry*, **17**(2), 385-389.

Buttery, R.G., Bomben, J.L., Guadagni, D.E. & Ling, L.C. (1971b) Some considerations of the volatilities of organic flavour compounds in foods. *Journal of Agricultural and Food Chemistry*, **19**(No 6), 1045-1048.

Chandrasekaran, S.K. & King, C.J. (1972) Determination of activity coefficients. *A.I.Ch.E. Journal*, **18**, 513-520.

Clarke, R.J. (1990) Physical properties of the volatile compounds of coffee. *Café, Cacao et Thè*, **XXXIV** (No 4), 285-294.

Cullere, L., Escudero, A., Campo, E., Cacho, J. & Ferreira, V. (2009) Multidimensional gas chromatography-mass spectrometry determination of 3-alkyl-2-methoxypyrazines in wine and must. A comparison of solid phase extraction and headspace solid phase extraction methods. *Journal of Chromatography A*, **1216**, 4040-4-45.

Ferreira, V., San Juan, F., Escudero, A., Cullere, L., Fernandez-Zurbano, P., Saenz-Navajas, M.P. & Cacho, J. (2009) Modeling quality of premium Spanish red wines from gas chromatography – Olfactometry data. *Journal of Agriculture and Food Chemistry*, **57**, 7490-7498.

Gretsche, C., Grandjean, G., Maering, M., Liardon, K. & Westfall, S. (1995) Determination of the partition coefficients of coffee volatiles using static head-space. In: *Proceedings of the 16th ASIC Colloquium (Kyoto)*, pp. 326-31, ASIC, Paris, France.

Hayasaka, Y., Baldock, G.A. & Pollnitz, A.P. (2005) Contributions of mass spectrometry in the Australian Wine Research institute to advances in knowledge of grape and wine constituents. *Australian Journal of Grape and Wine research*, **11**, 188-204.

Howard, K.L., Mike, J.H. & Riesen, R. (2005) Validation of a solid-phase microextraction method for headspace analysis of wine aroma components. *American Journal of Enology and Viticulture*, **56**, 37-45.

Martin, A., Carillo, F., Trillo, L.M. & Rosello, A. (2009) A quick method for obtaining partition factor of congeners in spirits. *European Food Research and Technology*, **229(4)**, 697-703.

Plutowska, B. & Wardencki, W. (2007) Aromagrams – Aromatic profiles in the appreciation of food quality. *Food Chemistry*, **101**, 845–872.

Plutowska, B. & Wardencki, W. (2008) Application of gas chromatography-olfactometry (GC-O) in analysis and quality assessment of alcoholic beverages – A review. *Food Chemistry*, **107**, 449–463.

Pollien, P. & Yeretzian, C. (2001) Measurement of partition coefficients. *Proceedings of the 19th ASIC Colloquium on Coffee*, CD-ROM, ASIC, Paris.

Wardencki, W., Biernacka, P., Chmiel, T. & Dymerski, T. (2009) Instrumental techniques for the assessment of food quality. *Proceedings of ECOpole*, **3**, 273–279.

Chapter 5

Wine Tasting Procedures and Overall Wine Flavour

5.1   Wine tasting

In Chapter 3 we discussed the individual components in a finished wine that mainly contribute to the colour and taste of wine, while in Chapter 4 we examined the volatile components, which are likely to contribute to the smell of the wine. However, it is the combination of the wine colour, taste and smell that we appreciate when appraising wines. The word flavour is usually employed to indicate the combination between smell (or odour) and taste. The word 'tasting' is often used to indicate that the flavour of the wine is being judged. Judging wine is generally in contrast to drinking wine, the former is often a more or less formal occasion, while the latter is a social event, during which we mainly consume wine because we enjoy doing so.

Our evaluation of the wine during a social gathering is influenced not just by the wine, or aspects immediately related to the wine such as its image, the label, the cork or the shape of the bottle, but also by many other factors such as our mood, the friends we are with, the atmosphere of the evening, the food we are eating, how relaxed we may feel, etc. This may explain why many people are convinced the wine they drank on holiday does taste quite different at home and is often disappointing. It is not that 'the wine did not travel', but the 'wine drinker is not travelling'. Thus it is not just the smell and taste of a wine which influences our choice and acceptance of wine, but a combination of psychological factors; product information, mood, anticipated sensory properties of the wine and psychological effects the wine may have (such as the effect of the alcohol in the wine) will all influence our perception and judgement. This illustrates the importance of wine tasting procedures when tasting wines, to ensure that the wine is being assessed and not the moment or the event.

Indeed wine tasting procedures are described and recommended by wine writers/experts (e.g. Robinson, 1993; Grainger, 2009). Methods for tasting wines used by scientists are standardized to give objective results, or as

Wine Flavour Chemistry, Second Edition. Jokie Bakker and Ronald J. Clarke.
© 2012 Blackwell Publishing Ltd. Published 2012 by Blackwell Publishing Ltd.

objective as can be, without bias. For more in-depth information, the reader is referred to Jackson (2008), and Jackson (2009) gives a detailed account of current knowledge of wine tasting. Some general objectives of wine tasting will be discussed in this chapter.

The tasting procedures for scientific purposes, appraisals by experts and judgements by amateurs can all be described in three major parts:

(1) Presentation of the sample to the person assessing the wine, including glass, sample size, sample information, temperature, light, no distractions.
(2) Accurately described task, comprising instruction and training regarding what the person tasting the wine has to do, including how to taste the wine, what attributes to the score, and what and how to score the wine.
(3) Procedure for collecting and analysing, interpreting and presenting the information.

Many influences on our sensory perception of taste and probably more so on smell, affect our overall judgement of food, and most likely also wine, over and above those environmental factors briefly mentioned above. In particular, our sense of smell is influenced by tasting techniques such as sniffing, hunger and hormonal changes. A brief overview on the mechanisms involved in smell perception is given by in Chapter 4 and on taste perception in Chapter 5.

The considerable investigations carried out on the physiological mechanisms involved in interpreting tongue sensations by the brain through the nervous system have been well described by Heath (1988). Less familiar are the concepts and mechanisms involved in the odour assessment, for example from wine, where the taster has to interpret the effect of up to 400 volatile compounds found in wine. Each compound is present in very small quantities, which by smelling from the glass or by drinking (even slurping) reaches the olfactory organ at the back of the nose. This results in a fairly complex cascade of biochemical reactions, briefly discussed in Chapter 4, giving signals that can be interpreted by the brain. The mechanism whereby the human nose detects volatile compounds for the brain to characterize its particular aroma characteristics has been the subject of several different theories, as discussed by Heath (1988). The most obvious is to relate 'smell' to chemical or molecular structure, and much progress has been made in this area of research but to date this has not proved sufficient to predict smell. Since the publication of the Nobel winning paper of Buck & Axel (1991) much progress has been made on odour perception, discussed briefly in Chapter 4, and has probably outdated an older theory (Turin, 1997), which assumed that molecular vibrations determine smell, so that the receptor cells in the nose behave as 'vibrational spectroscopes'. Guti et al. (2001) have given an overview of the various theories.

There are also numerous biological influences on our perception of food and therefore probably also wine. Some factors we may, to an extent, be aware of, such as hunger, and in the case of wine also the way we sniff (discussed below). Our sensory perception of wine is based on the integration of information about many of its aspects. The sensory properties of foods are conveyed to the brain by our senses: colour and surface structure are assessed by vision, structural information is gained by tactile and kinaesthetic senses, combined with hearing while the food is chewed; volatile compounds are perceived by the sense of smell, and the water-soluble chemicals by the sense of taste. The trigeminal nerve (which has endings in lips, oral and nasal cavity) registers temperature, touch and pain, such as in wine the burn of ethanol or the presence of carbon dioxide giving a prickling sensation. Research has shown that this nerve may also enhance the perception of certain volatile components (see Maruniak & Mackay-Sim, 1985). In all, while eating or drinking our senses collect an enormous amount of information, which is rapidly conveyed to our brain. Smell is probably the sense that gives us most enjoyment when eating or drinking. Wines we tend to drink for enjoyment, not for any nutritious reason. Hence the smells we detect from well made wines will make consuming them a pleasurable experience.

5.2 Wine tasting procedure

Although there are endless possible variations, once the wine has been selected and opened, there is generally a set order of events, ranging from the choice of glass to whether or not to expectorate the sample. It is important that for more formal wine assessments, the room is comfortable, free of smells and has northern daylight, in order to assess the wine colour. Of course, sample presentation, sample identification and the order of the samples need to be carefully controlled for formal assessments. Volatile aroma compounds are best perceived if present near or above the sensory threshold for detection and recognition, although synergistic effects need to be borne in mind (see Section 5.3). Some aspects relevant to wine tasting are briefly discussed below and summarized in Table 5.1

5.2.1 Tasting glass

In all wine tasting, a very important aspect is the use of an appropriate wine glass. The glass must be generous enough in size to enable it to be tilted sideways to inspect the colour at the rim of the glass and to swirl the wine to allow a better judgement of the aroma of the wine. The wine glass should have a stem that allows the taster to hold the glass, so that the wine is not unintentionally warmed up, or the clarity of the glass obscured. Such glasses tend to be tulip shaped, e.g. narrower at the top. The glass should be only part filled to give a generous amount of air-space above the glass from which the assessor can sniff the aromas escaped from the wine and present just above the wine in the glass. The glass has to be wide enough at the top to allow the nose to be comfortable above the wine, but not so wide that insufficient of the volatile compounds remain present in the air-space immediately above the wine.

The BS ISO glass has been especially designed to fulfill these requirements, however many other wine glasses having that general tulip shape can be used

Table 5.1 Aspects relevant in wine tasting.

Glass	Tulip shape, allowing appraisal of colour and flavour.
Serving	Comfortable, clean room, no smells, with adequate daylight for the appraisal of colour.
	Wine temperature: light wines cool, dark wines warmer, maximum room temperature.
	Decanting, only needed when wine have deposit or reductive aroma.
	Avoid oxygen for delicate older wines and white wines.
Sight	Judge appearance of wine (clear, no deposit).
	Colour – tilt glass, at meniscus indication of age in red wine (blueish, orange/brown)
	Depth of colour, indication of anthocyanin content, possibly related to other phenols
Smell	Swirl wine to help release aroma volatiles from wine
	Sniff (typically 200 ml in 0.4 seconds) for nasal aroma perception
	Adaptation, temporary reduced ability to smell, short exposure
Taste	Sample of about 30 ml in mouth, swirl round in mouth whilst sucking in air
	Retronasal perception of volatile aroma compounds
	Perception of mostly sour, sweet, bitter and astringency
	Perception of combination of taste and aroma compounds (flavour, bouquet)
	Adaptation to taste components
	Role of bubbles
	Perception of stringency, mouthfeel, time needed to perceive and recover
	Expectorate or swallow
Interactions	Psychological factors, such as expectation
	Compounds, for example sugar may influence fruity perception
	Compounds having both smell and taste, such as acetic acid
	Interactions between aroma compounds, giving suppression or enhancement effects
	Effect of food on wine perception
Appraisal	Hedonic like or dislike
	Descriptive, use vocabulary to describe sensory properties
	Value for money

for assessing wines. Reports in the popular press seemed to suggest some effects of the phenolic composition as a result of the presentation of a red wine in different shapes of glass (*New Scientist*, 31 August 2002, p. 23); however, the information given was inaccurate and in the short time span the wine is in the glass it is likely that the main effect is the release of the volatile compounds rather than the chemistry of the phenols.

5.2.2 Serving

Serving the wine is also still a topic shrouded in some mystique. Of course the wine to be tasted from the glass needs to be at the 'right' temperature and wine books recommend temperatures considered best for wines (e.g. Robinson, 1995). Generally, the lighter the wine, the lower the serving temperature, for example light white wines are served the coldest, light red wines are slightly cooled and full red wines are served at room temperature. Breathing the wine is rarely necessary, and if the wine is left in the bottle after extracting the cork the access of oxygen is probably highly ineffective since wine presents a relatively small surface area for oxygen diffusion. Many modern white wines are unlikely to benefit from oxygen at all.

Decanting tends to impart more oxygen into the wine than when used directly from the bottle and is best only done carefully with wines that have a sediment in the bottle. Clear wines do not need decanting and it is thought that aged wines or very fresh young wines (especially white ones) can easily lose their aroma as a result of reactions with oxygen. The room in which the wine is tasted should be normal room temperature, free of strong smells, which could interfere with our assessment of the wine. The colour of the wines should be assessed in natural northern daylight; in particular, our perception of the red colour quality can be affected by lighting. Special daylight tube-lighting is usually fitted in professional or scientific tasting rooms, so any variation of daylight can be controlled.

5.2.3 Visual

The assessment of the quality and quantity of colour as well as the clarity of the wine is done entirely by eye, usually at the beginning of the tasting. The glass is angled against a white background and the wine should be clear, unless the wine is drawn directly from a wine barrel. The colour can indicate the grape variety and age of the wine, although the information is not precise or conclusive. A red wine with a deep-red intensity suggests Cabernet Sauvignon or Syrah wines, generally grapes from a warm region and/or wines made with long maceration. A light red wine suggests a cool climate and grapes such as Pinot Noir.

The colour quality of the wines at the rim of the glass gives an indication of the amount of ageing in particular a red wine has received. Young wines are blue or more purple at the rim but as the wine matures this changes to brick red, orange–brown or even yellow. A comparable colour at the edge of the glass of two different wines does not necessarily indicate the same age, since the colour depends on other factors, such as the aeration the wine has received during maturation and the maturation temperature. White wines can also develop a considerably brown colour, such as Sherry that has been stored under oxidative conditions. The colour of a wine can also affect our perception of wine quality.

It is thought that even a very crude indication of the alcohol content can be obtained. In such an appraisal, colourless streams of wine running down the side of the glass after the wine has been swirled around (often referred to as tears or legs) is taken as a measure of the alcohol content; presumably, differences in viscosity and surface tension of the wine assessed give different effects. However, although wine books often refer to such possibilities, there seems to be no research data confirming that such judgements can indeed be made. Very sweet wines probably have a higher viscosity from their sugar content and after swirling it may take longer for the wine to run down the glass. Again, there is no research data to confirm this speculative statement. Pure ethyl alcohol is, however, known to have a much lower surface tension (22.3 dynes cm^{-1} at 20°C) than pure water (728 dynes cm^{-1} at 20°C) but the viscosity of both pure alcohol and water are the same (1.1 cp at 20°C). However, the viscosity of alcohol–water mixtures is generally higher than that of either (e.g. 3 cp for a 40% solution).

Jackson (2008) quotes research from Neogi (1985), who attributed the formation of tears to a physical process. After swirling, water droplets form tears slowly sliding down the sides of the glass, counteracted by evaporative losses of ethanol on the sides of the glass, which offset the action of gravity pulling the drops down by pulling wine up from the glass.

5.2.4 Smell

Next, our sense of smell will be used. There are two routes for the volatile substances to reach the olfactory organ, which is placed just behind the top of the nose where smells are somehow being recognized, followed by signals to communicate this information to our brain in order to identify the smell. The two routes are referred to as nasal and retronasal. Nasal means that volatile compounds will reach the olfactory organ through the nostrils of the nose during the period of *nosing* the wine from the glass. Retronasal means the smells escape from the wine in the mouth and travel via the back of the mouth to the olfactory organ.

Nosing is the traditional sniffing of the air-space above the glass of wine, usually before any sample is placed in the mouth. Sniffing serves a useful purpose. Under normal breathing conditions, only 5% of the respired air reaches the olfactory organ, while during sniffing this increases to about 20% (see Maruniak & Mackay-Sim, 1985), hence only a fraction of the odour compounds released from the wine into the air will reach the olfactory organ. Thus sniffing presents a much more efficient sample of volatile compounds than normal breathing would have done. Studies have shown wide variations among sniffing patterns between people but individual patterns were remarkably consistent over time. The average sniff of subjects was 200 ml in 0.4 s, giving an inhalation of 30 L^{-1} min. The individual sniffing techniques do not appear to affect performance in sensory perception, hence people seem to have their own individual sniffing technique that helps them to get the most smell from the wine.

Retronasal perception means wine is in the mouth when the flavours are released for perception. There will be a temperature change, with the wine warming to body temperature. Also the wine will to an extent be diluted with saliva. Genovese *et al.* (2009) found that dilution with saliva in model systems significantly affected the flavours released from the wines tested. In white wines diluted with saliva, the release of esters and fusel alcohols was reduced by 32–88%, whilst the release of 2-phenylethanol increased 27% and furfural increased 155%. These differences are large enough to explain the difference between nasal and ortonasal perception. The effect was less marked on red wine, showing only a decrease of some esters by 22–51%. The authors suggested polyphenols in wines binding to the proteins in saliva inhibit the activity of the salivary enzymes.

Interestingly, we 'adapt' to a certain smell, which means that during prolonged exposure to a certain smell we tend to no longer notice it. In sensory terms, adaptation is always a temporary change in sensitivity. For example, if you sit in a badly ventilated smelly room, you probably do not notice it yourself, at least not after a while. However, after a break outside this room and coming in from outside, you probably do. Thus, when tasting a wine, the first sniff will give information about many smells in the wine to the brain via the olfactory system. However, we can maximize our performance by not constantly sniffing the wine but by sniffing and drinking at intervals, giving our senses a chance to recover. Prolonged sniffing can lead to adaptation and a 30–60 second interval is thought to be sufficient for adaptation to disappear. There is also cross-adaptation, whereby a compound with a similar chemical structure inhibits the perception of another.

The physical laws controlling the release of volatile compounds have been fully discussed in Chapter 4. Generally, the volatile compounds in wines will be readily released while we are tasting. The concentration of each compound released into the air depends on its physical properties. Since wines contain usually between 11 and 15% alcohol by volume, alcohol will be the main influence on the 'behaviour' of the volatile compounds, determining what amounts of each volatile compound in the wine will travel into the air-space. Swirling the wine in the glass will aid the otherwise slow and diffusion-controlled process to reach maximum concentration of the volatile compounds present in the wine in the air-space (equilibrium will only be apparent, since the volatile compounds will continue to escape from the space above the open glass). Thus, swirling the wine before nosing it in the glass will help release more volatiles from the wine into the air-space above it and gives us the opportunity to have a good smell of the wine when nosing. Similarly, gently warming the wine will help release more volatiles and indeed may affect the balance of volatile compounds in the air-space. A wine served very cold might have a different 'nose' once it has warmed up to room temperature. A very cold wine will release only the most volatile compounds, while as the temperature increases relatively less volatile compounds are increasingly released from the wine into the air-space. Apparently, no data is available on times to reach equilibrium or maximum concentration. The time is probably of the order of minutes. Calculations are possible, but not reported, on the basis of mass transfer coefficients, including the known diffusion coefficients of volatile compounds (Dj) in water–alcohol, which are of the order of $1 \times 10^{-5}\,cm^2\,s^{-1}$ at 20°C.

5.2.5 Flavour

Next, our senses of smell and taste are used to assess the flavour of the wine. Firstly three of the taste sensations (sweet, sour, bitter) from non-volatile substances are perceived on the tongue. Many scientific, and even some popular wine books covering taste, include a sketch of the tongue showing the distribution of the sensitivity to the basic tastes. Sweet is placed at the tip of the tongue, bitter at the base, although the placements of salty and sour vary. These maps are wrong (Bartoshuk, 1993). According to Bartoshuk, the so-called tongue map has become an enduring scientific myth, based on some erroneous interpretations of research data. Data supports the view that all four classic tastes can be perceived on the sites of the tongue where the taste receptors are located, and although the intensity of taste varies over the tongue, this in no way implies that taste qualities are localized on the tongue as suggested by the maps. Heath (1988) agrees, though also states that the central region is relatively insensitive to all.

There is a temporal element in the perception of taste. Sweetness and sourness are rapidly perceived. We tend to adapt readily to sweetness and wine acids and phenols reduce the perception of sweetness. Adaptation to sourness is thought to be slower, and may cause a lingering aftertaste. To reach the maximum intensity of bitterness can take 10-15 seconds, and it takes time to disappear after swallowing or expectoration. Astringency is usually the last attribute to be detected, it takes 15 seconds to reach the maximum intensity and even longer to disappear. This information illustrates the importance of allowing time between assessing samples, so the senses can recover and overlapping effects of the samples are avoided.

Wine flavour is perceived once the wine is in the mouth and is a combination of three of the taste sensations from non-volatile substances perceived on the tongue, and the aroma (or smell) sensation from volatile substances perceived by the olfactory organs behind the nose. A generous portion of the wine has to be put in the mouth, e.g. up to about 30 ml. Once the wine is in the mouth, the wine is warmed up, moved around in the mouth and there is the option of noisily sucking air through the mouth. All these actions help the volatile compounds to escape from the wine and to travel retronasally via the back of the mouth to the olfactory organ. In physical terms, increased temperature will enhance the diffusion coefficients, affecting the partition coefficients, of the volatile compounds. Bubbles in the wine will increase the surface area from which the volatile compounds can escape. It is impossible to predict whether any such differences are large enough to give sensory differences between retronasal and nasal. Volatile compounds detected during nosing are often described separately, and may or may not be similar or identical to those detected from the palate. Again, research data is lacking.

It is important to allow sufficient time to assess the sample, since the quality of aroma often changes during sampling. One explanation for this may be that olfactory adaptation may allow the perception of other compounds.

5.2.6 Interactions

There are many possible interactions, between aroma compounds, between aroma and taste compounds, or between aroma or taste compounds and the matrix, possibly leading to enhancement or suppression of perception. Such interactions make the interpretation of wine flavour difficult to predict, even when considerable analytical data are available. Scientific debate continues regarding the extent of interaction between taste and smell and numerous sensory interactions have been determined, see reviews (Delwiche, 2004; Palaskova et al. 2008). It is interesting to note that interactions even occur at sub-threshold level, for example when a subject is given a sub-threshold concentration of an odour compound together with a sub-threshold concentration of a taste compound, subjects are able to detect the combination. This cross modal summation of selected smell and taste compounds at sub-threshold concentrations demonstrates that central neural integration of smell and taste signals is occurring. Examples of interaction causing suppression also exist, for example, the addition of sugar to fruit juice reduced the perception of sourness and bitterness, increased the perceived sweetness but also increased ratings for fruity ratings. There have been studies to determine whether the presence of, for example, sugar influences our perception of certain smells in wines, such as fruitiness. There are documented physical effects, such as at high concentrations sugar increases the partition coefficients (see Chapter 4).

There is likely to be an effect of ethanol. Le Berre et al. (2007) studied the effect of ethanol on the physico-chemical and perceptual interactions of a woody aroma (whisky lactone) and a fruity aroma (isoamyl acetate). Whisky lactone volatility was less in the presence of alcohol. Numerous interactions between the three compounds were determined, such as a masking of woody odour by the fruit odour and both odours masked the so-called alcohol odour, showing the importance of perceptual interactions even when tasting a very simple model wine.

There are other known interactions between aroma compounds; a well known example is an early study on esters in wines contributing to the fermentation aroma (see Francis & Newton, 2005). Six esters all present above threshold level were added in typical wine concentrations to a deodorized wine. However, there were no differences in odour intensity of this mixture when one of the esters was absent in the mixture, hence it was concluded that the absence of one ester was masked by the odours of the other esters. Other interactions reported are masking of the bell pepper aroma in Cabernet wines by fruity aromas, fruity aromas in red wines enhanced by C13-norisoprenoids (see the review by Palaskova et al. 2008). Volatile compounds with a similar character also are thought to enhance each other, for example two compounds present below threshold value are perceived by an additional effect.

However, cognitive associations can increase the flavour intensity scores (i.e. we expect something sweet to be fruity and therefore we think it to be more fruity than it really is). It would be interesting to study whether the sugar in sweet wines affects our perception of the flavour of the wine. Considering the large numbers of flavour compounds analysed in wines (see Chapter 4), the interactions in our sensory perception make it even more difficult to determine the contribution individual compounds make to the overall wine flavour.

In addition, there are compounds that we can taste and also smell. For example, acetic acid prevalent in vinegar has both a taste (sour) and a smell (vinegar). In wood-aged mature wines (Madeira) the acetic acid content may be high enough to contribute to both taste and smell.

The flavour or taste of drinks can, to an extent, be affected by compounds present in foods consumed with the drink – these are usually referred to as flavour modifiers or flavour enhancers. For example, it has long been said that chewing artichokes spoils the flavour of fine wine and makes water taste sweet (see Moulton, 1982). This sweetening effect lasts 4–5 minutes, although not everyone experiences this effect. Monosodium glutamate is perhaps a better-known flavour enhancer. Research in simple taste solutions showed that the threshold for sourness and bitterness were both lowered but there is no research data available for wines.

A recent paper by Bastian *et al*. (2010) reported interaction between food and wine. The authors reported that eating cheddar cheese just before Shiraz wine tasting reduced the flavour length and astringency of the wine, whilst the tannins were perceived as silkier. Most sensory studies are on single food or drink items, in particular with wine where there are so many well established opinions on food and wine combinations, there is scope for further research studies.

5.2.7 Astringency

The moving around of the wine in the mouth also helps to assess mouthfeel, which contributes significantly to our perception and enjoyment of wines. In particular, red wines are often referred to as 'full bodied' or 'robust', and would be described as lacking 'body' and 'backbone', without adequate astringency. This is not a taste but a sensation perceived in most parts of the mouth. This sensation is usually referred to as astringency. The mechanism of astringency perception is still debated in scientific research and probably involves phenolic compounds interacting with salivary proteins (Chapter 3). The sensation tends to be a drying, puckering experience in the mouth. The accepted definition of astringency is the complex of sensations due to shrinking, drawing or puckering of the epithelium as a result of exposure to substances such as alums or polyphenols. Astringency perception is not confined to any particular region of the tongue or mouth and gives a diffuse feeling of extreme dryness and roughness. Some astringency is needed for wines to have adequate mouthfeel. An excess is unpleasant and such wines need to mature to soften.

5.2.8 Judging the wine

To avoid having the sense of judgement affected by imbibing alcohol during multiple tastings, expert wine tasters usually 'spit out' (expectorate) the wine sample after the assessment. Many wine books also comment on the desirability of 'length' or 'finish', which means that the desirable smells and tastes should linger in the mouth after swallowing or expectorating the wine. The comment 'short', in this context, indicates that the wine flavour disappears pretty well as soon as the wine leaves the mouth. Wines with 'good length' are generally more highly valued by expert tasters.

There are many different tasting terms but a confusing number of definitions of these terms. No consensus seems to exist in the use of terms like bouquet, aroma, etc. and different wine writers may use these terms with different meanings. The term 'aroma' is most commonly used to describe the smell of the wine derived from the grapes, while 'bouquet' tends to refer to the smell of the wine formed as part of the development during maturation. In fact, many different terms are used in sensory analysis, the science of tasting food and drink and many of these terms are also applicable to wines. Some terms are well defined in, for example, the *'British Standard Glossary of Terms related to sensory analysis of food'* (1992). Even so, there seems to remain some confusion, specifically with terms like bouquet, aroma, flavour, nose, where even scientists seem to give definitions whereby some terms have multiple meanings. Some terms relevant to wine tasting defined by the *British Standard Glossary of Terms* are:

- Hedonic – Relating to like or dislike.
- Sensory – Related to the sense organs.
- Odour – Organoleptic attribute perceptible by the olfactory organ on sniffing certain volatile substances.
- Aroma – An odour with a pleasant connotation.
- Flavour – Complex combination of the olfactory, gustatory (taste) and trigeminal sensations perceived during tasting. The flavour may be influenced by tactile, thermal, painful and/or kinaesthetic effects.
- Taste:
  (1) Sensations perceived via the taste organs when stimulated by certain soluble substances.
  (2) Sense of taste.
  (3) Attribute of products including taste sensations.

There are many lists of wine descriptors, further discussed in Section 5.5.1. For the researcher, it is important to tailor a selected set of tasting terms to the wine samples to be assessed and train the panel appropriately. However, the amateur can experiment with the information on taste terms and interestingly research has shown that having an adequate vocabulary to describe the wines helps towards becoming a more discriminating wine taster (Hughson & Boakes, 2002).

It is not really surprising that sensory information contains usually a lot of variation. There are differences in sensitivity between people, interactions between and within samples, a general lack of adequate terms to describe sensory perceptions and a wide array of information in one sample has to be assessed.

5.2.9 Reasons for wine tasting

There is a great difference between the sort of tastings that can be done and the information that can be obtained from such exercises. The rigidity of the tasting and the recording of the information depend entirely on the purpose of the tasting. Wine tasting can be done purely for the experience and left just as mental judgements at the time. However, if the information is needed for publication in magazine or newspaper articles, some notes are usually made. In the scientific tasting procedures used to analyse wine, much effort is put into designing the experiment so that bias is taken out as much as possible and the collected data can be analysed using statistical analysis, so that calculated values have a known error term.

Sensory analysis

Why do people taste wine, other than for pleasure? The answer is relatively simple. There is an abundance of instrumental measurements available, many used to ensure the wine complies with legal and safety restrictions. Many chemical constituents can be measured instrumentally but the overall impression of the 'taste' of wine cannot generally be predicted from instrumental analysis. Hence tasting remains a very important part of the tests performed on wine and over the years tasting has also grown into a relatively new branch of science that more commonly is referred to as sensory analysis. Over recent decades this branch of science has made valuable contributions to our understanding of wines and the effect of various changes of production on the sensory properties of wines.

Tasting forms a valuable assessment of the wine's sensory properties and is used at many stages before the wine is enjoyed at the dinner table. The producer will taste the wine during various stages of production, from the vineyard (are the grapes ripe?), just after wine-making (young wine) and at various other stages of the maturation until the wine is sold. The brokers, the merchants, the négociants/shippers will also taste the wine to assess whether the wine is what they expect it to be. Official bodies for certifying wines with *Appellation Contrôlée'* type quality denominations will also taste wine to check whether it is typical for the region. Competitive tastings have a rôle in comparing wines and making selections regarding quality judgements. The consumers will ultimately taste the wine and decide whether they like it enough to ever buy another bottle.

Quality tastings

Many of the wine tastings are done to determine the quality of the wine, especially the tastings by expert wine tasters. A pertinent question is what this quality actually is. In its broadest sense, 'quality' means fitness for its intended use. In many instances the wine drinkers confuse the often rather ill-defined word 'quality' with 'like'; however, these words have very different meanings in the world of assessing drinks. For wine 'quality' is an integrated response to the sensory properties of the wine based on the expectation one has of a particular wine based on previous experience. Thus, for example, young red Bordeaux wines intended for maturation before drinking are likely to be deeply coloured and with a fairly high tannin content. Quality is an individual and subjective response, dependent on the person, based on the person's experience, expectation and preference.

The word 'like' should denote how much we like or dislike a wine, a personal preference in other words, and should be independent of the 'quality' assessment of the wine. However, in practice it is not easy to keep our 'like' or 'dislike' of a wine from interfering with the 'quality' assessments. Hence most experienced wine judges will have differences in opinion about wine quality. Many quality assessments are typically to test whether the wine meets the specification set. In wine terms that may mean that the wine should conform to the expected regional distinctiveness and standards. This is of interest to a wine buyer when assessing whether to buy the wine. Or, possibly, the wine buyer may be looking for a more innovative wine, having some 'artistic merit', in order to market a wine with certain characteristics in a special promotion.

Even though 'quality' assessments are likely to contain an element of personal preference, as discussed above, the opinion of the wine expert is generally well respected. Despite the shortcomings of such expert judgements, there is generally a remarkable agreement on many attributes of the wines. Many quality judgements are made by experts within a sub-group, such as Australian red wine made from Cabernet Sauvignon. This appears to be quite a satisfactory basis for judgement. The price of wine appears to be a good arbiter, since the quality of the wine, no doubt in conjunction with its image, allows, for example, Château Lafite of 1990 to command currently £500–750 for a bottle; the price has rocketed over the decade since 2000. However, although these wines are attributed with having a high quality, this does not mean that all consumers like an expensive wine more than a well made cheaper one, as is sometimes illustrated in TV programmes, where the most expensive wine is not necessarily always the most well liked.

Identifying wines by tasting

Identifying a wine 100% correctly in a blind tasting is very difficult. These 'blind' tastings shown on TV wine and food programmes have excellent enter-tainment value, but to identify a wine correctly on tasting every time is likely to lead to many embarrassing mistakes. In fact, the chemical identification of wines to ensure they come from a certain area, grape and quality standard, the so-called authenticity of wine, forms part of a branch of science that uses highly sophisticated analysis to try and establish wine authenticity, outside the scope of this book but refer to reviews (Isci *et al.*, 2009). Using sensory methods for authenticity tests cannot be relied upon.

Experienced experts, exposed to tasting many different wines each day, seem to be able to identify wines quite well and there is a strategy to use (Broadbent, 1979). However, with an ever increasing number of good quality wines being generally available, the task becomes ever larger. Even experts tend to think that 'One glance at the label is worth 50 years of experience'. However, with some training and thought it is possible to analyse the wine while tasting and pick up the characteristics of the wines (colour, smells, taste, etc.). It may be possible to determine the grape variety and, combined with aspects of the style of the wine, one can make an intelligent guess about the possible origin of the wine. In fact, a study by Hughson & Boakes (2002) compared the performance of testers in a group of wine experts and a group of novices to wine. They found that a short list of appropriate variety-relevant descriptors enhanced performance. In addition, differences in performance were attributed in part to the lack of vocabulary and information on varietal types that experts employ in tasting tasks.

Sensory analyses used in research

For research purposes, the scientist hopes, usually, to answer a very precisely defined question; typically only one variable is changed, and the effect is compared with the normal so-called control wine. For example, the effect of grape maturity on the wine may be tested by picking grapes from the same vineyard at two different picking dates. The word quality would be defined in specific terms, for example, by testing if there is a difference in fruitiness between the two resulting wines.

In such a research experiment, a panel of people is used to perform sensory assessments, rather than just using one or two experts. Using a larger number of tasters allows the collected data to be pooled and analysed using statistical methods. Such panels are often specially selected to ensure their taste and smell sensitivity. In addition, they are trained, depending on the tasks they are asked to perform. For example, for wine tasting they would be trained to assess wines and are taught to recognize certain smells or tastes in wines. The sensory experiment has to be carefully controlled to eliminate any bias and to ensure that the experiment can be repeated so as to give comparable results. Samples are usually presented blind, with a random number, in identical glasses. The experimental design also needs to be compatible with the use of statistical analysis.

Consumer tasting

There are many different ways to taste foods and drinks under controlled conditions, which are used in addressing research questions. One important style of tasting is hedonic – this means, do you like it? For example, a hedonic question can be asked to compare two wines to determine how many people prefer one to another wine and possibly get some pointers for the reason why (too sweet, too bitter, etc.). Currently these sorts of tasting are often done using 'normal consumers' (not trained) and the methods used form an important tool of market researchers, for example, trying to establish in advance how successful a new wine may be. Depending on the information the researcher wants to collect, wines can be presented blind or in their normal packaging.

Analytical tasting

Another style of tasting is strictly analytical and the person assessing the wine does not allow his or her like or dislike to interfere with the description of the wine. For example, a person assessing Australian red wines based on Cabernet Sauvignon might normally prefer 'fizzy' white wines from Italy. There are three main types of analytical tasting:

(1) Difference testing. The first step is often to determine whether samples are different, established in an experiment whereby the samples are presented totally blind. A form of triangle test is often used, by selecting the odd one out in a set of three, or by matching one or two samples with a given standard. There is a good chance of guessing the answer correctly in such tests, hence there are statistical tables that should be consulted to check whether the sample difference is significant or whether chance would have given the same result. Therefore tests have been developed to reduce the so-called statistical odds, for example by selecting wines which differ from three others in a set of five samples.

(2) Descriptive analysis. Typically a well-trained taste panel uses words to describe the differences between the wines accurately. Such a sensory panel is used like an instrument and trained in the use of terms to describe the sensory properties and scoring the properties. A scoring system is used so that numeric data can be collected and subjected to statistical analysis. Typically, a so-called scalar rating is used for example, using a line marked with low and high at the extremities. The panelists mark on the line the quantity of each defined attribute, for example the amount of blackcurrant smell in the wine. Very detailed sensory information can be obtained and this data can be matched with, for example, chemical analysis related to the compounds that impart smell and taste to the wine. For example, the effect of picking date of the grapes can be correlated to differences in the smell and chemical composition of wines. This sort of tasting analysis is very informative and permits analytical evaluation of the qualitative and quantitative differences between the tested wines.

(3) Ranking tests. To determine threshold concentrations (the concentration of a compound a person detects) a ranking test is sometimes performed. Samples of increasing concentration of the test compound are presented and the person is asked to rank the series according to their strength.

Of course the presentation is very important and the assessor has to be aware that if the strongest sample has been assessed first, time may be required for the senses to recover in order to perceive the weaker samples.

One example of a scientific experiment involved a trained panel to describe the properties of 24 Bordeaux wines from four different communes, whereby the panelists discriminated between the samples primarily on the basis of astringency and bitterness (Noble *et al.*, 1983). Eight Masters of Wine (highly trained and very knowledgeable experts in tasting of wines) described the quality and the properties of the wines. The Masters of Wine were consistent in the specific descriptions of the wines; however, the quality scores reflected, more than anything, the difference in interpretation of quality among the Masters of Wine.

5.2.10   Wine tasting information and analysis

Essential information about wine and tasting should be written down to record more than just a memory of a good wine. Many wine experts also have some sort of scoring system, such as a 20-point system, with high numbers given for high quality wines. The following information may usefully be recorded (as suggested by Broadbent, 1979), some of it simply copied from the bottle, but also some comments based on one's own experience:

- date of tasting;
- name of wine, country, district, vineyard, quality denomination;
- year of vintage;
- bottler;
- price;
- description of appearance (colour, depth, clarity, bubbles);
- description of its nose (aroma, bouquet);
- description of its taste (component parts, finish); and
- general conclusions (maturity, quality, value).

Whether this information is sufficient or far too much rather depends on what you want to know. If the question of interest is whether the wine is liked or not, the information recorded needs to be quite different from when the question of interest is what the precise sensory properties of the wine are. The methods of tasting and recording tend to be tailored. There is a great difference between the type of information recorded in research and the quality evaluations typically used by experts, as the expert tasting usually has a different aim to that in research.

Statistical analysis

Numerical data recorded by a panel of trained tasters or a large group of consumers is usually too unwieldy by itself for the experimenter to draw any sensible conclusions without subjecting this data to statistical analysis. The development of computing power has allowed the rapid and cheap use of older statistical techniques but it has also led to the development of very powerful new statistical analysis. The large data sets generated by sensory analysis can be handled in several ways, and the experimental design needs to be compatible with the use of statistical analysis. Many detailed textbooks are available, usually covering both the methodology of the actual tasting procedure and the statistical analysis (Piggott, 1984 & 1986; O'Mahoney, 1985; Jackson, 2009; see also Section 5.7.3). Only some general information will be given here.

Statistical analysis forms an important and integral part of tasting with taste panels. When consumer tests are carried out to find out which wine the consumer likes best, the experimenter's main interest is in the differences between the consumers and especially what their differences may be due to; for example, preferring sweet wines to dry ones, or preferring the bouquet of one wine to another one. Data from trained taste panels usually addresses a more analytical question about the samples; for example, is there a difference between these three wines, or what are the flavour characteristics that differentiate these wines? Even this data usually shows large differences between people, attributed to the differences in sensitivity between the assessors, even though the taste panel has been specially selected and trained. Commonly, each assessor tastes the same sample in duplicate or triplicate (all presented blind, so the person does not know the identity of the sample). Presentation order of the samples forms part of the experimental design.

A first step is often analysis of variance, to determine whether there are differences between the samples and whether the individual panelist's assessments are reproducible. The means of the various terms used to describe a wine are calculated, together with some information regarding the likelihood of there being real differences, or whether any differences are just due to 'noise'. Comparisons of a few wines for their mean scores (and errors) can be done on a *spider* plot (Fig. 5.1).

These large sets of data can now be manipulated using specialist statistical packages on computers. Usually, the statistician first decides whether the data 'makes sense', which means tests are done to ascertain that all panelists are judging samples more or less in the same way. Any so-called outlier data may indicate that the person had not understood the task, or some other problem. Any interference in the data is usually also tested for, to make sure the particular attribute tested is not dependent on another one. This should be known and the information will form part of the interpretation of the results. The most appropriate test will be chosen for the particular question the experimenter is addressing with the collected data set. The last stage is to draw conclusions from the analysis of the results, attached with a calculated *uncertainty* or *error* (which is present in all scientific data).

The advancement in computers has allowed the development of so-called multivariate statistics required to analyse large and complex data sets. Essentially, all the data is projected in a multidimensional space (where there are more than three orthogonal possible, somehow). The cut is determined,

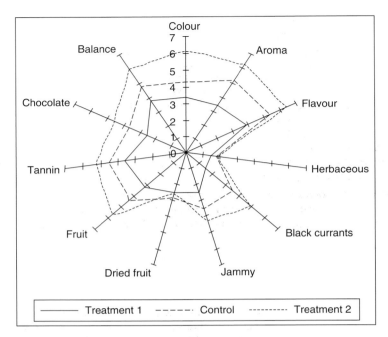

Figure 5.1 Spider plot showing the mean values of three wines assessed for intensity of 11 attributes on a 0–10 scale (seven assessors, duplicate tasting).

which represents on two or three orthogonal axes the most variation. If one is lucky, 50%, or even more, of the variation can be explained by the first two axes, and this may be enough to answer the questions asked. Otherwise, the information on a third, or possibly even a fourth, axis many need to be examined. Data from descriptive analysis can be represented graphically, which is usually easier to examine than large tables. The development of statistical techniques, including the so-called multidimensional statistical analysis, has enabled the handling of sensory data and has greatly helped to progress sensory testing.

5.3 Factors influencing sensory perception

Numerous factors that have nothing to do with our sense of smell or taste can influence our perception, as has been briefly touched on in Section 5.1. However, our sense of taste and smell are very sensitive and robust, illustrating their biological importance. Without these senses, our enjoyment of food and drink would be rather limited and our motivation to eat and drink a varied diet would be diminished. Damage to taste or smell is relatively rare and the nerve cells of the taste organs on the tongue and the smell organs located behind the nose are constantly renewed. When people say 'I cannot taste anything', they usually have a temporarily reduced sense of smell, commonly a result of a severe cold, causing congestion.

Although it may seem straightforward that we identify tastes and smells with relative ease, many factors affect our ability to taste and smell, many of which are relevant to our perception of wine. These influences can be physical, chemical, biological and even psychological. Temperature is an obvious physical influence and it is generally assumed that we perceive tastes best when food or drink is at mouth temperature. It seems that the perception of bitterness in red wines increases as the temperature of the wine decreases but little research data is available. The presence of other compounds can also influence our perception, a well known example is adding sugar to reduce the sourness of a drink and thus, residual sugar in wine will reduce our perception of its sourness and also bitterness. The pH (a measure of the amount of acid in a liquid) is an example of a chemical influence and can affect the perception of taste – one effect could be directly on the receptor proteins.

Numerous taste components exert more than one taste or sensory effect, for example alcohol gives a sweet taste but also creates a burning sensation and gives in wines the effect possibly referred to as 'weight'. Another example is acetic acid, which has both a taste (sour) and a smell (vinegary). A biological example is the interaction of wine with the production of saliva. Components in wine stimulate the amount of saliva produced in our mouth; the extra saliva then dilutes the wine. In addition, the protein present in saliva can bind with some of the components in the wine, thus possibly changing our perception of the wine.

5.3.1 Threshold and sensitivity

The threshold concentration for taste (defined as the concentration at which a compound is correctly defined by its taste, for example sweetness) can be determined using single sets of dilutions. There is also a detection threshold, at which concentration a person detects the presence of a compound without being able to recognize it. The detection threshold is usually lower than the recognition threshold.

Studies have been carried out in grouping people according to their taste sensitivity, and people are divided into Supertasters (those very sensitive to the used test compounds), Tasters (those less sensitive), and Nontasters (those not perceiving the test compounds). Supertasters tend to be more sensitive to other compounds also, for example certain sweeteners. One fascinating study, investigating people with a great preference for sweetness, showed that, generally, people with a 'very sweet tooth' were Nontasters (Bartoshuk, 1993). However, not all studies produced consistent results. Surprisingly, the differences in taste sensitivity recorded are small in comparison with the physiological differences observed.

Our sense of smell is extremely sensitive and discriminating, with certain compounds being detected in parts per trillion in the air. Threshold concentration and detection thresholds can be measured, although it is more difficult than these determinations for taste. There can be about a hundredfold variation in thresholds for smells between people. Surprisingly, even the same person fluctuates considerably, dependent on, for example, time of day and hunger (Stevens *et al.*, 1988). Threshold levels for smell/odour are discussed in more detail in Chapter 4.

When comparing what we smell with others during wine tasting, it has to be borne in mind that some people cannot perceive certain smells, or only at extremely elevated concentrations. It is rare for someone to be unable to smell anything at all; mostly there are certain groups of compounds for which a person is insensitive, a so-called specific anosmia (Amoore, 1977). Several diseases, such as Alzheimer's, Parkinson's and HIV are believed to be associated with loss of sensitivity for smells or even specific anosmias. Another sensitivity problem discussed above is that we tend to smell differences and hence suffer from adaptation (always a temporary change in sensitivity) when exposed to the same smell for some time. Hence, we tend to smell best if the smell we are presented with is different from the background smell. We 'adapt' to a certain smell, which means that after prolonged exposure to a particular smell we tend to no longer notice it; hence the importance of swirling and sniffing wine at intervals.

Finally, another sensory phenomenon is masking, whereby 'stronger'-smelling volatile compounds mask the perception of other volatile compounds. Although there seems to be no data available, masking probably contributes to the wine lover's observation that the perception of wine changes over time, when possibly some masking volatiles have diminished or larger concentrations of the masked volatile compound have been released. The above discussion shows that, besides our personal preference for certain tastes and smells, there are numerous chemical and biological processes that will affect our perception of wine.

5.3.2 Vocabulary

Although people are generally very sensitive to smells and can recognize very many different ones, it tends to be difficult to describe the different smells. A first problem is 'finding the words'. The smell is familiar but out of context it is not always easy to remember the word. In addition, even when you recognize cheddar cheese correctly, one is sometimes hard pushed to describe it any other way than 'cheesy' or 'cheddar-like', which does not convey much information. Wine experts have developed sometimes quite extravagant sets of words to describe a wine but there does not seem a ready-made vocabulary to draw on to describe the smells of wine.

5.4 Balance of taste sensations in wine

The interactive flavour effect of the different taste sensations in wine – bitterness, acidity and sweetness – discussed above, leading to the desirable situation of 'balance', is also the subject of many wine experts and writers. In particular, Peynaud (1986) and Ribéreau-Gayon (1978) have given the concept of balance a formal and semi-mathematical interpretation in a *suppleness index*.

The first factor is the equilibrium of basic tastes, where they state that acid and bitter tastes should be balanced by the sweet taste: sweet taste ↔ acid taste + bitter taste.

They especially note the important contribution that alcohol content makes to sweetness. This equation explains why red wines rich in tannins and thus bitterness, cannot tolerate as high a level of acidity as can white wines, which are low in tannins. The inevitable high acidity of wine, compared to other fermented drinks such as beer, is tolerable because it is counterbalanced by the sweet taste of the alcohol. Sweet wines containing residual sugars need the high acidity to balance the sweet taste. Although sweet taste is the only one, which is pleasant by itself, an excess in wine is considered 'cloying'.

Wines, especially reds, must be 'supple', meaning that they should not have either an excessively bitter or acid taste. A numerical suppleness index has been defined (Equation 5.1), in which a value of 5 was regarded as balanced. An increase in this index (in mixed units) corresponds to an increase in the sensation 'softness' and 'fullness of body' characterization (highly desirable in red wines), whereas a decrease corresponds to 'firmness' or 'thinness'. However, Ribéreau-Gayon recognizes the limitations of such an index, which is valid only within a certain compositional range.

$$\text{Suppleness index} = \text{Alcohol (degrees GL)} - [\text{Acidity (g L}^{-1}\text{ H}_2\text{SO}_4) + \text{Tannin (g L}^{-1})] \tag{5.1}$$

Other factors, especially in red wines, Ribéreau-Gayon considers as important. 'Hardness' (decreasing suppleness) depends upon additional compositional elements, notably acetic acid and ethyl acetate. These compounds in excessive quantities indicate bacterial spoilage and give wines particular organoleptic characteristics reminiscent of vinegar and nail varnish, generally considered undesirable in wines and reducing their quality. Even when present below the organoleptic threshold (of the order of 1 g L^{-1} for acetic acid and 150 mg L^{-1} for ethyl acetate in wines) these compounds, especially the latter, intervene in sensory evaluation, particularly of the aftertaste, reinforcing impressions of 'hardness' and 'burning'. He also notes the importance of other features of tannins, according to their structure in a wide range of complexity and their capacity for change through ageing, from strong astringent characteristics to 'smooth', 'supple' and 'velvety', as described in Chapter 3.

5.5 Wine aromas

Wine is primarily described according to its *bouquet*, and to the odour/aroma element of its flavour on tasting. Peynaud (1986) prefers to use the term 'bouquet' rather than odour of the wine being nosed, referring to the smell arising from the surface of the wine being swirled around in a suitable glass vessel, for wines which have been aged, either 'in-cask' or 'in-bottle' (so-called 'oxidative' or 'reductive' bouquet respectively), as already described in Chapters 3 and 4. There is the implication here that, unless the wine has been aged, it will have very little aroma perceptible on nosing. Certainly, the concentration of the volatile compounds will depend on the ageing of the wine, influencing the quality of the smell, but there is no scientific evidence regarding its quality. All wines will have *palate-aroma*, a term used by Peynaud (1986) referring to the retronasal perception already described, so that bouquet will also form part of this aroma/odour assessment. The exact distinction is difficult to define; palate aroma (retronasal) will certainly be more intense than 'nose' (nasal) aroma as a larger amount of the same vapour (in compositional terms, approximately) will reach the olfactory organ via the back of the mouth in drinking (and slurping) rather than merely 'sniffing' for the vapour to reach the same organ, where the odour characteristic is actually perceived. Of course, in drinking, a proportion of the aroma/odour will be taken up directly from the nostrils also. There is a comparable sensory discrimination in coffee tasting, where 'brew aroma' is referred to, when 'sniffing' from the surface of a cup of coffee. However, a different physical situation relates to sniffing a dry roast and ground coffee (and instant coffee), where the volatile compounds released in amount/composition will be somewhat different from those expected to be released when drinking an aqueous solution, due to partition coefficient factors (see Chapter 4).

To consider wine aroma alone, without reference to basic tastes, a large vocabulary of terms has been generated to describe these organoleptic sensations. This necessarily uses words borrowed from the smelling and tasting of other odoriferous substances, often using other food (especially fruit) or drink, or other smells (flowers), or smells associated with some chemical processes. Descriptions are essentially by analogy, though may refer to the odour of known single pure chemical substances.

These vocabularies have undoubtedly created 'wine-speak' and jargon, as so amusingly criticized some time ago by Simon Jenkins (*The Times* article, September 1993), then quoting from British Rail Intercity wine list the following description: 'Hardy's Nottage Hill Cabernet Sauvignon: another Aussie stunner – loads of colour, loads of flavour, but the flavour is soft, not rasping, and the fruit is deep plums and blackcurrant, with just a hint of spice'. There are no prizes for guessing the identity of this particular wine writer. Interestingly, a correspondent to *The Times*, noting this Jenkins article, said that she expected wines to taste of grapes, not these and other exotic fruits; this letter indicated a massive misconception about the nature of wine. Serious scientific wine authorities use numerous terms, e.g. Peynaud (1986) in his book *The Taste of Wine* gives a vocabulary of some 200 different words, whilst Broadbent (*Wine Tasting*, 1979) lists 120 words commonly used of wine, but also has 12 pages more of complex ones. Clarke (*Wine Fact Finder*, 1987) is content with some 31 for the ordinary reader (some referring to taste rather than the volatile component). A later book by Clark (*Introducing Wine*, 2003) adopted a different approach for tasting terms. Fruity flavours were used to group red and white wines separately, whilst 50 non-fruit terms were listed to describe wine. Some of the terms, such as astringent, were related to the mouth drying effect of tannin in wines, whilst other terms were more suggesting an intensity measure, such as meaty, defined as a heavy red wine with solid chunky flavours. Wine books all seem to propose their own set of terms to describe the sensory properties of wines. Possibly this diversity in approaches towards describing wines illustrates the difficulty of the task. Increasingly wine writers have web sites publishing their tasting notes, for example, Jancis Robinson.

Wines can be praised for their 'fruitiness' in wine writers' descriptions, though it is important to describe which kind of fruit flavour they are referring to. Nearly all these fruit flavours are generated during the fermentation; whilst only those best described as 'grapey' (also fruity with floral notes) were originally present in the grapes and remained largely unchanged during fermentation, as described in Chapter 4. Changes in all these flavours occur on ageing. Noted is the wine writer's use of the word 'aromatic', which probably means the same as 'grapey' since the responsible constituents are terpenes present at different levels in different grapes, as opposed to neutral, as in Chardonnay (without terpenes).

5.5.1 Odour/aroma classification

Odour classification and vocabularies have much in common with those used in the perfumery industry, where perhaps they first originated. One such system with eight main groups, proposed by Peynaud (1986) for the wine industry, is derived from the foregoing and is shown in Table 5.2.

Peynaud (1986) believed that all the odour/flavour terms used in wine tasting can be placed in one or more of the groups. Each group in Table 5.2 consists of terms that can be divided into sub-groups. These terms have separate names, which are those used in practice in the wine trade and industry and other specialist organizations concerned with aroma or flavour, such as in perfumery or coffee.

Thus, the Fruity Group (no. 5 in Table 5.2), of particular interest to the wine trade and industry, can be divided into sub-groups carrying terms as follows:

- Citrus – lemon, limes, grapefruit (yellow fruit), orange (orange).
- Berry – 'grapey', strawberry, blackcurrants, gooseberries, cherries, etc.
- Tree fruit – apricot, plum, etc.
- Tropical fruit – kiwi, lychees, passion fruit, banana, pineapple, etc.
- Dried and other fruit – figs, prunes, raisins, etc.

In practice, the flavour term used in the vernacular or in botanical parlance, such as 'blackcurranty' in wines from Cabernet Sauvignon grapes, will be used first, which can then be allocated to an appropriate sub-group/main group. Similarly, flavour terms can also be slotted in other relevant odour groups.

Table 5.2   Classification of odour sensations.

Type		Origin
1	Floral	arising from the flower parts of plants
2	Woody	associated with hard vegetable matter
3	Rustic/Vegetal	associated with soft vegetable matter
4	Balsamic	associated with resinous substances
5	Fruity	associated with the crushed fruits of plants
6	Animal	associated with the odour of animals
7	Empyreumatic	arising from heating/roasting processes
8	Chemical	associated with synthetic chemical manufacture
9	Spicy	associated with aromatic spices and herbs
10	Etherish	associated with ether substances (slightly sickly)

Some common terms may be appropriate to one or more groups, thus 'oaky', whilst others such as 'soapy', 'rancid', may be difficult to fit in.

These tiers, essential descriptors classified at three different levels, can be graphically illustrated as an Aroma Wheel, developed by Noble *et al.* in 1987 – a modified version for wine aroma terminology has been quoted in books by wine experts (Robinson, 1995), presumably indicating the use of such a system in the tastings of wine experts. Aroma Wheels have been described for coffee and for beer (Meilgaard & Peppard, 1986). Like all classifications schemes, overlaps and omissions can be found, but even so they will be found to serve well.

It is not often possible to characterize a particular wine aroma with a particular pure chemical compound. As can be seen from Table 4.7 in Chapter 4, many of these compounds have been described with a range of odour/flavour sensations, even in more than one category, and often have differing flavour impressions, dependent upon their concentrations in aqueous solution or in the air-space above. However, we can attribute their sensory effect from their presence in wines, arising as from very dilute aqueous solutions, in most cases approaching infinite dilution.

From Tables 5.2 and 5.3, some groups of chemical compounds with the olfactory sensations can be identified (described in Tables 5.5 and 5.6) for wines from specific grape varieties. Table 5.4 lists known or believed associations from the descriptions given in Tables 4.14, 4.15, 4.16, 4.17, 4.18, 4.21, 4.22, 4.23, 4.24, 4.25, 4.26, 4.27 and 4.28 (Chapter 4), though the listing is incomplete, owing to a lack of information.

5.5.2   Aroma/odour characteristics of wines from particular grape varieties

It is useful to set out the odour/flavour characteristics of wines from the main grape varieties commercially used, as described by leading wine writers and indeed by lay-persons with experience. Wines from some ten main grape varieties at 100% usage were so listed by Goolden (1991), upon which Tables 5.5

Table 5.3 Flavour/odour terms in groups/sub-groups/common names.

Group/ sub-group		
Number	**Name**	**Common names used for wines**
1	Floral	Roses, geraniums, violets, etc.
2	Woody	(A term widely used in the coffee industry for green/ roasted coffee: only occasionally used for wine).
3	Rustic/vegetal	Fresh – herbaceous, potatoes, peas, green bell peppers, eucalyptus, etc. 'Grassy' is often used equivalently to 'herbaceous' as a flavour defect. 'Earthy' is often synonymous. Canned/cooked – cabbage, asparagus, etc. Dried – hay, tobacco, tea, leather saddles. 'Rotten' – 'sulfurous'.
4	Balsamic	(The term resinous is often a synonym).
5	Fruity	Citrus – (yellow fruit) lemon, limes, grapefruit; (orange) orange. Berry – 'grapey', strawberry, blackcurrants, gooseberries, cherries, etc. Tree fruit – apricot, plum, etc. Tropical fruit – kiwi, lychýes, passion fruit, banana, etc. Dried and other fruit – figs, prunes, raisins, etc.
6	Animal	The originating animal is quoted, e.g. 'goaty'.
7	Empyreumatic	'Toasted', as from charred barrels used in ageing. Caramel, 'smoky'.
8	Chemical	Kerosene, bottle-age (a catch-all term, with sub-divisions relating to the petroleum industry, creosote and phenols).
9	Spicy	Cloves, pepper, liquorice but specific note often not quoted.
10	Ethereal	Buttery, sweetish, caramel-like, cake-like, vanilla.

and 5.6 were based, although now extended to include many more grape varieties. The second column provides the odour terms used, with those in italic indicating a wide acceptance. The third column places them into relevant groups/sub-groups according to the principles set out in Section 5.5.1. The fourth column distinguishes, where possible, particular aromas/odours as whether primary (P), i.e. arising from original grape (varietal), secondary (S), arising from the fermentation (sometimes from the pre-fermentation stages), or tertiary (T), arising from maturation, in-cask ageing [T (oak) when oak barrels are used], or in-bottle ageing. Finally, other relevant comments are also included.

Notably the main characterizing flavour descriptions in Tables 5.5 and 5.6 command a considerable degree of unanimity amongst wine writers/experts. Emphasis on one or more of the characteristics may be different, as can be expected on account of the different area origins of the same variety. Flavours will differ according to the degree of ripeness of the grapes when picked and

Table 5.4 Relationship of flavour/odour terms in wines to specific chemical compounds.

1	Floral	Terpenes: linalool (rose), α-terpineol (lily-of-the-valley), cis-Rose oxide (geranium), etc. Phenyl alcohol. Complex ketones such as β-ionone (violets), β-damascenone (Z isomer), undecanone.
2	Woody	Higher unsaturated alkyl aldehydes [e.g. (E)-nonenal]. Oak lactones (oaky), see Chapter 4, Table 4.16.
3	Rustic/Vegetal	Particular alkyl aldehydes and alcohols; thus 1-hexanal and 1-hexanol, and trans-2-hexenal in high dilution ('grassiness'). The methoxypyrazines, in particular 2-methoxy-3-isopropyl- in low dilutions (potatoes, peas; also green bell peppers). Some thio-compounds (cooked vegetable), benzenemethanethiol (gunflint, smoke), 2-methyl-3-furanthiol (cooked meat), 2-furanemethanethiol (roast coffee), see Chapter 4, Table 4.28. Complex lactones (oak lactones – 'coconutty') see Chapter 4, Table 4.26.
4	Balsamic	Alkyl hydrocarbons o-cresol ('medicinal').
5	Fruity	Certain fruity (mainly tree and bush fruit) flavours. Alkyl ethyl and other alkyl esters from C_4–C_8 carboxylic acids, and benzoic acid; acetates of higher alkyl alcohols (see Chapter 4, Tables 4.11 and 4.12). Thio-ketones, 4-methyl-4-mercapto-pentan-2-one (tropical fruit – kiwi). Branched esters ethyl 2-, 3- and 4-methylpentanoate and cyclic ester ethyl cyclohexanoate (sweet, fruity).
6	Animal	Butanoic acid ('goaty').
7	Empyreumatic	–
8	Chemical	TND 1,1,6-trimethyl-1,2-dihydronaphtalene (kerosine, bottle age). Phenols: 2-ethyl- and 2-ethyl-guaiacol ('smoky/phenolic'). p-cresol ('tarry/smoky'). guaiacol ('smoky/woody').
9	Spicy	Isoeugenol (cloves). Sotolon, or 3-hydroxy-4,5-dimethyl-2(H)-furan-2-one ('spicy'). Rotundone (peppery).
10	Ethereal	Diacetyl or butan-1,2-dione ('buttery'). Vanillin ('sweet, creamy' vanilla). HDMF or 4-hydroxy-2,5-dimethyl-2(H)-furan-3-one ('sweetish-caramel').

of course, on the mode of vinification and the length and type of ageing. However, the flavour descriptions given by wine experts/writers are often not clearly defined in respect of grape ripeness and particularly of the type/ length of ageing. It is also sometimes difficult to secure wines from one truly single variety in the commercial market, especially in Europe. Many wines in France (except Alsace), though conforming to Appellation Contrôlée

Table 5.5 Grape varieties used for red wines, with described flavour characteristics relating to odour/aroma, for further information (P, S, T) see text.

Grape variety	Main odour characteristics	Odour group	Aroma type
Cabernet Sauvignon	*Blackcurrants*	Fruit/berry	S
	Violets (esp. French)	Floral	S, T
	Cigar box	Woody	T (oak)
	Cedar wood	Woody	T (oak)
	Green capsicum or bell pepper (esp. Chile and California)	Vegetal	P
	Cake	Etherish	S, T
	Rosehip	Floral?	–

Other comments: 'Herbaceous/vegetal' flavour when unripe, inimical to 'blackcurranty'. Minty ('spicy'). 'Tobacco' from new oak ageing

Merlot	*Plums*	Fruity/berry	S
	Redberries	Fruity/berry	S
	Dundee cake	Etherish	S, T
	Toffee caramel	Etherish	S, T
	Pencil shavings	Woody	T (oak)
	Minty	Spicy	P

Other comments: Damson. Cherry. Velvet (texture term). Cloves. Toasted nuts. Complete range of aromas

Gamay (Noir) (Fr.)	Iodine	Chemical	–
	Gym shoes	Chemical	–
	Tar	Chemical	–
	Cherries (*kirsch*)	Fruity/berry	S
	Jam	Fruity (mixed)	S
	Fruit kernel (bitter almond)	Fruity	S

Other comments: Rapid vinification as by carbonic maturation can result in strong banana/pear drop/boiled candy/nail polish remover notes. Peppery

Syrah (Fr.)	*Raspberries*	Fruity/berry	S
[Shiraz – Australian]	Fruit gums	Fruity/berry	S
	Pepper	Spicy	S, T
	Black liquorice (Australian)	Spicy	T (oak)
	Creosote	Chemical	S
	Leather and saddle soap	Etherish	T
	Goulash (going meaty on ageing)	Animal	T
	Tarwood smoke 'smoky'	Empyreumatic/chemical	S

Other comments: Special characteristics of Australian; mulberry, burning rubber, with regional Australia differences of flavour also 'sweaty saddles' (Australia, Hunter Valley). When young – carnations and violets. When aged in-bottle 'gamey', 'leathery', 'tobacco'.

Cabernet Franc	Raspberries/blackcurrants	Fruit/berry	S
	Sponge cake/crumbs	Etherish	–
	Grass and green leaves (blackcurrant)	Vegetal	S
	Herbaceous border	Vegetal	S

(Continued)

Table 5.5 (*Continued*)

Grape variety	Main odour characteristics	Odour group	Aroma type
Other comments: cherries, pencil shavings.			
Pinot Noir Fr.	*Strawberries*	Fruity/berry	S
[Spüt-Burgunder (G.)]	Cabbages	Vegetal	T
	Gamey meat	Animal	T
	Black cherries	Fruity/berry	S
	Rusty metal	Chemical	–
	Earth compost	Vegetal	–
	Soft (velvety) and jammy	Floral	–
Other comments: Variable complex characteristics Mature wines with flavours of wood smoke, rotting vegetables.			
Nebbiolo (It.)	Truffles	Spicy	–
	Prunes	Fruity	S
	Violets	Floral	S, T
	Liquorice	Spicy	T (oak)
	Chocolate	Empyreumatic/ chemical	–
Other comments: Requires long ageing to counteract high acidity/tannin. Tar and roses, cherries, damsons, leather			
Grenache Noir (Fr.)	Raspberries/strawberries	Fruity/berry	S
Garnacha Tinto (Sp.)	Peppery	Spicy	P
Other comments: roasted nuts, leather, honey on ageing.			
Zinfandel	*Berries* (brambles)	Fruity/berry	S
(California, USA)	Black pepper, cloves, cinnamon, oregano	Spicy	–
	Violets/roses	Floral	S, T
Other comments: increasing ripeness to blackcurrants, prunes and raisins. Unripe flavours of green beans, artichokes, eucalyptus, mint.			
Sangiovese (It.)	*Cherries*	Fruity (tree)	S
	Violets	Floral	S, T
	Liquorice	Spicy	T
Other comments: dense plumminess when fully ripe. Vanilla/spice from oak ageing.			
Tempranillo (Sp.)	*Strawberries*	Fruity/berries	S
[also Cencibel, Sp.]	Orange	Fruity/citrus	S
	Incense	Empyreumatic	T
	'Oaky'	Various	T (oak)
Other comments: Strawberries/plum jam when young; overripe produces 'figgy' and 'sweet'. Long oak ageing also produces 'tobacco', 'toasty/spicy'.			
Mourvèdre (Fr.)	Berries (raspberry/ blackberry)	Fruity/berry	S
	Liquorice	Spicy	T
	Tobacco spice (vanilla, pepper, cinnamon)	Spicy	T
	Carnation/violet	Floral	S, T
	Musk/gamey	Animal	T

Table 5.5 *(Continued)*

Grape variety	Main odour characteristics	Odour group	Aroma type
Other comments: 'farmyardy', then 'leathery' with age.			
Carignan (Fr.) (Mazuelo, Carineña, Sp.)	'Hot berries'	Rustic	
Other comments: by carbonic maceration, gives rustic/vegetal and fruity.			
Cinsaut (Fr.)	'Low' berry	Fruity/berry	S
Other comments: aromatic when young.			
Malbec (Fr.)	Gamey	Animal/Rustic	S
Other comments: associated with low acidity. Damson/violets (especially in Argentina). Raisins, damson skins, and tobacco (Cahors, France).			
Petit Verdot (Fr.)	Peppery	Spicy	
	Liquorice	Spicy	
	Plum	Fruity	S
Other comments: banana when young, violets later.			
Graciano (Sp.)	Liquorice	Spicy	T (oak)
Other comments: spicy, aromatic, intensely flavoured.			
Pinotage (S.A.)	Mulberry	Fruity (berry)	S
	Blackberry	Fruity (tree)	S
	Liquorice	Spicy	T (oak)
Other comments: often aged; 'toasted marshmallows'.			

requirements, are in fact blends of different varieties, often necessarily determined by the climatic conditions in a given vintage year. A particular label name may mean a somewhat different blend composition in different years, e.g. Châteauneuf-du-Pape, though the overall aim is to achieve a degree of uniformity over the years. Barr (1988) has discussed this issue in detail, though his observations may not command universal agreement in the wine-producing fraternity. This practice of variety blending is less common in the newer wine industries of the USA, in Australia and New Zealand, where wines made from a single variety are often marketed and labelled accordingly; for example, Cabernet Sauvignon or Chardonnay. Single varietal grape wines may be possible due to more predictable climatic conditions. Nevertheless, grapes may still be drawn from a very wide growing area. A similar situation arises in the roasted and instant coffee industries, where brand names are predominant, with products based on blends of different types of coffee beans; the exact composition will vary from time to time according to availability and cost. However, the marketing of single 100% coffee types, such as Kenya, Costa Rica and Colombia, has become more commonplace.

Wines from a specific area/grape variety(ies) for a certain year are, of course, continuously assessed by producers, merchants and wine experts/writers and

Table 5.6 Grape varieties for white wines with described odour/aroma characteristics.

Grape variety	Main odour characteristics	Odour group	Type aroma
Chardonnay (Fr.)	*Butter*	Etherish	T
	Peachy	Fruity/tree	S
	Honeysuckle	Floral	T
	Patisserie/cheesy	Etherish	T
	Clove (esp. Australian)	Spicy/floral	T
	Pineapple (esp. New World)	Fruity (tropical)	S
	Apple/peach/melon (esp. California)	Fruity (tree)	S
	Angelica	Spicy	–
	Vanilla	Etherish	T
Other comments: barrel fermentation/ageing aroma (T) e.g. toasted hazelnuts from Vosges oak (especially strong in Côte d'Or). Cedary. Chablis (wet straw, wet stones). Coconut (American Oak).			
Sauvignon Blanc (Fr.)	*Gooseberries*	Fruity/berry	S
	Asparagus (canned)	Vegetal	T
	Green apples	–	S
	Grassy/herbaceous	Vegetal	S, P
	Boxwood/broom	Vegetal	S
Other comments: sharp aromatic. 2-methoxy-3-isopropyl-pyrazine – not necessarily appreciated, from under-ripe grapes. Presence of thio-ketones for boxwood notes.			
Sémillon (Fr.)	Butter	Etherish	T
	Egg custard	Etherish	T
	Straw	Vegetal	–
	Toast (Australian)	Empyreumatic	T
	Mangoes/melon/fig	Fruity/tropical	S
	Apricots/peach	Fruity/tree	S
	Candle wax	–	S
Other comments: lanolin, fresh apples – butter, honeyed and smoky on ageing. Lacks positive aroma. Oaked Australian, picked ripe – apricots/mangoes, and vanilla from oak.			
Pinot Blanc (Fr.)	*Cream*	Etherish	–
(Weisburgunder, G.)	Citron pressé	Fruit (Citrus)	S
(Pinot Bianco, It.)	Lavender	Spicy	–
	Parma violets	Floral	S
	Canned lychées	Fruity (exotic)	S, P
	Bracken	Vegetal	–
Other comments: can be aged to good effect. Pear and Apple (Italy).			
Chenin Blanc (Fr.)	Grapefruit	Fruity (citrus)	S
	Damp straw	Vegetal	–
	Honey	Floral	T
	Wet wool	–	–
	Cheese	Etherish	–
	Kiwi fruit/*guava*	Fruity/tropical	S
	Concentrated orange juice	Fruity (citrus)	S
	Camellia blossoms	Floral	P
Other comments: associated with high acid content (popular in California). Greengages and angelica and quince (aged).			
Riesling (G.)	*Petrol*	Chemical	T
	Revving up aircraft	Chemical	T
	Nettles	Vegetal	–
	Honey	Floral	T

Table 5.6 (*Continued*)

Grape variety	Main odour characteristics	Odour group	Type aroma
	Sap in a snapped twig (grapey)	Floral/fruity	P
	Lime	Fruity	S
	Green apples	Fruity/tree	S
	Roses	Floral	P
	Truffles	–	–
Other comments: various flavours according to source. Good German Riesling flavour needs bottle ageing, with increasing times with increasing ripeness, from four to five years even for Kabinett (QmP wines).			
Gewürztraminer (G.) (Fr.)	Scented soap	Etherish	S
	Exotic fruit (lychees)	Fruity	S
	China tea	Vegetal	–
	Cinnamon	Spicy	P
	Smoky	Chemical	S
Other comments: rose floral.			
Muscat Blanc à Petit Grains (Fr.)	Grapey	Fruity/floral	P
Other comments: high terpene content.			
Müller-Thurgau (G.)	–	Fruity/floral	S
Other comments: clove (spicy). Low acid. Dull aromatic.			
Ugni Blanc (Fr.) (= Trebbiano, It:)	–	–	–
Other comments: neutral (tart acid). Used for brandy manufacture.			
Clairette (Fr.)	Neutral	–	–
Other comments: fully flabby, oxidizing easily.			
Traminer (G.)	–	–	
Other comments: Less aromatic/spicy than Gewürztraminer			
Aligoté (Fr.)	Buttermilk/soap	Etherish	S
	Pepper	Spicy	–
Other comments: tartish, for early consumption.			
Scheurebe (G)	Blackcurrants	Fruity/berry	S
	Rich grapefruit	Fruity/citrus	S
Other comments: must be ripe for use.			
Viognier (Fr.)	*Overripe apricot*	Fruity/tree	S
	Musky	Spicy	–
	Peach	Fruity/tree	S
Other comments: glacé fruit, ginger and floral character.			
Silvaner (G.)	Passion fruit/pomegranates	Fruit/exotic	S
Other comments: light, slightly tart and earthy. Tomatoes when aged.			
Pinot Gris (Fr.) Rülander (G.)	Smoky	Chemical	S
Pinot Gris Pinot Grigio (It.)	Musky	Spicy	P
	Honey	Spicy	–
Other comments: pear/apple (Oregon and NZ). Aromatic.			
Grüner Veltliner (Austria)	Fruity	Fruit	S
	Peppery	Spicy	–
Marsanne (Fr.)	Citrus	Fruit	S
Other comments: with age floral character, honeysuckle, jasmine.			

judged for quality, value for money, etc. Wines are necessarily identified by their producer, e.g. *Château* or *Domaine* in France, or by suppliers. Perhaps the most important assessment is the identification of the quality attributes of wines made in (exceptionally) 'good years' for a particular wine on a regional or more local basis. It can refer back to many years from the present. There is usually a fairly good consensus on these dates, at any given time; thus, for clarets from Bordeaux the outstanding years were 1945, 1961, 1982 and possibly 2000, whilst 1968 was considered a disaster (due to a very wet August). The growers proclaimed the product of the 2000 harvest to be the best for a decade, or even a century (*Times* report, 2 February 2001), and this early assessment seems to be borne out to an extent, as it is currently described as a vintage of 'great consistency and balance' (Robinson, website), although this may not be the terms to describe the best vintage for a century! The vintage of 2005 is described as 'textbook perfection in respects other than price'. Similarly, tables are published showing how a particular type of wine is developing with age and whether a particular vintage year has reached the maximum quality. All such information can only be obtained by studying the publications of reputable wine writers, e.g. Clarke & Rand (2001), Hugh Johnson's *Pocket Wine Book* (each year) or Jancis Robinson (website). Nevertheless, fashions of particular styles of wine change amongst the general public, if not also amongst the experts.

The next two sections reflect the kinds of difference of quality and flavour judgements given by wine experts when assessing wines from different sources, albeit the same variety of grape was used for vinification.

5.5.3 Variants in Cabernet Sauvignon wine flavour

A wine from a 100% grape variety such as a Cabernet Sauvignon, using the harvesting/ vinification processes typical for the different countries in which the grapes were grown, will inevitably give a range of different wines. However, one particular flavour characteristic, 'blackcurranty' (Table 5.5), appears to be present regardless of origin and thus in general, varietal. A woody flavour characteristic; 'cigar box', 'pencil shavings' or 'cedar' arises when the wine has been aged in-cask, together with vanilla. The presence or otherwise of the green bell pepper herbaceous note from 3 methoxy-2-isopropylpyrazine is evident in the descriptions.

Some of the differences noted in wines based on Cabernet Sauvignon grapes are described below. Thus Ribéreau-Gayon (from Peynaud, 1986) comments that:

> Cabernet Sauvignon grapes grown in soils that are too fertile, or grown in climates that are too hot, give wines that have a rather crude and more vegetal aromatic character, evoking a bruised leaf rather than a fruitiness.
>
> Ribéreau-Gayon (1986)

(This may, however, be due, as he also mentions, to the use of slightly unripe grapes.) He itemizes such differences by characterization, as 'cloves' from a Californian winery (Guadeloupe Valley), 'liquorice'; Rioja, Spain, 'seaweed';

France (Languedoc); 'industrial fumes'; Chile (Santiago); 'soot', Hungary (Egar); and states that 'some closely resemble the Médocean model, but mostly have defects paraded as a virtue'.

However, currently the Australians are seen to be at the cutting edge of New World wine. Formal wine training:

> ... is at a very high level and the reputation of Australian wines is at a pinnacle. With exports year-on-year increasing. Despite being a relatively small producer, it certainly exerts influence in world wine-marketing and receives positive write-ups for many of its wines.
>
> (Robinson, 1995)

The current water shortages limit the irrigation of vines in hotter regions, normally producing consistent quality grapes and the resulting production may well become affected.

Descriptions of differences noted by other wine writers are similarly indicative, though not so sarcastic and are quoted below, including reference also to the basic taste. Thus Clarke describes a Cabernet Sauvignon wine from the Médoc as:

> ... a dark tannic wine with a strong initial acid attack, and a stark pure black currant fruit. When aged in new oak, its black currant fruit combined with a cedar cigar-like perfume is stunning.
>
> Clarke (1989)

This description is similar to that given by Robinson (1995), who goes on to describe wines made in Eastern Europe as:

> ... less refined than Bordeaux counterparts, often recalling red fruits rather than black, and rarely showing any oak influence (other than occasionally, use of heavy-handed oak chippings).
>
> Robinson (1995)

McQuitty (1998) refers also to a 'grassy, herbaceous quality, reminiscent of green pepper or green olive', but also to 'gorgeous, smoky, cedar scents ... after two years of oak ageing', in Cabernet Sauvignon wines from Bordeaux, though usually blended with some Merlot and Cabernet Franc grape wines. Cheap Chilean Cabernet Sauvignon-dominated wines are described as 'relatively ripe, cassis-sweet and glycerine rich', whilst New World Cabernets are 'bigger, bolder and jammier than their European equivalents'. In particular, Californian Cabernet reeks of 'ripe, minty, fruit', and 'Australian versions all have eucalyptus well to the fore', along with 'a burly, peppery quality from the Shiraz (grape) with which they are often blended'.

5.5.4 Variants of Chardonnay wine flavour

Chardonnay is another internationally planted grape variety, originally associated primarily with the Burgundy area. Its most quoted aroma/flavour characteristics are given in Table 5.6 where it can be seen to have a number of fruity characteristics (apple, peach, melon). Many Chardonnay wines are barrel aged, so that other additional characteristics can be perceived, e.g. vanilla, together with 'buttery' as a descriptor, also 'toasted', 'nutty' from the oak lactones. Again, overall flavour characteristics do depend upon where the vine is grown and how the grapes are vinified. The grape is regarded as 'amiable' since the vines are easy to grow. Both Clarke (1989) and Robinson (1995) emphasize the flavour significance of ageing in oak barrels where practiced, together with fermentation also in oak barrels. The Chablis, Côte d'Or, especially Côte de Beaune (e.g. Meursault, Montrachet) areas in France represent the peaks of vinifying performance and are thus described as 'luscious, creamy honeyed yet totally dry, the rich ripe fruit intertwined with the scents of new oak' of Côte d'Or wines, and the 'gentle, nutty richness, steely acidity' of Chablis (Clarke, 1989). Máconnais Chardonnay wines are claimed to be somewhat different, being produced in stainless steel tanks, with little ageing and the wines are meant to be drunk young, 'having a slightly appley flavour, as well as something fat and yeasty' (Clarke, 1989).

McQuitty (1997 and 1998) distinguishes between Chardonnays from the more northerly and southerly regions of the Burgundy area. Thus wines from the areas in the cool northern climes are described as:

> ... distinctly herbaceous, leafy in style, green apple/lemon in the mix; Chablis, vegetal richness has an almost cheesy aspect, underpinned by a steely lemon acidity due to its heavy limestone soil.
>
> McQuitty (1997)

Further south, Chardonnay produces 'those ethereal, creamy, buttery peach and pineapple scents. With judicious oaking, these wines yield a fine, nutty quality'. Robinson (1995) refers to Chablis as tasting of 'wet stones, with some suggestion of very green fruit', and ...'after eight years in-bottle, they can develop much more complicated, often deliciously honey-like flavours'. The Côte d'Or wines she describes as 'hazelnuts, liquorice, butter and spice'.

Nowadays, Chardonnay is planted in vineyards all over the world. Chardonnay wines are produced in a number of other areas in France, e.g. the Loire and Languedoc (as a Vin de Pays). Chardonnay grapes have produced very successful wines in California, e.g. 'positive fruit salad of flavours with figs, melon, peaches and lychées all fighting for prominence, and some buttery oak to round it out'. 'Best are drier, steelier, with a slightly smoky, toasty taste from the partially charred oak barrels they mature in'. Australian Chardonnays have conquered the market and are described as having 'a simple fruitiness, perked up with (usually added) acid and often oak chips'. 'Not necessarily good for keeping inbottle, unless sufficiently acid, leaving a lemon-peel/citrus flavour'. New Zealand Chardonnay wines reflect the cooler climate and thus more acid, with one example described as 'a herby base with sweet spicy, hazelnut-scented oak' (McQuitty, 1998). Chardonnay has been planted in Italy (Alto Adige, Friuli) to give wines with a 'characteristic cream-like taste with a little nutmeg'.

5.5.5 Flavour description of some other commercial wines

At times, the flavour descriptions of certain specific commercial wines by wine writers, particularly during the 1980s and 1990s, have been quite disparaging. It has to be recognized also that certain wine types become popular for a period (to the UK consumer) and then become unfashionable. Quality and flavour descriptions are bound to be subjective. Certain wine types and vintages can clearly be hyped up for sound commercial reasons, as already described. Nevertheless, opinions that have been expressed may be of interest to the wine chemist.

For example the Liebfraumilch and Blue Nun brand from the Rheinhessen and Pfalz regions of Germany, both essentially branded wines produced in large quantity (for export only), were particularly subject to adverse criticism. Thus, Clarke (1986) wrote 'should be designated just "tafel wein", that is, low in alcohol, low in acid, from grapes which are not ripe, sugared up (before fermentation)'.

Blue Nun was re-launched in 1997, but McQuitty (*The Times*, 15 November 1997) commented that it had a 'weird, spicy scent and very thin palate. Still a lot of sugar there'. Jancis Robinson refers to Liebfraumilch as 'almost any medium dry, vaguely aromatic blend can qualify; Niersteiner Gutes Domtal made up of over-priced Müller-Thurgau grown miles away'. Goolden (1980) refers to Liebfraumilch with its 'tartness, hints of stale milk and sulphur'.

Other branded wines similarly disparaged by McQuitty (1997) were the French Piat d'Or (at one time heavily TV advertised with amusing commercials); and Vin du Pays d'Oc (white) described as 'oxidised and sulphury'. The grape juice from which 'this wine has been made must have been left open to the air. Very acid.' And the red. 'nasty perfume'. Mateus Rosé, the once very popular Portuguese wine, did not do much better, being described as 'gross, sweet with a jammy, dirty flavour, as if it had been made with the sediment of old Victoria plums'. Goolden (1980) refers to its former popularity but gives a reasonable verdict. Mouton Cadet (red 1995 vintage) was described (McQuitty, 1997) as 'dry, starkly grassy, thin and unpleasant', and the white 'sulphurous, rubbery, disagreeable'. McQuitty (1997–98) also wrote some negative comments about some Paul Masson Californian wines and some wines from Jacob's Creek, Australia.

Underlying these comments is the belief that good quality means the absence of any mouth-cloying and sickly characteristics, as is imparted by sugar and certain milk flavours but there should be some slight bitterness and acidity present. The concept of dryness is very powerful also with Martinis, Sherries and some other drinks, especially those normally to be drunk as aperitifs and along with savoury meals. Sweet wines are only considered acceptable after the sweet has been served.

Wine writers are bound to be accused of non-scientific assessments. However, their rôle is not intended to be scientific. As described above, wine writers have a lot of knowledge about the taste of wines and generally assess the wines in the light of their experience and expectation, and they do not usually hide their personal opinion of the wine. However, some of the science of taste and smell does create misunderstandings, although these also occur within the scientific community (see the earlier discussion on the tongue map).

The apparent misunderstanding of the rôle of the tongue in tasting is not, however, unusual. Goolden (1980) referring to a Muscadet: 'Get past the nose (sulphur) and into the *taste*, and in a good version (generally "*sur lie*") you can pick up the faint waft of flowers. There's a definite citrussy edge blended with the flavour of unripe melon'. Again, Clarke:

> Interrogate your *taste* buds more thoroughly and they give you the flavour of hay. But the Muscadet tradition is to leave the new wine undisturbed on the lees until bottling. This does two things for the taste; the wine picks up yeasty, salty flavours, and keeps its fresh prickly taste, because some of the CO_2 from fermentation is delayed.
>
> (Clarke, 1986)

Clarke & Rand (2001), in the glossary to their valuable encyclopaedic text on grape varieties, somewhat unhelpfully define: 'Flavour or aroma compounds – substances in wine that can be smelled or tasted'. Use, however, has occasionally been made of Jackson (1994, 2008) with reference to Sotolon in Vin Jaune but it is not clear what 'glucosyl-glucose compounds' are. Similarly, Robinson (1995) makes erudite technical reference to 2-methoxy-3-isobutyl pyrazine, but surprisingly refers to 'fruit' as the youthful combination of flavour (aroma) and body coming from the grapes rather than wine-making or ageing and 'esters' formed during fermentation or ageing, often intensely aromatic ('... nail polish remover smells strongly esterified'). In fact, it is the other way round, 'fruity' is from largely non-aromatic alkyl esters, whilst aromatic fruity or 'grapey' comes from terpenes in the grapes. Descriptions of flavour can lack scientific clarity. Thus (Clarke & Rand, 2001) 'oaked Australian Sémillon is different, being picked riper, has richer fruit flavours of greengages, apricots and mangoes, all mixed with the custardy vanilla of the oak'; though presumably unoaked Australian would have the same fruit flavours (secondary) and oaking is only adding vanilla (tertiary aroma).

5.5.6 Off-odours and taints

A number of olfactory sensations in wines are regarded as undesirable when they occur, as indicated in Table 5.7. An off-odour is an atypical flavour often associated with deterioration or transformation of the product (British Standard, 1992), for example, an excess of acetic acid caused by wine spoilage. A taint is a taste or odour foreign to the product (British Standard, 1992), for example the contamination of the wine with the musty odour attributed to 2,4,6-trichloroanisole, a compound associated with faulty corks. The quality of wine in recent decades has improved and faulty wines with off-odours or taints are relatively uncommon. In part, scientific research has given a much better understanding of the causes and formation of off-odours, and the industrial response of improved grape and wine handling (avoiding oxidation, improved fermentation, scrupulous hygiene, etc.) has eradicated many wine faults.

However, changes in winemaking culture as a result of changes in consumer preference for the required style of wine has lead to an increase in some off-flavour formation, for example volatile ethyl phenols. Wines can also suffer from unsightly hazes but again scientific understanding of these faults has led to hazy wines being virtually a problem of the past. Some of the more common taint problems are briefly discussed below, and for further information on taints and measures to prevent taint as much as possible refer to the review of Sefton & Simpson (2005), and Ribéreau-Gayon *et al.* (2006).

Cork taint

Cork taint is a serious economic problem, which can also have a negative effect on the reputation of a producer. It is difficult to ascertain the size of the problem, although estimates of affected bottles vary from 1-5%, (see Sefton & Simpson, 2005). Despite the developments in instrumental analytical methods, cork taint remains difficult to analyse and even the sensory assessment is not always easy, especially in wines that have been aged in wood.

Cork taint odour is generally attributed to 2,4,6-trichloroanisole, a stable chemical compound in wine. Other anisoles that give rise to cork taint odours include 2,3,4,6-tetrachloroanisole, 2,4,6-tribromoanisole and pentachloro-anisole (Grainger, 2009). The incidence of bromoanisole taint in wines appears to be increasing. However, the identification of the musty smell may be a cause of confusion and occasionally a musty smelling wine is wrongly attributed to a cork problem. For example further investigation shows that the musty smell is due to 1-octen-3-one (mushroom odour), which can be present in wine from grapes affected by powdery mildew, or the musty smell may be geosmin (earthy aroma), also attributed to mouldy grapes. It is estimated that about 15% of wines labeled with cork taint have musty smells due to another cause (see the review from Sefton & Simpson, 2005). Strong earthy aromas have also been detected in wine resulting due to 2-methoxy-3-*iso*-propyl-pyrazine, derived from mouldy barrels.

Cork taint is transferred from faulty corks into the wine after bottling, although incidences of this taint have been reported that could not be traced back to corks. Much work has been done to try and eradicate this cork problem, however, the musty smell as a result of a faulty cork remains a problem, in part because the sensory threshold is very low (Land, 1989). Land quotes that 50% of the population can detect 1mg of 2,4,6-trichloroanisole in 5 million gallons of water. Interestingly, ethanol has a significant effect on threshold values, in water Land quoted 3×10^{-8} ppm ($0.03\,\mathrm{mg\,L^{-1}}$) whereas in wine of about 11% ethanol the threshold is 1×10^{-5} ppm, a 333-fold increase in concentration.

Thresholds for detection are reported to be between 1.4 and $4.6\,\mathrm{ng\,L^{-1}}$, whilst recognition concentrations tend to be a little higher (see the review by Sefton & Simpson, 2005). Consumer rejection thresholds reported are

Table 5.7 Off-odours in wine.

Name of off-odour (taint or defect)	Chemical compound associated	Possible cause and formation	Threshold levels in wine
Oxidized (rough, chemical, Sherry-like, flat)	Acetaldehyde	Excess formation from vinification or during maturation under oxidative conditions	Only in free state above, say, 100 mg L^{-1} combined
Acescence (turning sour), odour (vinegary), taste (sour)	Acetic acid	Excess, by spoilage of aerobic growth of acetic acid bacteria, especially during ageing process in barrels.	Volatile acidity 0.6–0.9 g L^{-1} (H$_2$SO$_4$) equivalent or 0.72–1.1 g L^{-1} (acetic acid) white wines up to 1.2 for red wines
	Ethyl acetate	Reaction of excess acetic acid with ethanol.	0.15 g L^{-1} odour
Irritating burnt odour	Sulfur dioxide	Excess/improper addition	See Chapter 3
Sickly (buttery as a negative)	Di-acetyl	Excess above threshold from vinification/ageing	>3 mg L^{-1}.
Phenol taint	Vinyl-4-phenol	Vinification procedures.	720 μg L^{-1} for a 1/1 mixture in white wines.
	Ethyl-4-phenol	*Brettonomyces/Dekkera* Yeast activity	420 μg L^{-1} for a 10/1 in red wines
Geranium-like (penetrating)	2-Ethyl-hexa-3,5-diene, derived from 2,4-hexa-di-ol	Compound formed from sorbic acid, if present, by lactic acid bacteria	Highly odoriferous
Cork taint	Mainly 2,4,6-trichloroanisole	From corks containing moulds, sterilized by chloro compounds	4–10 μg L^{-1}
	2-methoxy-3,5-dimethylpyrazine	fungal must	2 ng L$^{-1 a}$
Reduction odour (rotten) eggs	H$_2$S	Vinification conditions, formed from yeasts	<16 μg L^{-1}

Aroma	Compound	Origin	Threshold
- onions - cooked cabbage - asparagus Herbaceous	mercaptans (ethyl/methyl) methionol dimethyl disulfide Leaf aldehydes and aldehydes	Oxidation of grape lipids on crushing – usually in unripe grapes. Excess amounts.	$<10\,\mu g\,L^{-1}$ $<1500\,\mu g\,L^{-1}$ $<2.5\,\mu g\,L^{-1}$
Bitter almond	Benzaldehyde	Derived from benzyl alcohol. Excess formation in resin-lined vats.	$<20\,mg\,L^{-1}$
Mousiness (disagreeable, reminiscent of mice)	Acetamide 2-acetyl-tetrahydropyridine 2-ethyltetrahydropyridine 2-acetylpyrroline	Action by *Brettanomyces* and *Lactobacillus*	Highly odoriferous, humans vary greatly in sensitivity (see text)

[a]Quoted by Sefton & Simpson (2005).

between 3.1 and 3.7 ng L^{-1}. However, this appears to depend on the methodology used and the experience of the consumer, and individual thresholds for experienced panelists have been reported to vary by a factor of 100, whilst inexperienced panelists varied by a factor of 10 000. One can only assume that human sensitivity for this compound varies and training will help with the recognition of the taint.

The presence of 2,4,6-trichloroanisole leaves the wine dull, devoid of fruity smells, often with a perceptible nasty chemically musty smell. Instrumentally, it is not easy to pick up the low concentrations of this compound able to spoil the flavour of the wine.

Recently different methoxypyrazines have been reported as taint compounds in wines derived from cork. This taint has been described as musty, earthy or mouldy. The taint compounds have been identified as 2-methoxy-3,5-dimethylpyrazine and 2-methoxy-3-isopropylpyrazine. (see the review by Sefton & Simpson, 2005). Sensory descriptors reported are fungal must, aldehydic, coffee, acrid and nutty taint. In water the sensory threshold has been determined as earthy at 0.4 ng L^{-1}. There are not much quantitative data on these musty taints in wine, presumably it is not easy to analyse and identify this taint which the human nose can detect at very low concentrations.

Pratt *et al.* (2009) reported a study on mousiness in wine using DNA fingerprinting techniques, in order to match the taint with the microbiological organism responsible for the taint. They found strong links between the microbial composition and some of the most common taints in wine, trichloroanisole was related to *Penicillium variabile* in most samples, but the accumulation of 2-methoxy-3,5-dimethylpyrazine in cork was complex and it is assumed that a number of micro-organisms are involved. These authors also discussed the possible degradation of taint compounds by micro-organisms, which may ever help to reduce the risk of wines becoming contaminated with these chemically stable taints.

Mousiness

A relatively rare but very unpleasant off-flavour is mousiness in wine. It has been described as a 'peculiarly disagreeable flavour in wine, which is closely resembling the smell of a residence of mice' as quoted in a recent review on mousy off-flavour (Snowdon *et al.*, 2006); some of their main points are summarized later in this section. Currently there is no method to remove the mousy off-flavour from the unpalatable wine. Its sensory perception is usually delayed on the palate and it cannot be perceived by sniffing, however, once perceived it can be very persistent. Human sensitivity varies and is thought to be genetically determined and ranges from very sensitive to anosmic. This variability must make it more difficult to deal with this off-flavour in the wine industry and wine appreciating circles. Since the off-flavour is very persistent and is reported to last up to ten minutes, various other ways for testing have been developed. One common one is to rub some wine on the back of the hand and sniff close to the skin. This method can easily be used by the amateur wine drinker.

Three chemical compounds have been identified to give a mousy off-flavour to wine: 2-ethylhydropyridine, 2-acetylhydropyridine and 2-acetylpyrroline. Contaminated wines tend to contain more than one of these compounds, although there is no scientific information regarding the individual contribution to mousiness of these compounds. The formation of these mousy off-flavours is under microbiological control and can be formed in wines infected with lactic acid bacteria or *Dekkera/Brettanomyces* yeast. The conditions leading to the formation of mousy off-flavour are not yet understood.

Ethylphenols

Raised levels of 4-ethylphenols in wines are associated with undesirable aromas describes as phenolic, leather, horse sweat, stable, varnish, etc. This problem, reviewed by Suarez *et al.* (2007) has become more prominent in aged red wines, in particular due to the current trend requiring ageing in wooden barrels, since it appears that certain organisms can thrive in old wood. Wines are subjected to ageing in wood without any physical clarification processes, which is thought to enhance the quality of the wine, but this also poses an increased risk of spoilage by slow growing yeasts such as *Brettanomyces bruxellensis*, *Brettanomyces anomalus* and *Saccharomyces baiili*, as well as some lactic acid bacteria, all capable of producing undesirable off-flavours. These organisms appear to be able to grow in the presence of ethanol and require minimal nutrients.

The sensory threshold for 4-ethylphenol is low (230 µg L^{-1}, quoted by Suarez *et al.*, 2007) and typical sensory descriptors are phenolic, leather, horse sweat, stable and varnish. The related 4-ethylguiacol gives is typically described as smoked or bacon and has a lower sensory threshold (47 µg L^{-1}, quoted by Suarez *et al.*, 2007), however, its negative impact on wines is thought to be less. Wines identified as having a high, medium and low typical character due to the growth of *Brettanomyces* have been reported to contain 3.0, 1.74 and 0.68 mg L^{-1} 4-ethylphenol. Variations in sensory threshold have been reported to depend on the type of wine. For further description of formation see Section 7.4.9.

5.6 Wine and food flavour

When wines are consumed with food, wine and food flavours are intimately mixed in a lunch or dinner and usually aligned with specific courses in the menu. Many food writers like to recommend also the wines that should accompany specific recipes of food. There have always been traditional guiding principles in this choice, in which the most obvious is the recommended drinking of full-bodied red wines with roasted red meat. Similarly, dry white wines should accompany white meats and fish; whilst sweeter wines should be provided with the dessert. In fact, the French preference for eating the cheese and biscuits before the 'pudding' lies in the idea of a suitable food to accompany remaining (red) wine not consumed during the main course. When only one wine is used during the meal, modern-day choice often seems to be red wine in preference to white, whatever the food.

These principles can be summarized by the concept of similarity in strength of flavour, so that a strong flavoured wine should accompany strong flavoured food and not be allowed to overwhelm a delicate tasting dish. More sophisticated relationships between food and drink, including non-flavour factors such as political ones, are described by food and wine writers and their chefs or sommeliers, in particular Simon (1946) and Johnson (1979). They feature in menus of public and State occasions, as exampled below:

State Banquet at Buckingham Palace in honour of the French President 21 March 1939

The Food	The Wines
Consommé Quennelles aux trois couleurs	Sherry, 1865
(soup of fish balls with three colours)	Madeira sercial, 1834
Filet de truite saumonée roi Georges VI	Piesporter Goldtropfchen, 1924 (Middle Mosel, Riesling) Deidesheimer Kieselberg, 1921 (Rheinpfalz, dry white wine)
Rouennais à la gelée Reine Elizabeth; Garniture Buzanay Mignonette d'Agneau Royale; Petits pois à la française; Pommes nouvelles, rissolées au beurre Poussin Mercy-le-haut; Salade Elysée Asperges vertes; Sauce maltais Bombe L'Entente Cordiale; Corbeille Lorraine Cassolette Basillac	Perrier-Jouet, 1919 (Champagne) Château Haut-Brion, 1904 (Red, Graves) Château Yquem, 1921 (Sauternes, sweet white) Port (Royal Tawny, 1912) and Brandy (1815) (Adapted from Simon, 1946)

Of interest here is the expected appearance of French wines from prestigious vineyards and their age, i.e. some 35 and 18 years respectively. The German vineyards from which wines featured in this banquet are equally prestigious to this day, see also Section 6.3.2 on Tawny Port.

Modern recommendations for drinking particular wines with certain foods have been made characteristically by Johnson (1979) in great detail, a small selection of which is given in Table 5.8.

Popular wine books, web sites and wine retail outlets, even sometimes the back label of the bottle will all give information regarding what wines to

Table 5.8 Modern recommendations for wine and food.

First Course	
Asparagus	– white Burgundy or Chardonnay. Tavel Rosé
Consommé	– medium dry Sherry, Madeira
Croque Monsieur	– young red, e.g. Beaujolais
Fish	
Fish fingers (with tomato sauce)	– white Burgundy, Chassagne-Montrachet
Coquilles St. Jacques	– white German wines with cream sauces, otherwise Gewürztraminer or Hermitage Blanc
Plaice (fried or grilled)	– white Burgundy
Meat	
Roast beef	– fine red wine
Duck	– rather sweet Rhine wine (fat counter balance)
Steak (beef)	
Filet	– red of any kind
T-bone	– Barolo, Hermitage, Australian Cabernet
Ham	– fairly young red Burgundy
Cheese	
English (strong, acid)	– Ruby, Tawny or Vintage Port
	– red wine, e.g. Châteauneuf-du-Pape
Desserts	
Baked Alaska	– Sweet Champagne or Asti Spumante
Christmas Pudding	
Fruit salad	– None
Strawberries and cream	– Sauternes or Vouvray

Adapted from Johnson (1979).

consume with selected foods, usually based on the general principles discussed earlier, or occasionally the image the wine company likes to present about its wines. In sensory research there is a great awareness regarding interaction of the wine samples, the time required between samples tasting for assessors to judge all samples under the same conditions and avoidance of a tired palate. A recent research paper researched consumer preference for Shiraz wines and how it was influenced by tasting different cheddar cheese samples just before wine tasting (Bastian *et al.*, 2010). They reported that eating Cheddar cheese just before wine tasting reduced the flavour length and astringency of the wine, whilst the tannins were perceived as silkier. Possibly, this can be explained by the residual fat from the cheese in the mouth interacting with the wine volatiles and phenols. The most preferred wine and cheese combinations tended to contain the wine and cheese samples getting the highest quality and 'like' scores. In all pairs the wines had a stronger flavour than the accompanying cheese.

5.7 Aroma indices and statistical methods

Aroma indices described in the literature are essentially based on the ratio shown in Equation 5.2, which can be determined for each constituent. Such indices have also been described as Odour Activity Values in studies on roast coffee brews (Grosch, 2001), whilst so-called Flavour Units have been used in beer flavour studies (Meilgaard, 1982; Meilgaard & Peppard, 1986, in greater detail).

$$\text{Amount of constituent (say in ppm) / Threshold level of constituent (say also in ppm)} \quad (5.2)$$

As they are based upon perceived intensities, they are subject to Stevens' Law (Chapter 4, at the end of Section 4.1.3) so that there are theoretical shortcomings to the application of these concepts. The difficulty associated with trying to interpret flavour units may be the reason why few scientific papers have been published using flavour units. Considerable careful work is needed to establish reliable threshold levels of all the constituent compounds.

5.7.1 Flavour unit concept

As used by Meilgaard, Equation 5.2 is expressed in terms of a difference threshold (as opposed to absolute values) as described in Chapter 4, which depend upon the amount of the same compound already present in the beer.

In determining threshold levels in practice, Meilgaard & Peppard (1986) stress a number of important points: (1) due to the wide variation in sensitivity of perception amongst individual people, group-averaged data (i.e. from trained panels of some 12–15 members) is preferred; (2) due to the possible aroma perception effect of even very small amounts of some contaminant compounds (e.g. ppt) it is essential with most reference flavour substances (e.g. butadione) to prepare them in a very highly purified form, involving multiple distillations and absorption techniques, before use in tasting.

Meilgaard & Peppard (1986) recognized the limitations imposed by Stevens' equation but quoting from the findings of Teghsoonians (1973), in the restricted range of FU to 0.5–5.0 units in beer samples, n is around 1. The errors will then rarely exceed about ±20%, except in the case of ethanol, which causes direct solvents in the mouth.

Meilgaard (1982) regarded this concept and the resulting technique as useful in producing or giving best estimates of the flavour effect of changing the composition of beers in respect of particular constituents by changing relevant brewing practices.

The flavour significance of a particular substance in a given beer is assessed by knowledge of its threshold value and flavour unit in a so-called null beer, which had a composition not too different from that of the sample under test. The composition of the sample beer has to be known for all the compounds of flavour (odour/taste) interest and, of course, that of the null beer. This technique is used primarily for the prediction of flavour differences, from composition data.

For many substances, threshold values are known for a given null beer, as determined by numerous co-operative studies organized by the American Brewing Institute; thus purified isoamyl acetate (estery/fruity – banana) has a standardized difference threshold of 0.5–$1.7\,mg\,L^{-1}$ in a beer already containing 1–$3\,mg\,L^{-1}$. In a particular test situation, isoamyl acetate (actually a 25:75 mixture of 2-methyl-butyl and 3-methyl-butyl) was shown to have a threshold of $1.2\,mg\,L^{-1}$ in a null beer already containing $1.8\,mg\,L^{-1}$. In a sample beer containing $2.4\,mg\,L^{-1}$, it can then be calculated that sample beer had $2.4/1.2 = 2$ FU of this component, whilst the null beer has $1.8/1.2 = 1.5$ FU. The difference between the two is 0.5 FU, so that the sample beer was considered not likely to show any additional 'banana-like ester' impact over that in the null sample. There is also a small amount of a 2-methylpropyl acetate ester (with similar odour characteristic), in each (about $0.1\,mg\,L^{-1}$) but inclusion of this figure would not alter the conclusion. If the 'banana' esters content is added to the fusel oil content (i.e. C_3–C_5 alcohols + 2-phenylethanol) of each, there is, however, now a marked difference of 1 FU between the sample and null beer.

In relating flavour-to-flavour units, Meilgaard (1982) set out the following groups:

(1) Primary flavour constituents ('primary' used here in the sense of effective intensity, not origin). These are present at above 2 FU, so that removal of any such constituent in a given beverage would produce a decisive change of flavour.
(2) Secondary flavour constituents are those that are present between 0.5 and 2 FU; removal of any one constituent from this group would produce a minor change in flavour.
(3) Tertiary flavour constituents occur at levels of 0.1–0.5 FU where they cannot be perceived individually, so that removal of any one would not produce a perceptible flavour change.
(4) Background flavour constituents are those that are present below 0.1 FU, often the majority of constituents present.

In using this technique, Meilgaard (1982) also considered the question of interactions between different constituents in respect of different threshold values, by introducing an interaction factor (d_{int}) where combinations of individual constituents were being examined. Where flavour effects are additive, $d_{int} = 1$, but where synergism exists, d_{int} will be >1.00; and values less than 1.00 are possible. In general, he found that where constituents had similar flavours, such as the banana esters and the apple esters (ethyl hexanoate and ethyl octanoate) there was little apparent synergism. For octanoic acid and ethyl acetate, with such distinctive flavours, suppression was occurring (i.e. $d_{int} = 0.54$).

Only three components in beers, hop bitters, ethanol and carbon dioxide, were considered likely to produce FU values of about 2.0 in general. Off-flavour substances when present will come into this category. It is evident, however, that if we would have 20 similarly flavoured but different chemical compounds (thus 'fruity esters'), each with an individual level near 0.1 FU, their overall FU would be 2 or more.

Meilgaard & Peppard (1986) recognized the rough-and-ready nature of the calculations involved and stated that it could not be expected to cope with the more complex aspects of flavour, such as the effect of changing yeast strain or of oxidation, which affect many compounds and groups of compounds simultaneously. This work is of potential interest in the study of wine flavour, where the composition of the various groups of substances described in Chapter 4 shows that very few individual aroma compounds are likely to show high flavour unit values.

5.7.2 Odour activity unit

Grosch (2001) showed some very high ratios of the Aroma Index, which he called Odour Activity Value, based upon threshold values on an absolute basis in pure water.

The concentrations of aroma substances in coffee brews are often much higher than in wines, e.g. thus furaneol at 7.2 mg L^{-1} and 3-methylbutanal at 0.57. Some very high individual OAVs have been recorded, particularly amongst sulfur compounds, like 3-methyl-2-butan-1-thiol at 2000-fold. As in wine, β-damascenone is a recently discovered component at a low content level, though at some 2130 times the determined threshold in water. These two substances will clearly play an important rôle in the flavour of the brew. OAVs were not, however, used directly to determine which volatile compounds were determining flavour.

Grosch (2001) has described laboratory procedures to determine the really significant contributors in coffee brews. Roasted coffee has even more known volatile compounds than wines, over 800, including some 80 pyrazines generated by roasting. There is perhaps a lesser complexity, in that attention can be primarily directed to one roasting level (medium), two species, some four varieties/cultivars and two methods of green bean preparation, rather than the wide range of wines in the market place. Grosch and his colleagues (2001) examined distillates (by the Likens–Nickerson method, using ethyl ether or a low temperature extractant) and determined that there were only some 28 really important odoriferous substances present. They then prepared model aroma distillates and compared the aroma intensity with an actual coffee distillate (from medium roasted Colombian arabica coffee) by flavour panel testing. Omission experiments (i.e. by omitting one or more components from the model) together with other techniques enabled fine honing into a model aroma closely comparable with an authentic coffee aroma (both from a brew and the roasted coffee itself). Quantitative data on the content of each of these 28 compounds, together with their odour threshold values in water, was presented. All these compounds were tabulated according to six of the odour groups described earlier in this chapter. It is of interest that a number of these compounds are also the ones described as present in wines – the various furanones, alkanals, and even isobutyl-methoxy-pyrazine. Like wines, some of these compounds come through from the raw bean or grape but in coffee most are generated by the roasting process and in wine by the fermentation.

Interestingly, the earlier but reliable quantitative GC-MS work of Silwar *et al.* (1987) on roasted coffee volatile compounds showed that about 800 mg of coffee distillate was obtainable from 1 kg of a medium roast arabica coffee. This distillate was made up of some 38–45% of furans of all kinds (including sulfur derivatives), 25–30% pyrazines and only 3–5% aliphatic compounds, in the many hundreds of separately identified compounds; but no connection was attempted as to any relationship with odour characteristics or activity values. Grosch (2001) in his investigations shows some 500 mg distillate, containing 28 different compounds, screened out as being the important odorants for roast coffee, of which certainly 28% by weight can be seen to be furans of different kinds (including furanones and thiols). As already mentioned, most of these compounds had OAVs determined in an actual brew from the coffee of well over ten, and several compounds had very high OAVs. Grosch used absolute values in pure water for thresholds; no difference values in coffee brews appear available. Compounds such as the many thiazoles and oxazoles were not seen to be important, as also all of the simpler pyrazines, identified as being present and thought very important in many earlier studies.

A review by Francis & Newton (2005) considered recent publications with reliable identifications of volatile aroma compounds in wine, including quantification and threshold values. Based on this literature survey they compiled a list with compounds in wines all with OAV greater than one, consisting of the compounds listed in Table 5.9.

Of course, interactions between compounds in wines as well as inherent difficulties in determining accurate threshold values will influence these data. Only a subset of compounds may be present in wines but these compounds may typically contribute to the aroma of a wine.

Beverages, including wines and spirits, have been happy hunting grounds for chemists wishing to flex their analytical muscle to the full, whereas the odour significance of many of these compounds has been found subsequently to be minimal. No comparable work in narrowing down the really significant contributors to the flavour of each important wine, such as those from Cabernet Sauvignon and Chardonnay grapes, appears to have been published.

5.7.3 Multivariate and other statistical procedures

As has been noted in Section 5.1, the number of variables involved in assessing the flavour of wine can be very large, both in the elements describing the flavour of a wine and in the contributing factors to that flavour from grape variety, growing location and vinification practices. By the use of so-called multivariable statistical procedures and techniques, connections can be made between such sets of variables, see the review by Noble & Ebeler (2002).

Table 5.9   List of compounds in wines with an OAV greater than 1.

No.	Chemical groups	Compounds
7	Ethyl esters	Ethyl isobutyrate, Eethyl 2-methylbutyrate, ethyl isovalerate, ethylhexanoate, ethyl octanoate, ethyl decanoate
3	Acetates	Isoamyl acetate, phenylethyl acetate, ethyl acetate
2	Cinnamic esters	Ethyl dihydrocinnamate, *trans*-ethyl cinnamate
8	Acids	Isobutyric acid, isovaleric acid, acetic acid, butyric acid, propanoic acid, hexanoic acid, octanoic acid, decanoic acid
6	Alcohols	Isobutanol, isoamyl aclohol, 2-phenylethyl alcohol, methionol, 1-hexanol, (*Z*)-3-hexenol
4	Monoterpenes	Linalool, geraniol, *cis*-rose oxide, wine lactone
6	Phenols	Guiacol,4-ethylguiacol, eugenol, 4-vinylguiacol, 4-ethyl phenol, vannillin
7	Lactones	*cis*-Oak lactone, γ-nonalactone, γ-decalactone, γ-dodecalactone, 4-hydroxy-2,5-dimethyl-3-(2H)-furanone, (*Z*)-6-dodecanoic acid-γ-lactone, sotolon
2	Norisoprenoids	β-Damascenone, β-ionone
7	Sulfur compounds	3-Mercaptohexyl acetate, 4-mercaptomethyl pentan-2-one, 3-mercaptohexanol, 2-methyl-3-furanthiol, 3-methyl thio-1-propanol, benzenemethanethiol, dimethyl sulfide
6	Other	2,3-Butanedione, acetoin, 3-isobutyl-2-methoxypyrazine, acetaldehyde, phenyl acetaldehyde, 1,1-dioxyethane

Francis & Newton (2005).

These procedures have been used in the last few decades for a number of different beverages. There have recently been a number related to wines; for example, Vernin *et al.* (1993) have used statistical techniques (PCA) to classify Bandol Mourvèdre wines from various vintage years (1986–1988) on the basis of GC data from head-space analysis of some 40 different aroma compounds. In Vernin's work, the marked significance of such compounds as ethyl propanoate, isobutyl acetate, the methyl butanols, ethyl octanoate, diethyl succinnate is of interest. Multiple regression analysis is frequently the method of choice for prediction of specific flavour notes from measurements of volatile compound content. Such a technique has been described for the volatile compounds in tea by Togoni (1998); thus, the intensity of a 'sweet, floral note' could be predicted from compound concentrations,

$$= -0.0591[\text{pentanal}] - 0.671[\text{2-heptanone}] + 0.562[\text{linalool}]$$

$$+ 0.693[\text{2-phenyl ethanol}] + 0.0713[\text{Jasmine lactone}] - 0.134$$

for which reference to the original publication is necessary.

The use of statistics linking sensory data from trained taste panels and chemical analytical data from gas chromatography olfactometry are now increasingly linked together using statistical methods such regression analysis, partial least squares or principal components analysis trying to

Table 5.10 Compounds contributing significantly to quality of 25 Spanish red wines[a].

Compounds contributing fruity or sweet characteristics	Off-flavours contributing negatively to quality	Compounds which possibly suppress the fruity character of wines
Propyl acetate	3,5-Dimethyl-2- methoxypyrazine	Metionol
2,3-Butanedione	4-Ethylphenol	Metional
Isobutyl acetate	2,4,6-Trichloroanisol	(Z)-2-Nonenal
Ethyl butyrate	3-Ethylphenol	(E,E)-2,4-Decadienal
Ethyl 2-methylbutyrate	4-Ethylguiacol	3-Isopropyl-2-methoxypyrazine
2,3-Pentanedione	o-Cresol	Acetic acid
Ethyl 3-methylbutyrate		2-Methylisobomeol
Isoamyl acetate		
Ethyl 2-methylpentanoate		
Ethyl 3-methylpentanoate		
Ethyl 4-methylpentanoate		
Ethyl hexanoate		
β-Damascenone		
2,5-Dimethyl-4-hydroxy-3(2H)-furanone (Furaneol)		

[a]Information from Ferreira *et al.* (2009).

model the data in order to explain the the differences between the wines in chemical terms. Often the end result of the analyses is between two and four axis on which the data can be pictured showing the largest amount of variation between the samples. A predictive model can be the end result. Usually teams of scientists with an array of different skills need to work together, to ensure accurate chemical and sensory analytical data are collected on a well selected sample set. However, these methods have their limitations. The wines used for such an experiment need to have a wide range of sensory different properties and the resulting statistical model is only valid for wines with properties comparable to the ones in the original data set. However, without statistical methods it would be extremely difficult to draw any conclusions from such large and complicated data sets, and the application of statistical methods can lead to being able to unmask interesting information. Numerous studies have been done, usually confirming the significance of a relatively small number of volatile aroma compounds in their contribution to wine aroma of the sample set, some are briefly discussed by Francis & Newton (2005).

One such study for example (Ferreira *et al.*, 2009) used data obtained on 25 Spanish red wines analysed by descriptive analyses using a trained panel and data collected by quantitative gas chromatography olfactometry. Using correlation methods and partial least squares regression models, they found that the regression model containing three vectors explained 78% of the variation between the wines. Mainly 15 fruity and sweet smelling aroma compounds significantly contributed to the quality of this data set (Table 5.10).

The presence of any off-flavour had a significant negative effect on the overall quality. However, the presence of nine compounds with negative sensory attributes in very low concentrations also could significantly affect the perceived quality of a wine. The authors concluded that the presence of even very low concentrations of odorants with negative sensory attributes may strongly suppress the fruity characteristics of wines. Hence is may not just be the presence of positively contributing aroma volatiles, but the absence of any off-odours that exert important sensory quality attributes to wines.

Bibliography

Barr, A. (1988) *Wine Snobbery*. Faber and Faber, London.

Belitz, H.D. & Grosch, W. (1986) *Food Chemistry*, 2nd edn. Springer Verlag, Berlin.

Clarke, O. & Rand, M. (2001) *Grapes and Wines*, Webster, London.

Goolden, J. (1990) *Wine Tasting*, loc. cit.

Grainger, K. (2009) *Wine Quality Tasting and Selection*, Blackwell Publishing Ltd., Oxford.

Jackson, R.S. (2008) *Wine Science, Principles, Practice, Perception*. 2nd edn. Academic Press, San Diego.

Jackson, R.S. (2009) *Wine Tasting: A Professional Handbook*. 2nd edn. Academic Press, San Diego.

Labows, J.N. & Cagan, R.H. (1993) Complexity of flavour recognition and transduction. In: *Food Flavour and Safety; Molecular Analysis and Design* (eds. A.M. Spanier, H. Okai & M. Tamura), pp. 9-27. American Chemical Society, Washington.

O'Mahony, M. (1985) *Sensory Evaluation of Foods (Food Science and Technology 16)*. Marcel Dekker Inc., New York.

Rapp, A. & Marais, J. (1993) Changes in aroma substances during storage. In: *Shelf-Life Studies of Foods and Beverages* (ed. G. Charalambous), Developments in Food Science, 33, pp. 891-922. Elsevier, Amsterdam.

Ribéreau-Gayon, P., Glories, Y., Maujean, A. & Dubourdieu, D. (2006) *Handbook of Enology, Volume I, Wine and Wine Making*. John Wiley & Sons, Ltd, Chichester.

Robinson, J. (1993, 1995) *Jancis Robinson's Wine Course*. BBC Books, London.

Teranishi, R., Buttery, R.G., Stern, D.J. & Takeoka, G. (1991) Use of odour thresholds in aroma research. *Lebensmittel Wissenschaft und Technologie*, **24**, 1-5.

Vernin, G., Pascal-Mousselard, H., Metzger, J.L. & Parkonyi, C. (1993) Classification of wines originating from AOC bandol using multi-variate statistical methods. In: *Shelf-Life Studies of Foods and Beverages* (ed. G. Charalambous), pp. 975-990. Elsevier, Amsterdam.

References

Amoore, J.E. (1977) Specific anosmia and the concept of primary odours. *Chemical Senses and Flavour*, **2**, 267-281.

Barr, A. (1988) *Wine Snobbery*. Faber and Faber, London.

Bartoshuk, L. (1993) Genetic and pathological taste variation: what can we learn from animal models and human disease? In: *The Molecular Basis of Smell and Taste* (eds. D. Chadwick, J. March & J. Goode), pp. 251-267. John Wiley & Sons, Ltd, Chichester.

Bastian, S.E.P., Collins, C. & Johnson, T.E. (2010) Understanding consumer preferences for Shiraz wine and Cheddar cheese pairings. *Journal of Food Quality and preference* (IN PRESS)

British Standard Glossary of Terms relating to sensory analysis (1992). *BS.5098*, British Standards Institution, Chiswick, London.

Broadbent, M. (1979) *Wine Tasting*. Christies' Wine Publications, London.

Buck, L. & Axel, R. (1991) A novel multigene family may encode odorant receptors: A molecular basis for odor recognition. *Cell*, **65**, 175-187.

Clarke, O. (1986, 1989, 1998, 2001) *Wine Fact Finder*. Websters, London.

Clarke, O. (2003) *Introducing Wine*. Websters, London.

Clarke, O. & Rand, M. (2001) *Grapes and Wines*. Websters, London

Delwiche, J. (2004). The impact of perceptual interactions on perceived flavor. *Food Quality and Preference*, **15**, 137-146.

Ferreira, V., San Juan, F., Escudero, A., Cullere, L., Fernandez-Zurbano, P., Saenz-Navajas, M.P. & Cacho, J. (2009) Modeling quality of premium Spanish red wines from gas chromatography - Olfactometry data. *Journal of Agriculture and Food Chemistry*, **57**, 7490-7498.

Francis, I.L. & Newton, J.L. (2005) Determining wine aroma from compositional data. *Australian Journal of Grape and Wine Research*, **11**, 114-126.

Genovese, A., Piombino, P., Gambuti, A. & Moio, L. (2009) Simulation of retronasal aroma of white and red wine in a model mouth system. Investigating the influence of saliva on volatile compound concentrations. *Food Chemistry*, **114**, 100-107.

Grosch, W. (2001) Volatile compounds. In: *Coffee: Recent Developments* (eds. R.J. Clarke and O.G. Vitztham), pp. 68-89. Blackwell Publishing Ltd., Oxford.

Guti, H., Blehr, K. & Fritzler, R. (2001) Descriptors for structive-property correlation studies of odorants. In: *Aroma Active Compounds in Foods* (eds. M. Gundert & K.-H. Engel) ACS Symposis Series No. 794, pp. 93-108. American Chemical Society, Washington.

Heath, B. (1988) The physiology of flavour, taste and aroma perception. In: *Coffee Vol. 3 Physiology* (eds. R.J. Clarke & R. Macrae), pp. 141-170. Elsevier Applied Science, London.

Hughson, A.L. & Boakes, R.A. (2002) The knowing nose; the rôle of knowledge in wine expertise. *Journal of Food Quality and Preference*, **13**, 463-472.

Isci, B., Yildirim, H.K. & Altindisli, A. (2009) A review of the authentication of wine origin by molecular markers. *Journal of the Institute of Brewing*, **115**, 259-264.

Jackson, R.S. (2008) *Wine Science, Principles, Practice, Perception*. 2nd edn. Academic Press, San Diego.

Johnson, H. (1979) *Pocket Wine Book*, Mitchell Beazley, London.

Land, D.G. (1989) Taints - causes and prevention. In: *Distilled Beverage Flavour* (eds. J.R. Piggott & A. Paterson), pp. 17-32. Ellis Horwood, Ltd., Chichester.

Maruniak, J.A. & Mackay-Sim, A. (1985) The sense of smell. In: *Sensory Analysis of Foods* (ed. J.R. Piggott), pp. 23-57. Elsevier Applied Science, Barking.

McQuitty, J. (1997-98) Saturday articles on wine in *The Times* newspaper, London.

Meilgaard, M.C. (1982) Prediction of flavour differences between beers from their chemical composition. *Journal of Agricultural and Food Chemistry*, **30**, 1009-1017.

Meilgaard, M.C. & Peppard, T.L. (1986) The flavour of beer. In: *Developments in Food Science 33, Food Flavours, Part B, The Flavour of Beverages* (eds. I.D. Morton & A.J. Macleod), loc. cit., pp. 99-170.

Moulton, D.G. (1982) Sensory basis and perception of flavour. In: *Food Flavours, Part A, Introduction* (eds. I.D. Morton & A.J. Macleod), pp. 1-13. Elsevier, Amsterdam.

Noble, A.C., Williams, A.A. & Langron, S.P. (1983) Descriptive analysis and quality of Bordeaux wines. In: *Sensory Quality in Foods and Beverages* (eds. A.A. Williams & R.K. Atkin), pp. 324–334. Ellis Horwood, Ltd., Chichester.

Noble, A.C., Arnold, R.A., Buechsenstein, J., Leach, E.J., Schmidt, J.O. & Stern, P.M. (1987) Modification of a standardized system of wine aroma terminology. *American Journal of Enology and Viticulture*, **36**, 143–146.

Noble, A.C. & Ebeler, S.E. (2002) Use of multivariate statistics in understanding wine flavor. *Food Reviews International*, **18**, 1–21.

Palaskova, P., Herszage, J. & Ebeler, S.E. (2008) Wine flavor: chemistry in a glass. *Chemical Society Reviews*, **37**, 2478–2489.

Peynaud, E. (1986) *The Taste of Wine*. Macdonald, Orbis, London.

Piggott, J. (ed.) (1984) *Sensory Analysis of Foods*. Elsevier Applied Science, Barking.

Piggott, J. (ed.) (1986) *Statistical Procedures in Food Research*. Elsevier Applied Science, Barking.

Pratt, C., Ruiz-Rueda, O., Trias, R., Anticó, D.C., Capone, D., Sefton, M. & Bañeras, L. (2009) Molecular fingerprinting by PCR-denaturing gradient gel electrophoresis reveals differences in levels of microbial diversity for musty-earthy tainted corks. *Applied and Environmental Microbiology*, **75**(7), 1922–1931.

Ribéreau-Gayon, P. (1978) Wine flavour. In: *The Flavour of Foods and Beverages* (eds. G. Charalambous & G.E. Inglett), pp. 355–380. Academic Press, New York.

Ribéreau-Gayon, P., Glories, Y., Maujean, A. & Dubourdieu, D. (2006) *Handbook of Enology, Volume I, Wine and Wine Making*. John Wiley & Sons, Ltd, Chichester.

Robinson, J. (1993, 1995) *Jancis Robinson's Wine Course*. BBC Books, London.

Sefton, M.A. & Simpson, R.F. (2005) Compounds causing cork taint and the factors affecting their transfer from natural oak closures to wine – a review. *Australian Journal of Grape and Wine Research*, **11**, 226–240.

Silwar, R., Kempschoerer, M. & Tressl, R. (1987) Gas-chromatographic-mass spectrometry-quantitative determination of steam-volatile aroma constituents. *Chemistry Microbiolology Technology Lebensm.*, **10**, 176–87 [in German].

Snowdon, E.M., Bowyer, M.C., Grbin, P.R. & Bowyer, P.K. (2006) Mousy off-flavour: a review. *Journal of Agriculture and Food Chemistry*, **54**, 6464–6474.

Stevens, J.C., Cain, W.S. & Burke, R.J. (1988) Variability of olfactory thresholds. *Chemical Senses*, **13**, 643–653.

Suarez, R., Suarez-Lepe, J.A., Morata, A. & Calderon, F. (2007) The production of ethylphenols in wine by yeasts of the genera *Brettanomyces* and *Dekkera*: A review. *Food Chemistry*, **102**, 10–21.

Togoni, A. (1998) Tea flavour. *American Chemical Society Symposium Series (No. 173)*. American Chemical Society, Washington.

Chapter 6
Sherry, Port and Madeira

6.1 Introduction

The three classic fortified wines, Sherry, Port and Madeira, are very different in their sensory properties. The thing they have in common is that the grapes are grown in fairly hot regions, possibly giving young wines that are not necessarily desirable to drink as they are. In all three cases, the character of the finished wine is determined greatly by the maturation process. There are certain quality characteristics required for the grapes and there are specifications for the fermentation process for the production of the young wines. However, the wines are typically blended extensively during the maturation period and the finished wines tend to bear little recognizable resemblance to the original young wines, or even grapes. The processes used to make these wines are described in Chapter 1. Some of the chemistry relevant to the flavour and colour of these wines will be discussed in this chapter.

6.1.1 Sherry introduction

Sherry is the name of several related fortified wines made from grapes grown in Jerez de la Frontera, in the province Cadiz in the south of Spain. The hot climate of the region would be expected to produce white grapes that would turn into rather bland white table wines. Indeed the white table wine made from the Palomino grape, the main grape variety, is fairly neutral, lacking in acidity, without distinct varietal character and of little interest as a wine in its own right. However, as a result of a unique and complex method of maturation and blending, Sherry wines have evolved with great individuality and style. The neutral base wine, which forms the starting point of the Sherry manufacture, forms an excellent background for the delicate flavours produced as a result of the maturation and blending procedures.

The three main types of Sherry (Fino, Oloroso and Amontillado) are made from the base wine using different ageing techniques. In short, Sherry can be matured under *flor* (a layer of particular yeasts growing on top of the wine) to develop into Fino, which is pale yellow, dry and pungent. The wine can be also

Wine Flavour Chemistry, Second Edition. Jokie Bakker and Ronald J. Clarke.
© 2012 Blackwell Publishing Ltd. Published 2012 by Blackwell Publishing Ltd.

matured without *flor* yeasts to develop into Oloroso, which is usually dark brown, full bodied with a strong bouquet. A combination of *flor* maturation followed by a period of ageing without *flor*, results in Amontillado. This type of wine-making has also been adopted and modified by other wine-making regions but the discussion below will focus on these typical Sherry styles from Spain. The wines can be sold sweetened or dry and are usually fortified to 18%–20% v/v alcohol. A range of commercial products is manufactured, all of which are essentially based on the three main wine styles.

6.1.2 Port introduction

Port is a fortified wine made from grapes grown in the Douro region in Northern Portugal. It is a naturally sweet wine – part of the grape sugars are kept in the wine by arresting the fermentation process approximately halfway, by rapidly pressing the grapes and adding grape spirit (also referred to as brandy) to between 18% and 20% v/v alcohol. The fortifying spirit is not entirely neutral in character, and is thought to influence the maturation of the Port wine. Both red and White Ports are made, but White Ports only form a small part of the production.

There are two main styles of Port wine, Ruby and Tawny, both made from red grapes. Ruby Ports are red, full-bodied and often still quite fruity in character when the wines are ready to drink. Ruby Ports are aged in old wood, or larger tanks and do not usually have any wood-aged characteristics. Some special Ruby Ports (the so-called Vintage Ports) receive considerable bottle ageing, giving lighter red wines, with often a very fruity character, despite having aged for two decades or more. Tawny Ports are generally amber and have a typical flavour developed during prolonged ageing in old oak casks, which is best described as 'crisp, nutty, with an oaky note' and giving an impression of dryness. Within these two main styles the Port manufacturers make many products, each having its own sensory characteristics.

Although White Port is made in the same way as red and Ruby Ports, it is aged similarly to a young Ruby Port. Very little information has been published on White Ports, therefore this chapter is concerned only with red Ports.

6.1.3 Madeira introduction

In addition to Port and Sherry, Madeira is the third 'classic' fortified wine. The island of Madeira forms part of Portugal, lies in the Atlantic about 1000 km from the mainland of Portugal and about 750 km off the coast of North Africa. The wines are essentially shaped by their maturation, which involves heating the wine to up to 50°C, commonly referred to as the *estufagem*. This confers a strong and characteristic flavour on the wine. Despite this unusual method of maturation, leading no doubt to the formation of the caramel-like brown colour and distinctive flavour compounds typical for Madeira, it is only recently that scientific literature on the chemistry of these changes has been published, discussed later in this chapter.

There are both dry and sweet Madeira wines, which are drunk before and after dinner, respectively. The drier and lighter styles are Verdelho and Sercial, while the sweeter ones are Bual and Malmsey, named after the grapes from which the wines should be made. Madeira is thought to be one of the longest-living wines: the baking process, the high alcohol and the high acidity of the wine all contribute to its stability and hence, its keeping quality.

6.1.4 Comparisons between fortified wines

The three fortified wines are made by fermenting the crushed grapes, in many ways similar to the non-fortified table wines (Chapter 1). Various aspects of the general discussion in Chapters 3 and 4 regarding the basic taste and the volatile components of wine are also relevant to Sherry, Port and Madeira. Little specific published data on grape composition is available but the compounds are probably within the range of those reported in other hot climate wine-making regions. Volatile compounds formed during the fermentation are likely to be similar to those formed during the early part of table wine fermentations. However, since the maturation methods of these classic fortified wines change considerably the character of the young wines, the volatile compounds formed during fermentation are unlikely to contribute much to the flavour of the end-product. The flavour contribution of the grape to the wine is difficult to ascertain. There are very specific requirements for the grapes used for making each of the fortified wines and one can only assume that Sherry would be quite different in character had it not been made from the Palomino grape. For Port wines, many different grape varieties/cultivars are used, and there is evidence that the grape variety (or varieties) determines the character of the wine, even after the prolonged ageing process. Conversely, Madeira should be made from one of only four classic grape varieties/cultivars, and specific wine styles evolve from each variety, although a number of other varieties are also used, and in fact these, mainly Tinta Negra, account for about 88% of Madeira wine production (Elliot, 2010).

Many volatile compounds have been identified in wines, including the fortified wines. Often, non-quantified trace amounts of such compounds are reported but the presence of only small quantities of some minor volatile compounds may well characterize the typical flavour of the wine. The very different wine-making and maturation techniques of fortified wines mean that some volatile compounds seem to be characteristic of the fortified wines. Due to the similarity in volatile composition of many wines, there is also the view that quantitative differences among the volatile compounds are more important than qualitative differences.

Table wines and fortified wines differ in three ways, which are very likely to contribute significantly to the perceived flavour properties of these wines. Firstly, the ethyl alcohol level is raised to about 18–20% v/v, usually by fortification of either the fermented wine (Sherry), or by arresting the fermentation by the addition of ethyl alcohol, such as is customary in the production of Port wine. Secondly, many, but not all of the fortified wines are sweet. Port wines are all sweet and aged in the presence of a high sugar content. Sherry is aged dry, and can be sweetened before bottling. Madeira is aged dry, or sweet, depending on the wine characteristics required. Thirdly, fortified wines are all aged in ways that are not customary for table wines, usually involving a considerable amount of blending and in most cases, more oxidation than with table wines. Port wines can be aged for a minimum of three years to a decade or more in old oak casks to develop their character. Sherry undergoes numerous blending steps in the *solera* system (Chapter 1), while the wine is kept in oak butts, and the age of the wine is difficult to establish. In addition, the fermenting *flor* in Fino-style Sherry will determine the character of the wine. Madeira derives its unique flavour properties from the estufagem, or baking process.

6.1.5   Ethyl alcohol – sensory effect

The formation of alcohol by yeast fermentation of sugars in the must in fortified wine-making is similar to the fermentation procedure in table wine production and is discussed in Chapters 3 and 7. The higher concentration of ethanol in fortified wine is expected to affect the physical behaviour of the volatile compounds that contribute to the flavour of the wine. As discussed in Chapter 4, many such compounds become more soluble as the alcohol content in the water–alcohol mixture of wine is raised. This effect can be calculated by comparing the partition coefficient determined for the compound of interest at the two alcohol levels and is expected to show a lower partition coefficient at higher alcohol concentration. Thus, theoretically, many volatile compounds become less available for sensory perception when the ethanol content is increased to the level common for fortified wine. However, the higher ethanol content may assist in solubilizing compounds that have a limited solubility at lower ethanol levels, for example high molecular weight compounds. The increased amounts of such compounds dissolved in higher alcohol concentrations could contribute to the overall flavour of the wine. Thus, if we taste two almost identical wines, with the only difference being their ethanol content (for example 12% and 20% v/v), we would expect a difference between the wines, with the higher alcohol wine having less very small volatile molecules in its head-space (the air-space immediately above the liquid sample, where people take their sniff from, see Chapter 5).

However, it is difficult to predict whether sensory differences could be determined between such samples. Little information is available on the sensory effects of ethanol on the perception of flavour from alcoholic beverages (Bakker, 1995). To determine significant sensory differences in flavour, relatively large differences in flavour concentration in the head-space need to be present. In addition, the concentration of volatile compounds in the head-space of a glass is likely to vary when tasting a wine and the compounds are also perceived once they are released in the mouth via the retronasal route (Chapter 5). When tasting a wine, it is first sniffed from the glass, then some wine is taken into the mouth, where it is warmed gently, and many professional and amateur wine tasters will swirl the wine around in the mouth.

These processes will assist the volatile compounds to escape from the wine and are thought to enhance our perception of the wine volatile compounds. There may also be a time effect, with the most volatile compounds being perceived first, followed by lesser volatile compounds, for which it may take more time to reach the sensory threshold concentration (Chapter 4).

6.1.6 Ethyl alcohol – chemical effect

The fortifying spirit used for both Port and Sherry contains trace volatile compounds that could affect the quality of the wines. However, the more important influence will be the change in flavour development in the presence of the higher ethanol concentration. All three wines are stored for maturation in oak and there will be an increase in wood extractives because of the higher ethanol concentration. In table wines, many esters in wine derived from the fermentation tend to be present in excess and they hydrolyse until they are in equilibrium with their component acids and alcohols. New esters may be formed chemically, usually involving compounds present in high concentrations in wine. The equilibria of the formation of esters and acetals, compounds typically formed during wine maturation (Chapter 4), are likely to be increased due to the higher concentration of alcohol than in table wines.

6.1.7 Sweetness

Sucrose in solution increases the partition coefficients (Chapter 4) but there is no data regarding the effect of the typical sugar levels in wines. Some of the data suggest that sugar does not enhance the sensory flavour intensity of tested foods (Bakker, 1995). However, cognitive associations can give an increase in flavour intensity scores. There are some direct effects of sugar on the chemical processes, in particular the longer maturation times of sweet wines is expected to give some flavour compounds due to sugar degradation processes. In Port wine and Madeira there is evidence of this occurring (see Sections 6.3.9 and 6.4.7).

6.2 Sherry

There is much less information published on the chemistry of Sherry affecting the sensory properties than on other table wines, even though Sherry production (Chapter 1) is a much more complex process. Although all Sherry is made from one grape variety/cultivar, Palomino, the ageing under *flor* (layer of yeast on the top of the wine) or under oxidative conditions produces a remarkable range of wines with a definite recognizable character. Volatile compounds' production is influenced by the particular grape variety and its growing conditions, the primary fermentation process, maturation under *flor*, maturation without *flor*, storage in wood and the fractional blending system. Much of the knowledge of volatile compounds in Sherry can be attributed to the research of Webb and co-workers, as reviewed, for example, by Webb & Noble (1976). Some reviews describe detailed aspects of wine-making and focus on only the major changes in composition during maturation, without detailed information regarding the overall volatile composition (Goswell & Kunkee, 1977). A review of the different methods of Sherry production, including those used in America, Australia and South Africa, and the associated chemistry has been reported by Amerine *et al.* (1980). The current technology of fortified wine-making is discussed in detail by Reader & Dominguez (1994). Goswell (1986) has reviewed the microbiology of Sherry style production and Bakker (1993) has reviewed the composition of Sherry. The production of Sherry and its chemistry has been briefly reviewed recently by Jackson (2008). A general overview of flavour occurrence and formation in wine is given by Nykänen (1986).

6.2.1　Wine producers

Sherry has a history as long as its hometown Jerez de la Frontera, and English merchants have been involved in the Sherry production and trade for several centuries. Many of the Sherry companies were established in the first part of the nineteenth century although some date back to the eighteenth century. The bigger companies all have their own vineyards, but grapes are also grown by farmers. The wines are made and then matured in the *bodegas*, the large well-ventilated winery buildings in which the many oak casks with maturing wines are kept relatively cool. Prior to Spain joining the EU, larger companies exported their wines in bulk, and much of the preparation of the final blends and the bottling was done elsewhere, for example in England. These wines served the British market but were also exported. As part of the EU regulations, the wines are 'bottled at source' and the entire process of wine-making, blending and bottling is carried out in the Sherry region in Spain. There has been a considerable reorganization in the business operations as a result. Popular wine books list the Sherry companies, often including the main styles of wine they produce.

6.2.2　Commercial wine styles

Finished Sherries have very complex sensory characteristics. Fino Sherries are usually sold dry. Some fine, older wines can also be sold dry but most Sherries are sold sweetened. Fino has a pale straw colour, is very dry but without much acidity. It has a delicate, pungent bouquet and the alcoholic strength usually lies between 15.5 and 17% v/v. Manzanilla is a regional variation of the Fino style and is matured in the coastal town Sanlúcar de Barrameda. It is bone dry, with a clean and slightly bitter aftertaste, being slightly less full bodied than Fino. Manzanilla has a pale straw colour, much like a Fino and an alcoholic strength between 15.5 and 16.5% v/v. Fino and Manzanilla should both be drunk young and cold and opened bottles should be consumed within a day or so, since the flavour is not stable once the bottle has been opened. Amontillado is dry and clean, with a pungent aroma reminiscent of Fino but 'nuttier' and fuller bodied. It is amber and becomes darker with increasing age. The alcoholic strength is 17–18% v/v. Oloroso has a strong bouquet and is fuller bodied but has a less pungent odour than Fino or Amontillado. Even when Oloroso is dry, it has a slightly sweet aftertaste, possibly as a result of the wood extractives. Oloroso has the darkest colour, best described as dark gold and its colour intensity increases with age. Wines marketed as medium dry, medium, cream or pale cream Sherry usually are based on Oloroso wine, with an addition of Fino to lighten the colour and flavour and a small amount of Amontillado. Gonzalez Gordon (1972) describes the nomenclature used by Sherry tasters in Jerez. The descriptive terms are related to alcoholic strength, total acidity, the time the wine has matured under *flor* and the concentration of acetaldehyde that has developed. Other terms used are clean, dirty, soft, hard, full, empty, soft and dull.

6.2.3 Wine writers' comments

Although the wines are made from mostly one grape variety, the complex maturation procedures and the range of producers gives an array of different tastes and smells. Clarke (1996) comments that some of the commercial blends are too much aimed at mass acceptance of the wine, resulting in 'bland and forgettable wines'. However, he describes older Sherries as having 'positively painful intensity of flavour, mixing sweet and sour, rich and dry all at once and all Sherry, dry or sweet should have a bite to it'. Broadbent (1979) describes the three main Sherry styles. Fino is described as having a 'pale lemon straw colour, refined fresh *flor* aroma, dry light and fresh, with a long crisp finish'. Amontillado is described as 'deeper in colour, very slightly Fino reminiscent, richer and distinctly nutty, dry to medium dry'. Oloroso is described as a total contrast to Fino, 'deeper in colour, deep amber to warm amber brown, complete absence of *flor* tang, softer and sweeter on nose and palate, medium to full bodied'. The comment regarding sweeter on the nose presumably means smells associated with sweetness, as sugar itself does not have a sweet smell when in solution. Robinson (1995) describes Manzanilla and Fino as 'very pale, delicate, prancing, palate-reviving thoroughbreds', which are 'bone dry and tingling with life and zest'. The other two major styles, Amontillado and Oloroso, she describes as 'dark, nutty' wines which can 'thrill the palate, with its subtle shadings of mahogany and nuances that are direct and delicious results of extended ageing in oak'. Details of Sherry tasting can also be found on the web (for example wineanorak.com).

6.2.4 Grapes and must

The required base wine quality suitable for the successful establishment of *flor* is made from must of the Palomino grape with a specific gravity between 1.085 and 1.095, a phenolic compounds content of 300 to 600 mg L^{-1} (no method given but presumably expressed as gallic acid) and a total acidity between 3.5 and

4.5 g L^{-1} expressed as tartaric acid, equivalent to between 2 and 3 g L^{-1} expressed as a sulfuric acid (Goswell & Kunkee, 1977). However, titratable acidity is not a reliable indicator of acidity in Palomino must and a high pH is often associated with high contents of phenolic compounds (Reader & Dominguez, 1994), which are considered not desirable for a base Sherry to be aged under *flor*.

The grapes are crushed and pressed. The free run juice, containing less than 200 mg L^{-1} total phenolic compounds (presumably expressed as gallic acid) is most suitable for ageing under *flor* and is kept separate from the pressings containing higher concentrations of phenolic compounds. An addition of sulfur dioxide may be made during crushing but this is not always considered desirable and there is a trend trying to restrict its use. Sulfur dioxide inhibits the action of oxidative enzymes, thus reducing the onset of browning, but it increases the extraction of phenolic compounds from the grapes into the wine and prevents oxidation and precipitation of the phenolic compounds onto the grape solids. Its content present during fermentation is also believed to influence the development of wine aroma, since it forms complexes with carbonyls, such as acetaldehyde, ethyl 2-ketoglutarate and ethyl 4-ketobutyrate (Webb & Noble, 1976). When these complexes dissociate during maturation, the carbonyls become available to form volatile aroma compounds. When the acidity of the must is too low, it is adjusted by an addition of tartaric acid to the must.

6.2.5 Base wine

When the fermentation is finished, the dry wines have an alcoholic strength of 11–12% v/v. In November the malo-lactic fermentation, which converts the fairly sharp tasting malic acid into the softer tasting lactic acid and carbon dioxide, is completed by lactic acid bacteria naturally present in the wine. The wines are initially classified into two groups on the basis of their quality and fortified accordingly (Reader & Dominguez, 1994). Fino is made from the lighter dry wines, using mainly free run juice and some light pressings. This wine is pale yellow and has a low total of phenolic compounds (200 mg L^{-1}; the authors did not indicate in which compound the concentration is expressed but it is probably expressed as gallic acid), a good pungent aroma, a volatile acidity between 0.3 and 0.5 g L^{-1} (expressed as acetic acid), a pH between 3.1 and 3.4, contains less than 100 mg L^{-1} sulfur dioxide and is free of any bacterial spoilage. These wines usually develop *flor* spontaneously.

Oloroso is made from slightly darker wines, with higher concentrations of total phenolic compounds (up to 475 mg L^{-1}), which are less likely to develop and support *flor*. These base wines have a more vinous and full-bodied nose and a higher volatile acidity (0.7–0.9 g L^{-1}). Wines made from the higher pressings and subsequently with higher total phenolic compounds (over 550 mg L^{-1}) are classed as the lowest quality wines, although small quantities may after maturation be used for the Oloroso style. These wines are often exposed to higher temperatures during their maturation, occasionally outdoors in the sun, where they develop into dark Oloroso styles. Fino wines are fortified to 15.5% v/v alcohol, while Oloroso wines are fortified to 18.5% v/v alcohol.

6.2.6 Maturation

Wines are stored in seasoned oak casks (so-called Sherry butts), with a capacity of 500–600 L, in *bodegas*, which are tall, well-ventilated buildings, designed to stay relatively cool. The wine can either be matured under *flor*, to develop the typical Fino character, or they can be matured without *flor* to develop Oloroso wines. There are some major changes in volatile compounds content, either as a result of the biological ageing under *flor*, or due to the oxidative changes during maturation without *flor*. Since the concentration changes involved are relatively high, they probably impact on the typical flavour developed in the Sherry styles. These alterations in volatile compound content can directly be related to the process of maturation.

There are also minor changes in the type of volatile compounds during maturation of Sherry. Some compounds have been identified that are thought to contribute greatly to the flavour of Sherry, although the typical flavour is assumed to be due to the combined effect of several aromatic volatile compounds. The Solera system, the fractional blending system typical for Sherry, is described in Chapter 1. Variation in composition and resulting flavour of the wines between the different Sherry butts is very large; hence the fractional blending system seems essential to obtain a reliable supply of wine with similar flavour characteristics.

6.2.7 Maturation changes under *flor*

The *flor* micro-organisms, consisting of yeasts, occur in the Sherry *bodegas*, and provided the base Sherry has the required composition, usually a wrinkly film of yeast on the surface of the wine, the so-called *flor*, will establish on maturing Fino. The wines are fortified after the fermentation to 15.5% v/v alcohol (Chapter 1) with neutral spirit, which is not thought to influence the flavour of the wine. However, the resulting higher alcohol content inhibits film-forming acetic acid bacteria, which would rapidly spoil the wine. These wines are stored in 600 L butts that are kept about 80% full (content about 500 L) to maintain a high surface-to-volume ratio and stored between 15 and 20°C. Due to the temperature dependence of these organisms, there is a seasonal variation in *flor* yeast activity; the *flor* is most active between February and June and then declines until October time, when activity increases again. The dry storage conditions will also cause evaporation of water through the wooden butts, and over a four-year maturation period of Fino wines a 15% reduction of the initial volume has been estimated (Martinez de la Ossa *et al.*, 1987b), resulting in a concentration of the wine. Much smaller evaporative losses of alcohol (0.2–0.3% v/v) have also been quoted (Reader & Dominguez, 1994). Presumably the evaporative losses will depend on the storage temperature and the humidity and vary accordingly. The cooler and more humid climate at the coast, where the Manzanillas are matured, is presumed to influence the yeast metabolism, thus resulting in differences between

Manzanillas and Finos. The higher humidity is also thought to reduce the evaporative losses during maturation.

Both the origin and the taxonomy of the yeasts growing in the *flor* are still being researched (Reader & Dominguez, 1994). Strains from several species have been identified, including yeasts species belonging to the *Saccharomyces* species. Other authors suggest that *Saccharomyces fermentati* or *S. beticus* should now be known as *Torulaspora delbrueckii* (Kunkee & Bisson, 1993). These *flor* yeasts are physiologically different from the fermenting yeasts that dominate during the fermentation. *S. beticus* is the dominant yeast in younger wines, while *S. montuliensis* is dominant in older wines. There seems to be a strong relationship between wine composition and yeasts. For example, Mesa *et al.* (2000) used molecular techniques to characterize the yeasts in the *flor* and showed that small differences in the season and base wine resulted in the selection of different yeast genotypes. *Flor* yeast strains vary in their synthesis of volatile compounds, such as esters, higher alcohols and terpenes (Esteve-Zarzoso *et al.*, 2001). The strains of *S. cerevisiae* identified in *flor* appear to be unique, for example they have a greater resistance to the toxicity of ethanol and acetaldehyde, and are quite different from the strains found during wine fermentation. There is still the question whether there is a succession of yeast strains involved in *flor* maturation or whether there is one dominant strain.

The *flor* needs oxygen; the blending system in the solera, requiring regular movements of wine, should ensure an adequate supply of nutrients and oxygen for the *flor* to thrive. The *flor* yeasts are in a respiratory mode, requiring the absence of glucose and concentrations of oxygen greater than that found dissolved even in a saturated wine to oxidize ethanol (Kunkee & Bisson, 1993). Hence these *flor* yeasts only grow as a film on the surface of the wine, unless the oxygen concentration in the wine is increased under pressure (Goswell & Kunkee, 1977). Therefore all wine technology books stress the importance of partly filled butts to ensure an adequate surface-to-liquid ratio to ensure a sufficient supply of oxygen for the *flor* to thrive. The layer of *flor*, which actively consumes oxygen, protects the bulk of the Sherry from the uptake of oxygen and prevents browning due to the oxidation of phenolic compounds – hence Fino maintains a pale yellow wine colour.

Essentially, the *flor* yeasts cause a second fermentation. The energy and carbon sources come from ethanol and other alcohols, while oxygen is taken up from the head-space. *Flor* results in numerous biochemical changes, affects the chemistry during maturation and determines the character of the wine. There are reductions in alcohol, glycerol and volatile acidity, all used as a carbon sources for *flor* yeast growth. Martinez de la Ossa *et al.* (1987a) reported reductions in glycerol from 7 to 0.2 g L^{-1}, which may help to explain the very dry sensory perception of the Fino wines, which lack any hint of sweetness that may have been imparted by the presence of glycerol. They also reported the precipitation of potassium bitartrate during maturation, which reduces the tartrate concentration from 3.30 to 2.20 g L^{-1}, contributing to the reduction of total acidity from 5.3 to 4.1 g L^{-1} and thereby, resulting in a slight increase in pH. Martinez de la Ossa *et al.* (1987b) have reported a considerable reduction of volatile acidity due to the *flor* growth, from 0.39 to 0.24 g L^{-1} (expressed as tartaric acid), and a reduction of 1.6% v/v in the ethanol content. There was also an increase in the acetaldehyde concentration, from 96 to 286 mg L^{-1}, with the greatest increase occurring during the early part of the maturation. Increases have been reported to normally 260–360 mg L^{-1} but even higher concentrations can be formed. Acetaldehyde makes an important odour contribution to the typical oxidized, somewhat apple-like nose of Fino *flor* Sherries, but is also a precursor for compounds such as acetoin and 1,1-diethoxyethane and can react with certain phenolic compounds and alcohols, generating other volatile compounds, some of which are thought to be typical for Fino Sherry (see later in this section).

Possible autolysis (spontaneous rupture of cells that releases their contents) of old yeast cells that form part of the sediment on the bottom of the cask may well contribute to the typical flavour. However, little firm evidence to date is available on these more speculative flavour formation possibilities.

Interestingly, the word *flor* may not have the same meaning in all scientific textbooks. Evidently it was first used by Pasteur and, for example, Ribéreau-Gayon (2006) uses the term *flor* for spoilage of wine caused by a strain of yeast referred to as *Candida mycoderma*, which grows on the surface of a wine and oxidizes ethanol into carbon dioxide and water, leaving the wine flat, watery and turbid. A second more common meaning of *flor*, also used by Ribéreau-Gayon (2006), is the layer of micro-organisms growing in Fino Sherry but also on the yellow wines from Jura.

6.2.8 Maturation changes without *flor*

Fortification of young Oloroso wines to 18.5% v/v inactivates any *flor* yeast and prevents *flor* from growing or being formed. The casks are kept 95% full and their storage temperature is less critical than for the Fino wines. To produce a good Oloroso the neutral base wine needs to contain sufficient oxidizable phenolic compounds and is therefore not suitable for Fino wines. Storage under oxidative conditions results in the dark golden colour of the Oloroso, attributed to oxidation of phenolic compounds. The higher alcohol concentration, often combined with a higher storage temperature, may lead to increased extraction of phenolic compounds from the wood during maturation of Olorosos, explaining the higher concentrations of phenolic compounds. Typical maturation of Oloroso takes about eight years.

Although the wines are fermented to dryness, glycerol formed during alcoholic fermentation (7–9 g L^{-1}) is thought to give a hint of sweetness on taste. A considerable loss in volume during maturation and usually an increase in alcoholic strength has been reported in Oloroso wines, resulting in a higher concentration of non-volatile compounds (Martinez de la Ossa *et al.*, 1987b); the longer maturation time than with Fino makes the effect more pronounced. These authors estimated a 30–40% reduction of the initial volume during the 12 years maturation of Oloroso wine. Therefore some changes in concentration of compounds were thought to be due to this concentration effect, such as increases in alcohols (methanol, n-propanol, n-butanol, iso-butanol). However, they observed oxidation of acetaldehyde into acetic acid, resulting in increases in volatile acidity (0.21 to 0.74 g L^{-1} expressed as tartaric acid) and the formation of the ester ethyl acetate (from 170 to 280 mg L^{-1}), in excess of the increase due to the concentration effect. They speculated that other oxidation and esterification processes may also occur during ageing, contributing the aroma of Oloroso wines.

6.2.9 Maturation with and without *flor*

Amontillado Sherry is made by maturation with *flor*, followed by maturation without *flor*. During initial maturation under *flor*, the wines develop all the characteristics of a Fino. However, when wine is not refreshed with additions of younger wines in the solera system, or when wine has reached a considerable age, the wine may start to lose the *flor*. Further fortification to about 17.5% v/v alcohol is usual, to protect the wine against spoilage by acetic acid bacteria and prevent any further development of *flor*. The wine is matured in a second solera system in casks 95% full and the ageing processes change, with oxidation of the wine changing the pale yellow colour to amber and dark gold, together with the development of a nutty, complex flavour typical for Amontillado.

During the maturation in the solera under *flor* these wines develop like Fino Sherry, with reductions in alcohol, glycerol and volatile acidity. In the second solera system during the oxidative maturation, there are increases in volatile acidity (0.21 to 0.74 g L^{-1} expressed as tartaric acid) and glycerol (0.14 to 4.37 g L^{-1}), in part because of evaporative losses of water (Martinez de la Ossa *et al.*, 1987b). Changes in the second more oxidative solera system were similar to those observed for Oloroso wines, although there were quantitative differences as a result of the prior biological maturation. Acetaldehyde was lost (from 288 to 186 mg L^{-1}), which is attributed to oxidation to acetic acid, with an accompanying increase in ethyl acetate (from 60 to 190 mg L^{-1}) (Martinez de la Ossa *et al.*, 1987a).

6.2.10 Volatile compounds

Numerous publications report volatile compounds in Sherry and there are a number of compiled lists of qualitative data. Over the last decade more quantitative data has become available but small quantities of some minor volatile compounds may well characterize the typical flavour of the wine. A useful compilation of more than 130 volatile compounds identified in Sherry and Sherry-type wines has been collated by Webb & Noble (1976), and also reported by Montedoro & Bertuccioli (1986). In 1989, Maarse & Visscher listed 307 volatile compounds identified in Sherries. The compounds listed consisted of 28 alcohols, one hydrocarbon, 19 carbonyls, 47 acids, 65 esters, 16 lactones, 67 bases, four sulfur compounds, 13 acetals, one ether, 14 amides, 17 phenols, five furans, three coumarins, four dioxolanes and two dioxanes. Many of these compounds have also been identified in standard table wines also (Chapter 4) and are not especially typical for Sherry. Similarly, some of the changes during maturation are expected to be not unlike those described for table wines, as described for example, by Rapp & Mandery (1986) (see also Chapter 4). The focus in the literature tends to be on *flor* Sherry, with little specific information being available on Oloroso or Amontillado Sherry. Presumably, Oloroso wines are not very different from white wines aged under oxidative conditions in oak casks and some of the flavour compounds derived from the wood (such as oak lactone, see below) would be expected to be present in Oloroso wines. Maarse & Visscher (1989) do not differentiate between the Sherry styles in their compilation of volatile compounds. During maturation under *flor*, there tend to be increases in aliphatic and aromatic acids, terpenes and carbonyls.

Biological *flor* ageing does appear to give rise to numerous compounds during this stage. Brock *et al.* (1984) have made a qualitative study on the formation of volatile compounds during three months ageing of Palomino base wine, using a submerged *flor* (a variant American system whereby the *flor* is grown submerged in the wine). They reported the formation of 36 new compounds, which included 14 acetals, two acids, three alcohols, four carbonyls, four esters, one lactam, four lactones and four nitrogen-containing compounds. The authors suggested that many of these compounds may contribute to the characteristic aroma associated with submerged *flor* Sherry and may only be formed in *flor* Sherry.

Another study using Fino wines made from Pedro Ximenez grapes matured using *flor* in Montilla-Moriles (southern Spain) compared the composition of the wines before and after *flor* maturation (Moyano *et al.*, 2009). There were no details regarding the differences between *flor* ageing in this region, compared with Jerez, although the grapes used for the wine are different. These authors reported significant changes in composition, summarized in Table 6.1, showing significant increases in aldehydes, esters, acids and lactones as a result of *flor* ageing. The main odour active compounds identified by these authors are also listed in the Table 6.1.

As a result of the biological changes in wines matured under *flor*, further reactions of the relatively large amounts of small molecular weight compounds, such as acetaldehyde and glycerol, lead to the formation of new aromatic volatile compounds. Acetaldehyde is the most abundant aldehyde in *flor* Sherry. Glycerol can arise from several biochemical pathways and is a natural constituent in wine. During biological maturation, glycerol tends to decrease in concentration, whereas acetaldehyde increases.

The accumulation of acetaldehyde in *flor* Sherry gives the oxidized odour or aroma, or 'nose'. Four isomeric acetals formed by glycerol (and other polyols) and acetaldehyde were present in sufficient quantities to be detected and identified in Fino Sherry (Muller *et al.*, 1978). The presence of all four acetals (*cis*- and *trans*-5-hydroxy-2-methyl-1,3-dioxane, and *cis*- and *trans*-4-hydroxymethyl-2-methyl-1,3-dioxolane) is indicative of chemical equilibration reactions, rather than the enzymatic formation. These compounds are thought to be typically formed in Sherry type wines and contribute a certain

Table 6.1 Changes in volatile composition as a result of ageing under *flor* of fine wines made from Pedro Ximenez grapes.

Compounds	Concentration mg L^{-1}		Compounds with Odour Activity Values > 1
	Young wine	Fino	
Total acetaldehyde and derivatives	65	245	Acetaldehyde, 1,1-diethoxyethane, acetoin
Total alcohols	521	576	Isobutanol, isoamyl alcohols, furfuryl alcohol, phenethyl alcohol, methionol
Total esters	250	565	Ethyl acetate, isoamyl acetate, phenethyl acetate, ethyl butanoate, ethyl hexanoate, ethyl lactate, ethyl octanoate, ethyl decanoate, ethyl 2-furanoate
Total acids	1.8	4.8	
Total aldehydes	0	2.8	
Total lactones	4.9	26	(*Z*)-Oak lactone
Total volatile phenols	0	0.29	4-Ethylguiacol

Data adapted from Moyano *et al.* (2009).

pungency. The acetal 1,1-diethoxyethane (Webb & Noble, 1976) is considered to be formed in sufficient quantities to give a green note to the wine (Jackson, 2008), although this compound has also been described as fruity (Arctander, 1967). Moyano *et al.* (2009) confirmed the importance of this compound in *flor* Sherries, and contributing to the aroma (see Table 6.1). Aliphatic acetals tend to have green or fruity characteristics, while cyclic acetals tend to be pungent.

Lactones and the associated oxo and hydroxy compounds are important compounds, since they are thought to be the most significant volatile compounds contributing to the typical Sherry aroma. They are formed during maturation under *flor*. The biosynthesis of some lactones has been elucidated (Chapter 4), for example glutamic acid additions to maturing Sherry under *flor* enhance the concentration of some γ-lactones (Wurz *et al.*, 1988). Solerone (4-hydroxy-5-ketohexanoic acid lactone) has been reported to contribute an aroma of Sherry (Brock *et al.*, 1984), occasionally referred to as the Sherry lactone. Another lactone, sotolon (4,5-dimethyl-3-hydroxy-2(5*H*)-furanone) has also been isolated from Vin Jaune, a wine-like Sherry, made in the Jura in France (Dubois *et al.*, 1976). The authors proposed the formation of this compound by aldol condensation of pyruvic acid and α-keto-butyric acid. The same compound has been identified in *flor* Sherries (Martin *et al.*, 1990); it is present in sufficient concentrations (22–72 µg L^{-1}, mean 41) to contribute its toasty, spicy odour to the characteristic aroma of *flor* Sherry. These authors confirmed the presence of sotolon in Vin Jaune (75–143 µg L^{-1}, mean 115), and could not determine any sotolon in either non-*flor* Sherry or in red or white table wines. Sotolon has also been identified in wines made from botrytized grapes, in sufficient concentrations to be above the sensory threshold (2.5 ppb) (Rapp & Mandery, 1986). Numerous other lactones have been determined in Sherry, such as 4-hydroxybutanoic acid lactone, 4-hydroxydecanoic γ-lactone, δ-5-hydroxydecanoic acid lactone, γ-butyrolactone and various substituted forms (see Brock *et al.*, 1984), some of these are also found in table wines (Maarse & Visscher, 1989; see also Chapter 4). There is insufficient information regarding the sensory contribution that this array of lactones makes to Sherry. Oak-lactone, sometimes referred to as the whisky lactone (3-methyl-γ-octalactone), has also been found in Sherry (Chapter 4); it is extracted from wood during maturation of wines and spirits. It occurs in *cis* (*E*) and the *trans* (*Z*) forms, the *trans* form has a sensory threshold approximately ten times lower (0.067 ppm) than that of the *cis* form (0.79 ppm) (Maga, 1989; see also Chapter 4, Table 4.21). It contributes an oaky, woody character to the wine, as well as its particular character to many oaked table wines and also to whisky and brandy. Moyano *et al.* (2009) determined both *cis* and *trans* oak lactone in 'Fino-style wine' aged under *flor*, and *trans* lactones was identified well above threshold concentration (Table 6.1).

Due to the complexity of the Sherry maturation, laboratory soleras have been set up to study the volatile compounds in a controlled environment. For example, Criddle *et al.* (1983) analysed over 100 compounds in Sherry maturing under *flor*. An investigation on the formation of volatile compounds under *flor* in a model laboratory study published both quantitative and qualitative data on the changes in the wine (Begoña Cortes *et al.*, 1999). A number of the changes in volatile compounds reported are shown in Table 6.2, along with the initial content, the content when the *flor* yeast layer was fully formed after about 20 days and the content of compounds at the end of the experiment. Some data quoted by Moyano *et al.* (2009) have been added.

Some compounds tended to increase in the early part of maturation, while others increased after the *flor* was fully formed, see also Table 6.1. As expected, the data show increases in acetaldehyde, 1,1-dioxyethane and acetoin content. Most alcohols increased in concentration. The acetates of the higher alcohols decreased, except propyl acetate. Acetate esters usually decrease on maturation, accompanied by a decrease in the fresh and fruity character that these acetates impart to the wine (Rapp & Mandery, 1986; see also Chapter 4). The most abundant esters, ethyl acetate and ethyl lactate, increased in the early part of the Fino maturation, followed by a decrease in concentration. As

Table 6.2 Contents of volatile compounds in Fino Sherry.

Compound	Units	Initial concentration[a]	Whole film[a]	250 days maturation[a]	Threshold in water[b]	OAV after 250 days[c]
Carbonyls						
Ethanal (acetaldehyde) (total)	mg L^{-1}	84.8	133	146	4–120 ppb	>1
1,1-Diethoxyethane (diethyl acetal)	mg L^{-1}	22.1	75.4	69.7	1ppm[d]	
3-Hydroxy-butan-2-one (acetoin)	mg L^{-1}	1.7	5.7	48.6	30ppm[d]	
Alcohols						
Propan-1-ol (propyl-alcohol)	mg L^{-1}	13.6	12.3	14.8	9–40 ppm	<1
2-Methyl-propan-1-ol (isobutyl-alcohol)	mg L^{-1}	67.1	58.3	102	3.2 ppm	>1
3-Methyl-butan-1-ol (isoamyl alcohol)	mg L^{-1}	381	361	387	1ppm	>1
Phenyl-2-ethan-1-ol (β-phenyl ethyl alcohol)	mg L^{-1}	82.1	87.5	102	10 ppm[d]	
Propan-2-ol (sec-propyl alcohol)	mg L^{-1}	2.4	2.7			
Butan-1-ol (n-butyl alcohol)	mg L^{-1}	5.3	4.5	5.8	0.5 ppm	>1
Butan-2-ol (sec-butyl alcohol)	mg L^{-1}	1.1	1.9	1.2	1000 ppm[d]	
3-Methyl-pentan-1-ol	μg L^{-1}	117	114	144	50 ppm[d]	
4-Methyl-pentan-1-ol (isohexylalcohol)	μg L^{-1}	58.3	57.5	51	50 ppm[d]	
Hexan-1-ol (hexyl alcohol)	mg L^{-1}	2.3	2.3	1.7	0.5 ppm	>1
(*E*)-Hex-3-en-1-ol (*trans*-3-hexenol)	μg L^{-1}	80.8	79.8	74		
(*Z*)-Hex-3-en-1-ol (*cis*-3-hexenol, the leaf alcohol)	μg L^{-1}	70.8	70.6	78.9		
Phenyl methanol (benzyl alcohol)	μg L^{-1}	45.2	47.7	46	5.5 ppm flavour in water	<1
Esters acetic						
Propyl ethanoate (propyl acetate)	μg L^{-1}	41.7	47.1	74.5	3–11ppm	>1
2-Methyl propyl ethanoate (isobutyl acetate)	μg L^{-1}	24.9	21.1			
3-Methyl butyl ethanoate (isoamyl acetate)	μg L^{-1}	885	673	191	0.03 ppm[d]	
2-Phenyl ethyl ethanoate (β-phenyl ethyl acetate)	μg L^{-1}	228	223	103		
Ethyl ethanoate (ethyl acetate)	mg L^{-1}	36.8	42.3	15.8	6–60 ppm	<1
Ethyl mono-hydroxy propanoate (ethyl lactate)	mg L^{-1}	16.4	20.6	12.2	50–200 ppm	<1

		Beginning of maturation	After formation	After 250 days	Threshold	OAV
Acids						
Butanoic acid (butyric acid)	mg L⁻¹	2.4	2.1	7.5	6.8 ppm	<1
2-Methyl propionic acid (iso butyric acid)	mg L⁻¹	2.2	2.2	22.1	20 ppm[d]	>1
3-Methyl butanoic acid (iso-valeric acid)	mg L⁻¹	1.5	1.7	5.5	1.6 ppm	>1
Hexanoic acid (n-caproic acid)	mg L⁻¹	1.6	1.8	1.5	5.4	<1
Octanoic acid (caprylic acid)	mg L⁻¹	1.6	1.6	0.05	5.8	<1
Decanoic acid (capric acid)	mg L⁻¹	0.35	0.37	0.07	3.5	<1
Esters (others)						
Ethyl propanoate (ethyl propionate)	µg L⁻¹	109	154	433	9–45 ppm	<1
Ethyl-1-oxo propanoate (ethyl-pyruvate)	µg L⁻¹	201	138	81.3		<1
Ethyl 2-methyl propanoate	µg L⁻¹	41.6	28.9	351	0.01 ppm	<1
Ethyl butanoate (ethyl butyrate)	µg L⁻¹	172	193	392	1–450 ppb	>1
Ethyl 3-hydroxy butanoate	µg L⁻¹	446	473	747	67 ppm[d]	<1
Diethyl butandioate (diethyl succinate)	mg L⁻¹	0.8	1.2	6.1	100 ppm[d]	<1
Diethyl hydroxy butandioate (diethyl malate)	mg L⁻¹	0.8	1.1	4.1	760 ppm[d]	<1
Ethyl hexanoate (ethyl caproate)	µg L⁻¹	123	104	160	36 ppb	>1
Ethyl octanoate (ethyl caprylate)	µg L⁻¹	39.1	47.1	162	8–12 ppm	>1
Furanones						
Dihydro-3(H)furan-2-one (γ-butyrolactone)	mg L⁻¹	10.3	12.8	29.4	high	<1
Dihydro-3-hydroxy-4, 5-dimethyl-3(H)furan-2-one (pantolactone)	mg L⁻¹	0.47	0.69	3.22		
(E)-Dihydro-5-butyl-4-methyl-3(H)furan-2-one (oak lactone)	mg L⁻¹	0.22	0.22	0.04	490 µg L⁻¹ (in wine)	<1
Miscellaneous						
Linalool (see Chapter 4)	µg L⁻¹	9.4	11.6	32.2	6 ppb	>1
Citronellol (see Chapter 4)	mg L⁻¹	1.2	0.5	0.28	0.3 ppb	>1
3-Ethoxy-propan-1-ol	mg L⁻¹	0.25	0.28	0.49		
3-(Methylthio)-propan-1-ol (methionol)	mg L⁻¹	3.2	3.3	3.0	low, ppb	
4-(Prop-2-enyl)phenol (eugenol)	µg L⁻¹	129	230	347	6 ppb	>1

[a] Data extracted from Begoña Cortes *et al.* (1999). Measurements at the beginning of maturation, after formation of entire *flor* yeast film and after 250 days of maturation. [b] Data for the threshold values are extracted from tables in Chapter 4. [c] The Odour Activity Value (OAV) is the approximate ratio of the content after 250 days divided by the sensory threshold value. [d] Values quoted by Moyano *et al.* (2009).

Table 6.3 Concentration ranges in ng L^{-1} of ethyl 2-, 3- and 4-methylpentanoate, and ethyl cyclohexanoate in various commercial Sherry styles.

Wine type Threshold (ng L^{-1})	ethyl-2-mp 3	ethyl-3-mp 8	ethyl-4-mp 10	cyclohexanoate 1
3 Fino Sherries	2.8–18	112–514	748–1356	0–13
2 Cream Sherries	17–26	48–180	376–1439	4.7–36
1 Pale cream Sherry	9.6	18	142	0
4 Pedro Ximenez Sherries	0–1066	0–518	110–972	0–63

Data adapted from Campo *et al.* (2007).

expected the ethyl esters of other acids increased in content, except for ethyl pyruvate, presumably because there is a decrease in pyruvic acid. Two of the lactones identified in this study, γ-butyrolactone and pantalone, increased, in particular after the *flor* had been formed. There was also an increase in linalool, a monoterpene. The terpene alcohols linalool, nerolidol and farnesol are produced by *flor* yeasts (Fagan *et al.*, 1981) and may contribute floral notes to the wine.

Four branched esters have been analysed and quantified in a range of wines, with very low thresholds (see Chapter 4) and they are typical in many wines, including fortified wines (Campo *et al.*, 2007). They suggested that ethyl 2-methylpentanoate, ethyl 3-methylpentanoate, ethyl 4-methylpentanoate and ethyl cyclohexanoate are formed by esterification reactions with ethanol and the corresponding acids formed by micro-organisms. Their concentrations are especially high in aged fortified wines, such as Sherry and can reach concentrations well above threshold concentrations (Table 6.3) and their OAV values are well above 1 and very high for ethyl 4-methylpentanoate. These compounds are thought to contribute sweet fruity notes.

A detailed gas chromatography olfactometry study elucidated the compounds contributing to the distinctive aroma of Fino Sherry and Pedro Ximenez wines (Campo *et al.*, 2008). They confirmed that the high levels of acetaldehyde significantly contributed to the Fino aroma, and reported that diacetyl, ethyl esters of branched aliphatic acids with 4, 5 or 6 carbon atoms, 4-ethylguiacol and sotolon all contribute to the distinctive Fino character. The esters are formed by slow esterification of ethanol with acids formed by the *flor* yeast (see also Chapter 7). 4-Ethylguiacol is formed by yeasts (see Chapter 7). Sotolon was also identified as being present well above threshold (15 μg L^{-1}) concentration, and is formed by aldol condensation between acetaldehyde and 2-ketobutaric acid, a reaction mediated by *flor* yeasts.

Regarding the Pedro Ximenez wines, Campo *et al.* (2008) listed 3-methylbutanal, furfural, β-damascenone, sotolon, ethyl cyclohexanoate, phenylacetaldehyde and methional. They suggested that sotolon, with concentration up to 540 μg L^{-1}, was formed as part of sugar degradation, as has also been suggested for the formation of this compound in Madeira and Port wines. Madeira, Port wines and Pedro Ximenez wines are all aged under oxidative conditions in the presence of high sugar levels, so it is not surprising that the authors suggested that flavour formation in these wines is comparable. The authors quoted that the development of phenyl acetaldehyde and methional are dependent on the dissolved oxygen in wines, in Pedro Ximenez wines their concentrations were 68 and 20 µg L^{-1} respectively, whiles in Port the authors quoted values of 78 and 17 µg L^{-1}. The PX wines also contained high levels of the branched esters and cyclohexanoate, confirming findings from Campo *et al.* (2007) discussed above. β-Damascenone seems specific for Pedro Ximenez wines, averaging at 10 µg L^{-1} but a maximum of 21.7 µg L^{-1} was analysed. It may well contribute to the raisin notes of Pedro Ximenez wines. In lower levels this compound is also present in wines, where it is thought to act as an aroma enhancer (see Chapter 4).

Amontillado wines have been aged biochemically and chemically, although generally they are aged for a shorter time under *flor* than Fino Sherry. Zea *et al.* (2008) did a study determining which volatile aroma compounds are typical for Amontilado wines. They concluded that the flor maturation has a strong influence on the aroma of Amontillado, and there are many similarities in aromas of Fino and Oloroso style wines, however, sotolon has the highest impact on Amontillado and Oloroso wines, whilst acetaldehyde, ethyl acetate and eugenol were typical for *flor* aged Fino wines.

6.2.11 Changes during maturation in phenolic compound content

Concentrations determined in young Palomino Fino wines at the beginning and the end of fermentation showed that procyanindin B1, epicatechin, caftaric acid, *cis* and *trans* p-coutaric acid and to a lesser extent caffeic acid increased during fermentation (Benitez *et al.*, 2005). Data are given in Table 6.4. Interestingly, these authors also investigated the effect of de-stemming on the phenolic composition but there were no significant differences in phenolic content.

The phenolic compounds in Sherry can form a substrate for oxidative reactions, influencing the organoleptic properties of the product. Flavonols (quercetin, kaempferol and isorhamnetin) have been identified in maturing Finos but not in maturing Amontillados and Olorosos by Estrella *et al.* (1987), who postulated that their absence from the latter could be due to polymerization reactions whereby, under the oxidative conditions prevalent during maturation, they may contribute to the formation of coloured compounds. The Fino style wines were selected to be low in phenolic compounds but all wines are kept in wood at higher levels of alcohol than typical table wines, Oloroso wines being kept with the highest alcohol content.

The breakdown of lignin by ethanol gives small phenolic compounds in the wine, some of which may be further modified by yeast metabolism (Chapter 3). Estrella *et al.* (1986) studied non-flavonoid phenol compounds (cinnamic acids, benzoic acids, phenolic aldehydes and coumarins) during ageing in eight

Table 6.4 Phenolic contents in fermenting Palomino must and young wines.

Group	Compound	Concentration mg L^{-1} 2 days after inoculation	in young wine
Hydroxybenzoic aids	Gallic acid	3.7–8.3	2.4–10.5
	Syringic acid	0.9–2.2	1.2–1.8
Hydroxycinnamic Acids and Esters	Caftaric acid	22.7–33.1	37.7–40.9
	2-S-Glutathionyl caftaric acid	4.4–9.8	7.4–10.1
	cis p-Coutaric acid	1.8–3.9	3.5–4.2
	trans p-Coutaric acid	4.9–8.8	9.0–10.1
	Fertaric acid	0.3–0.7	0.6–0.9
	Caffeic acid	1.5–5.1	2.8–5.1
	trans p-Coumaric acid	0.4–1.2	0.3–0.6
	Ferulic acid	0.2–0.5	0.3–0.6
Flavan-3-ols	Procyanindin B1	0.9–2.2	1.6–3.4
	Catechin	5.0–6.9	1.6–3.4
	Procyanindin B2	5.8–8.6	1.7–4.5
	Epicatechin	2.2–4.6	2.0–6.4

Data adapted from Benitez et al. (2005).

Table 6.5 Contents of non-flavonoid phenols in Sherry[a].

Non-flavonoid phenols	Wine style	Concentration (mg L^{-1}) Minimum	Maximum	Mean
Cinnamic acids	Fino	8	60	16
	Amontillado	19	82	47
	Oloroso	12	138	52
Benzoic acids	Fino	6	242	80
	Amontillado	34	293	138
	Oloroso	32	392	122
Phenolic aldehydes	Fino	4	46	22
	Amontillado	10	123	55
	Oloroso	11	147	76
Coumarins	Fino	9	94	43
	Amontillado	14	115	53
	Oloroso	15	112	54

[a]Calculated from Estrella et al. (1986).

Sherry solera systems. They reported a steady increase in phenolic compounds but there were also considerable differences between soleras for the same Sherry style, attributed to different bodegas in which the studies were carried out. The minimum, maximum and mean concentrations of these four groups of non-flavonoid phenolic compounds calculated from their data are shown in

Table 6.5. These data show that Fino contained the lowest concentrations but the differences between Oloroso and Amontillado are small. The higher concentration of alcohol in maturing Oloroso and Amontillado may have contributed to the higher extraction of these phenolic compounds from the oak casks. In addition, water loss through evaporation, fractional blending and, in Oloroso and Amontillado, the oxidative maturation conditions will have contributed to the range of values.

6.3 Port wine

The production of Port is described in Chapter 1. There is relatively little scientific information on Port wine. The wine style originates from the Douro region in northern Portugal, but this style of sweet fortified wine is sufficiently popular for New World wine makers to make these wines in Australia, California and South Africa. The extent of its production varies with the local popularity of the drink. The discussion below focuses on the Portuguese Port wines. Most Port wines are red and due to the slightly different composition of the young Port wines compared with table wines, the chemistry underlying the colour changes are typical for these wines. The manufacturing procedures for the Port styles are all very similar. Besides the climate and the geological influences, both the choice of fruit and the maturation parameters determine to a great extent the chemical composition and the sensory attributes of the final product and will be discussed next.

The choice of young wine, which can be made from different varieties/cultivars produced in different parts of the extensive Douro region, coupled with a period of maturation in old oak casks (often with a capacity of 500–600 L, referred to as pipes) ranging from 3- 20 or more years, gives wines with a spectrum of different colours and flavours. Several reviews deal mainly with the technology of wine-making (Goswell & Kunkee, 1977; Goswell, 1986; Reader & Dominguez, 1994) and Bakker (1993) has reviewed the composition of Port. A brief review on the production and chemistry of Port wine is also given by Jackson (2008).

6.3.1 Port wine producers

The grapes are either grown by farmers or by the companies actually making the wines. Since traditionally these companies were also responsible for exporting the wines, they tend to be referred to as 'shippers'. Grapes grown by farmers can be made into wine in small, often quite elementary, wineries on the farm (usually referred to as *quinta*), as a rule under the supervision of the shipper who has agreed to buy the wines from the farmer. There seems to be a trend for the smaller farms to sell the grapes rather than the wines to the shipper, who will then make wine at their own winery. However, some of the bigger quintas produce their own wines and some of these wines are nowadays marketed as quinta wines. Some shippers also have vineyards of their own, so for part of their wine production they can be entirely in control of the quality of the wines.

The wines are made in the Douro valley, although as a result of the rugged terrain and many small roads there is still quite some transporting to be done to deliver all the grapes to the shipper's winery. The wines are usually matured in Port lodges downstream the river Douro, in Vila Nova de Gaia, near the quay-side of the river Douro. Oporto is just on the other side of the river. Most wine writers give lists of Port shippers or Port lodges, some with notes on the company and their wines. Many Port lodges were established in the early nineteenth century, although some date back to the seventeenth century, and quite a few have British origins.

6.3.2 Commercial Port wine styles

All finished Ports are sweet, as a result of arresting the fermentation by the addition of grape spirit, resulting in a residual sugar concentration of between 80 and 120 g L^{-1} and an alcohol concentration of about 20% v/v. There are three main styles of Port, i.e. Ruby, Tawny and Vintage. Within each basic style are a number of different quality categories. The maturation time of these categories is prescribed by the Portuguese authorities. Most Ruby and Tawny Ports are blends of different grape varieties/cultivars and of wines made in different years, which allow the shipper to maintain consistently its unique style, despite seasonal variation. Any indication of age on the label is a typical average age of the wine. Ruby Ports are drunk after about three to five years maturation in old oak casks and have a fruity character, a fairly deep red colour and plenty of phenolic compounds to give a full body to the wine.

Vintage Port has the year of the harvest date on the label and is wine all from one year. The wine is made in exceptionally good years, about three times in a decade and these wines command high prices at wine auctions. It is not treated before bottling and bottled at about two years after its harvest. It needs a decade or more of bottle-ageing before it is considered ready to drink. It will have thrown a considerable deposit. Hence the wine needs careful handling, ensuring that the sediment is not disturbed when removing the cork and it will need decanting before drinking to take the wine off the deposit in the bottle. Vintage Port is the premium Port wine from selected years in which the climatic conditions were ideally suited to produce excellent wine, usually from top ranking vineyards. The wine is matured in wood for two years, followed by a considerable period (often a number of decades) of bottle-ageing, during which it develops a different character to those wines solely matured in casks. The wines remain fruity and with a red colour, even after some of the colour intensity has been lost.

There are also Ruby Port wines that aim to have some of the Vintage characteristics, made in years of not quite Vintage quality. These wines will have received a combination of wood and bottle ageing, for example 'late bottled Vintage'. Such Ports are made from wine of one specific year (usually not a Vintage year) and are bottled after four to six years of maturation in wood. Because of their longer wood-ageing period, they can be drunk much younger than Vintage Ports. Another example is wine with Vintage Character, usually a blend of several years. A currently popular style is 'single quinta Vintage', which is wine from one farm and one year, although not usually a year 'declared' by the shipper as Vintage. It can be the wine that would be declared as Vintage in Vintage years and it tends to be released on the market ready to drink.

Tawny Ports are aged ten years or more, even up to 30 years, in old oak casks and develop a nutty, raisin like, even slightly oaky character. The phenolic compounds tend to soften during the prolonged maturation and the wine gains an orange–brown Tawny colour. These quality Tawnies generally have an indication of age on the bottle and may contain wines from a number of different years. Some younger Tawnies of a more commercial style may have matured considerably shorter, using lighter wines as a base; however, these wines lack the complexity of truly aged Tawny Ports. Some shippers produce blends of real wood-aged Tawny Ports from a single vintage, referred to as 'Colheita' Ports.

6.3.3   Wine writers' comments

There are many descriptions of the wines and because of the different styles and the individuality the many Port shippers impose on the wines, there is no doubt that there are many flavours and tastes in these wines. Clarke (1996) considers that all Ports will have a degree of sweetness and fieriness, the latter designed to be reduced during maturation but all Ports will need some 'bite' to balance their sweetness. According to Clarke:

> The best Ports have a peppery background to a rich fruit, which is both plummy and raisiny, getting a slight chocolate sweetness as they become older, and managing to mix perfumes as incompatible as fresh mountain flowers, cough mixture and old leather ...
>
> Clarke (1996)

whilst he considers that 'Tawnies have a gently brown sugar softness, with some bite to balance the sweetness'. Broadbent (1979) gives separate defini-tions of taste for Ruby, Tawny and Vintage Ports. According to him 'rubies should be full and true Ruby in colour, with a fruity, peppery "nose", not unlike a young "Vintage Port", invariably sweet, full and fruity, often strappy and hefty'. Some of these tasting terms are not typical for the description of drinks and terms such as 'strappy' and 'hefty' are not defined and presumably indicate some mouthfeel or astringency sensation. In contrast 'Tawnies should have an amber-Tawny hue, soft and nutty, sweet and harmonious, extended flavour and fine aftertaste'. Young Vintage wines are described as 'deep red, very purple, peppery, alcoholic and unyielding, very sweet, full bodied, fruity and slightly rasping'. Robinson (1995) describes the great majority of Ruby Ports as 'vigorous, juicy stuff'. She refers to cheap Tawny Ports as wines 'lighter and browner' than the Ruby Ports. The sensory credits are given to Tawny Ports that have aged for one or more decades in old oak casks. She describes real aged Tawny as having an 'alluring light shaded jewel-like Tawny colour' and continues to comment that 'my most hedonistic Port drinking experiences have been with 20-year old [Tawny] Ports, which taste as good served chilled in the heat of the Douro summer as they do next to the fireside in a British winter'.

6.3.4 Grapes and must

Historically the wines were made from up to 60 varieties and many of these are still in production (Goswell & Kunkee, 1977). Clarke & Rand (2001) suggest that only five of these varieties/cultivars are currently recommended, Touriga Nacional, Touriga Franca (the new name for Touriga Francesa, renamed in 2001), Tinta Roriz, Tinto Cão and Tinta Barroca, while the other varieties/cultivars are permitted but not used in great quantities.

The grapes are harvested when the specific gravity of the must is between 1.090 and 1.100, with a typical total acidity between 0.39 and 0.60 g per 100 ml (expressed as tartaric acid), a pH between 3.3 and 3.7 (in warmer regions the pH can be as high as 4.0) and a phenolic compounds content between 0.4 and 0.6 g L^{-1} (Goswell & Kunkee, 1977). When the acidity of the must is too low it is adjusted by an addition of tartaric acid. A study on 95 Port wines made from 16 different grape varieties/cultivars grown at five different sites over a six-year period showed a wide range of grape maturity, with specific gravities from 1.071 to 1.119, dependent on variety and climatic conditions. The pH values of the must ranged from 3.27 to 3.90, resulting in Ports with a pH of 0.29 ± 0.10 units higher than the must pH (Bakker *et al.*, 1986a).

To make wines that can mature for many years, even decades, it is important for the wines to have a sufficient amount of colour, due to the anthocyanins, and other tannins. The grape varieties/cultivars differ in their phenol and anthocyanin content, hence the wine maker will select the grapes he or she uses to make wines destined to mature and develop into either Ruby or Tawny Ports. In most vineyards, a mixture of varieties/cultivars is planted, making it difficult to pick all grapes at optimum maturity, which is one reason to plant new vineyards in blocks of the same variety.

In sixteen *Vitis vinifera* grape varieties/cultivars used to produce Ports, all the anthocyanins are 3-glucosides (Bakker & Timberlake, 1985a). In all but three varieties/ cultivars the anthocyanins are located in the skins and anthocyanins based on malvidin generally predominate. Malvidin 3-glucoside is the major pigment (33–94%), followed by malvidin 3-*p*-coumarylglucoside (1–51%) and malvidin 3-acetylglucoside (1–18%). Peonidin 3-glucoside (1–39%) is prominent in four varieties/cultivars, but delphinidin 3-glucoside (1–13%), petunidin 3-glucoside (2–12%) and cyanidin 3-glucoside (trace–6%) are present at low concentrations. These seven anthocyanins formed usually account for 90% or more of the anthocyanin composition. Although the same anthocyanins could be found in all varieties/ cultivars, the authors found that the ratio malvidin 3-acetylglucoside/total malvidin 3-glucosides (the latter being defined as the sum of the percentages malvidin 3-glucoside, malvidin 3-acetylglucoside and malvidin 3-*p*-coumarylglucoside) was characteristic of variety, independent of site and a useful aid to identify the grape varieties/cultivars. The structural formulae and some of the chemical properties of these compounds are described in Chapter 3.

6.3.5 Fermentation and base Port wine

The grapes are picked, crushed in the presence of sulfur dioxide, normally between 50 and 150 mg L⁻¹ and fermented to approximately half its sugar content before stopping the fermentation by adding fortifying spirit to bring the level of ethanol to 18% v/v. The young fortified Ports may still contain a residual concentration of sulfur dioxide (20-100 mg L⁻¹) used during crushing, which will influence the ageing mechanisms occurring in the young Ports. Young wines are usually full bodied, rather high in phenolic compounds, deep red, fruity and with a hint of grape spirit perceptible on the nose.

The short maceration on the skins means the wine maker has to ensure a sufficient extraction of the coloured anthocyanins in a very short time, unlike in red wine-making where the fermenting must as well as the wine can be left in contact with the skins if so desired. Numerous techniques in Port wine-making are tried and used to maximize extraction (for more discussion, see Reader & Dominguez, 1994). Fermentation can take place in a traditional *lagar*, which is an open stone trough, of varying sizes, but typically about 3 x 3 m, usually filled with approximately 0.5 m grapes, which can be 'worked' by foot treading or a mechanical pumping method. The fermentation can also take place in tanks, which are many times higher than wide; there is an opening at the top of the tank, used for filling, and the bottom can usually be opened to empty the tank. Two methods in current use for Port fermentation, the traditional open *lagar* fermentation, including treading and fermentation in a tank with pumping-over have been compared (Bakker *et al.*, 1996), using grapes of a single variety, picked in the same vineyard and on the same day. Their results showed that the type of fermentation vessel and extraction method used during Port fermentation had a very little effect on the characteristics monitored during fermentation, such as yeast growth, sugar depletion, alcohol formation and the metabolism of amino acids. The *lagar* method extracted a little more coloured pigments and phenolic compounds than the tank method but as the wines matured these analytical differences became insignificant. After three years maturation the sensory quality of the finished wines was not dependent on the method of production and the analytical differences between the wines was minimal.

A biochemical method of enhancing extraction has also been experimented with. The use of commercial pectolytic enzymes, Vinozym G and Lafase H.E., to extract pigments during the short processing of crushed grape mash prior to fortification to make Port wine has been tested. The pectolytic enzyme preparations were used to evaluate the effect on colour extraction during the short processing time (Bakker *et al.*, 1999). Results showed that both enzyme preparations enhanced colour extraction during vinification and gave darker wines than the control, without any apparent sensory disadvantages for their suitability as Ruby Ports. The wines underwent similar changes during maturation but significant differences in colour were maintained after 15 months maturation. There was also an indication of enhanced fruity character of wine made with pectolytic enzymes but this observation requires further investigation.

The short fermentation period on the skin makes an efficient extraction of anthocyanins (the coloured phenolic compounds located usually in the skins of the grapes) essential. In young single-variety Ports made over three different years from up to 16 varieties/cultivars the total contents of anthocyanins ranged from 143 to 1080 mg L^{-1}, with an average of 330 mg L^{-1} (Bakker & Timberlake, 1985b). The varieties/cultivars Touriga Nacional and Touriga Francesa (now known as Touriga Franca) tended to give high contents, while Mourisco gave low contents of anthocyanins in the resulting Ports. The distribution of anthocyanins in the young Port wines differs from those determined in the grape skins. The relative amounts of malvidin 3-glucoside were higher in Ports than in grape skins, indicating some hydrolysis of the acylated malvidin 3-*p*-coumarylglucoside during fermentation. According to Singleton (1992) extraction of phenolic compounds and anthocyanins into wine during the wine-making process is usually only about 40% of the concentration in the grape, with 60% or more of these compounds being 'lost' by incomplete extraction, adsorption, precipitation with solids or proteins, conversion into non-phenolic compounds (for example by oxidation to quinones) or polymerization to an insoluble condition.

The climatic conditions also seem to exert quite an influence on the accumulation of anthocyanins (Bakker & Timberlake, 1985b). These authors reported that the concentrations of total anthocyanins in young Ports made from single varieties/cultivars from the same site over three years varied two-fold. This variation was as great as that in the same variety from different sites in any one year. Varieties/cultivars contained consistently more anthocyanins when grown in the upper Corgo (relatively hotter region) than when grown in the lower Corgo (relatively cooler region).

The effect of seasonal variation on the colour of Port wines was investigated in up to seven harvests, and the effects of five different sites were examined in 95 young Port wines (Bakker *et al.*, 1986a). Immediately after processing up to sixteen single variety Port wines were analysed for colour, pigment (a measure of total colour potential independent of the pH of the Port) and phenol content. The authors reported that seasonal variation over six harvests was two-fold for pigments (in agreement with the two-fold variation in anthocyanin concentration, discussed above) and 1.6-fold for phenolic compounds. The differences between the varieties/cultivars were much greater than the effect of season, there was a 12-fold variation between the pigment contents of the Ports, but only a 3.6-fold variation between the phenol contents. These results show that even in this relatively warm wine-making region the seasonal variation will give wines with greatly different colours, presumably explaining why Vintage years are usually declared only two or three times in a decade.

6.3.6 Port wine compared to red table wine

The are two significant differences between fortified Port wine and table wine, in addition to their higher alcohol content discussed above. Firstly, Port wines are sweet, and during their lengthy maturation time breakdown products of the residual grape sugars may contribute to the volatile aroma compounds of the mature wine. Secondly, young Port wines tend to have higher acetaldehyde compounds than red table wines, influencing both the colour and flavour development of these wines. Especially, this second difference is thought to be relevant to the character of Port, particularly after maturation. Acetaldehyde (ethanal) is an important aroma compound in wine (see Chapter 4) and in Port wines it contributes also to the chemistry of the changes occurring during maturation. Its wide-ranging rôle in wine-making has been reviewed (Liu & Pilone, 2000). Acetaldehyde in Port wines is derived from the fermentation, the fortifying spirit, and formed during maturation.

During the fermentation the fermenting yeasts produce acetaldehyde in the must and the highest amount of acetaldehyde is formed when the yeast action is in its most vigorous phase (Whiting & Coggins, 1960; see Nykänen, 1986). Therefore, by terminating the fermentation at halfway, the acetaldehyde concentration in young Port wine remains higher than when the wine would have been allowed to ferment to dryness. This is thought to be one reason why the acetaldehyde concentration is higher in sweet fortified wines than in dry table wine. The amounts of acetaldehyde formed by yeasts also depend on the type of yeast and unpleasantly high concentrations can be formed in wines (more than $600\,mg\,L^{-1}$, see Liu & Pilone, 2000). The acetaldehyde is in part derived from the fortifying spirit itself, which is not neutral in character, and contains appreciable amounts of acetaldehyde. During maturation of Port wines under fairly oxidative conditions, further acetaldehyde may be formed by oxidation of ethanol (Wildenradt & Singleton, 1974) and data on Port maturation suggest that acetaldehyde is formed during maturation (Bakker & Timberlake, 1986).

Acetaldehyde binds reversibly but strongly, in equimolar concentrations to sulfur dioxide, forming the acetaldehyde-bisulfite complex ($CH_3CH(OH)SO_3^-$) (see Chapter 3). In table wine the usually low content of acetaldehyde tends to be bound to any sulfur dioxide present, the latter being added as a processing aid during various stages of Port wine and table wine-making. In freshly made Port wines the molar concentration of acetaldehyde is usually in excess over sulfur dioxide, hence there is both 'free' and 'bound' acetaldehyde in Port, giving by addition 'total' acetaldehyde (Bakker & Timberlake, 1986). Only free acetaldehyde participates in the polymerization reactions occurring during the maturation. The presence of free acetaldehyde (thus not bound to sulfur dioxide) in Port wine forms an important contrast to table wine, contributing to the different character of the wines after maturation. However, as the sulfur dioxide concentration decreases due to oxidation reactions, bound acetaldehyde is gradually released as free acetaldehyde. The average total acetaldehyde content of 55 young Port wines is reported to be 127.3 ± 25.6 mg L^{-1}, while the free acetaldehyde content is 65.1 ± 24.2 mg L^{-1} (Bakker & Timberlake, 1986). Figures for red table wines indicate a range of total acetaldehyde from 34 to 94 mg L^{-1}, although the author stated that the free acetaldehyde would be fairly low due to binding with sulfur dioxide (see Lykänen, 1986).

6.3.7   Maturation

Young Ports are usually sweet, intensely red with a purplish tinge, high in tannins and they taste sweet but are harsh and astringent due to tannins. Their smell is fruity and reminiscent of the fortifying spirit. The colour, aroma and flavour of young Ports are due to compounds from the grape, the fermentation and the spirit. These wines need maturation to develop the complex sensory attributes typical for the various Port styles. During maturation, they become browner, changing from a deep red with a purple edge at the rim of the glass, to a brick red or even amber Tawny colour, depending largely on the length of the maturation. Maturation times of three to five years generally lead to the development of Ruby styles, while a considerably longer maturation period (up to 20 years) in wood is needed to produce some of the Tawny styles. The chemical processes underlying the changes occurring during maturation are complex and by no means all understood. Some changes are more typical for Port wines but many of the reactions occurring during the maturation of red table wines are likely to be relevant during Port maturation also. Accompanying the colour changes are alterations in perceived astringency due to modifications in the phenolic compounds. The rather harsh character of the young wines is lost, and the tannins become 'softer'.

At the same time, the bouquet of the Port changes from its youthful fruity aroma to that of a typical Port. The final aroma depends on the length of maturation and whether the Port has matured in wood (usual for most Port wine styles) or mostly in-bottle (Vintage Port style). In the data on volatile composition of Port wines collected to date no terpenes have been reported (see discussion below), and the fruity part of the aroma of young Ports is, presumably, due to esters, in addition to compounds formed in wines with high sugar and alcohol content, matured under oxidative conditions (see Section 6.3.9).

6.3.8   Colour changes during maturation

Young wines are usually left in wooden (usually old oak) vats for two or three months after vinification. During this time yeast cells, suspended solids from the grapes and excess tartaric acid not soluble in the wine settle at the bottom of the vessel. The wine then receives its first racking and gets taken off the debris at the bottom of the vessel and put into a clean one. There is usually a certain amount of aeration involved in such racking procedures, giving the wine the oxygen that is thought to be involved in the colour changes occurring during the maturation of wines (Singleton, 1992). Unlike table wines, in which oxidation rapidly leads to detrimental effects on quality, the maturation of Port wine is thought to benefit from a limited regime of oxidation over a prolonged period of time.

The colour of young Port wines is mostly due to the red monomeric anthocyanins, but during maturation these anthocyanins are gradually 'lost'. Polymerization reactions between anthocyanins and other phenolic compounds lead to the formation of larger so-called oligomeric or polymeric molecules, in which the anthocyanins are used as building blocks. In Port wines two different polymerization reactions are thought to occur simultaneously, although the extent of each reaction depends on the chemical concentrations, in particular of acetaldehyde. These reactions coincide with the observed losses of anthocyanins and the qualitative and quantitative changes in colour during maturation.

During the first months of maturation the Port wine becomes more intensely coloured, a change typical for Port wine, and referred to as 'closing up'. The increase in colour can be up to 80%, measured as colour density [defined as the sum of the colour intensity measured spectrophoto-metrically at 520 nm (the red region) and 420 nm (the brown region)], depending on the free acetaldehyde concentration in the young Port wines (Bakker & Timberlake, 1986). This initial increase in colour during Port ageing is attributed to the formation of acetaldehyde-containing oligomeric pigments that are more coloured at Port pH than the anthocyanins from which they are derived (Bakker & Timberlake, 1986). Other phenolic compounds (mostly flavan-3-ols) are also involved in these reactions, leading to the formation of acetaldehyde-bridged polymers between anthocyanins and other phenolic compounds. These authors suggested that anthocyanins react strongly with free acetaldehyde but the oligomers formed become less reactive with increasing size. When the wine reached its maximum colour, at 'closing up', the larger oligomers are thought to become insoluble and start to precipitate. However, while the colour remains on this plateau, for a few months at most, the formation of new polymers and the loss of older and larger polymers by precipitation are thought to be in equilibrium. Gradually the precipitation process is faster than the formation of new oligomers and the colour of the Port becomes less intense. In addition, the polymers which remain soluble could become less coloured by incorporation of other colourless phenolic compounds. These changes in visible colour, viz. an increase in intensity followed by a plateau before a decrease in intensity, are accompanied by a measurable loss of monomeric anthocyanins.

The rate of acetaldehyde-induced polymerization reactions is governed by the free acetaldehyde concentration. During maturation free acetaldehyde is liberated from the acetaldehyde–bisulfite complex by oxidation of sulfur dioxide (Bakker & Timberlake, 1986). This process ceases to be important when all sulfur dioxide has been oxidized into sulfate. These authors also found that additional acetaldehyde was formed by oxidation of ethanol, confirming a previous observation (Wildenradt & Singleton, 1974). Small amounts of acetaldehyde were produced during maturation, presumably by coupled oxidation of ethanol, even in the presence of low concentrations of sulfur dioxide. The normal procedure of racking Ports during maturation is thought to give sufficient aeration for these oxidative polymerization reactions. Direct condensation of anthocyanins with other phenolic compounds also occurs but this process is slower than the condensation involving acetaldehyde. The extent of both reactions in Port wines is thought to depend on the free acetaldehyde concentration (Bakker & Timberlake, 1986), with the formation of acetaldehyde-bridged polymers being more prominent at higher concentrations.

Studies in model solutions have confirmed the occurrence of these reactions and their accompanying colour changes (Bakker *et al.*, 1993). These authors confirmed the existence of these dimers by determining the mass of these dimers, consisting of malvidin 3-glucoside linked to catechin linked by an acetaldehyde bridge. They used, at the time, the still fairly new technique of fast-atom bombardment mass spectrometry (FABMS), which allowed the accurate mass determination of large non-volatile compounds. Model solutions mimicking Port wines (containing free acetaldehyde as well as anthocyanin and flavan-3-ol) also showed large increases in colour during the early part of storage, concurring with the rapid loss of monomeric anthocyanins. Further studies on model Port wines showed that added sulfur dioxide slowed the loss of anthocyanins and the accompanying colour changes (Picinelli *et al.*, 1994). The mechanism of formation of acetaldehyde-containing oligomeric pigments in wine is probably similar to that described in model systems, with the formation of complexes containing an acetaldehyde link between anthocyanin and flavan-3-ol.

Colour changes during Port wine maturation are accompanied by losses in measurable anthocyanins (Bakker *et al.*, 1986b), which can be determined by HPLC. Just after fermentation polymeric pigments contribute between 22% and 30% of the colour, indicating that the chemical changes in colour composition have already started during the short fermentation (Bakker, 1986). After 16 weeks, more or less coinciding with the 'closing-up' colour plateau in Port wines, 63–66% of the colour in normal Ports is due to polymers, while in a Port containing a high concentration of acetaldehyde already 91% of the colour was due to polymers. After 46 weeks of maturation, between 78 and 95% of the colour was due to polymeric pigments, with the extent of polymerization depending on the concentration of free acetaldehyde. The losses of anthocyanins are logarithmic with time and the two acylated-anthocyanins (malvidin 3-acetylglucoside and malvidin 3-coumarylglucoside) are lost faster than malvidin 3-glucoside.

Monomeric anthocyanins are lost, mostly by participation in these polymerization reactions. The colour changes in maturing wines can be measured accurately and even relatively easily seen. However, many other reactions also occur, although the extent of each one in maturing wines remains very difficult to quantify. For example, monomeric anthocyanins are usually bleached by sulfur dioxide but as the wine matures it tends to become more resistant to any bleaching (see also Chapter 3). The polymeric compounds formed due to chemical changes during maturation are more resistant to bleaching by sulfur dioxide. Little is known about the colour properties of the polymerized anthocyanins, regarding both the quantity and quality of colour on a molecular basis, so the relative contribution of polymeric compounds to wine colour is difficult to assess. Interestingly, Bakker et al. (1993) determined in model Port wine solutions that maturation in the presence of high levels of acetaldehyde gave relatively less brown solutions than ageing of model wines in the absence of acetaldehyde, indicating that the different polymerization mechanisms may lead to qualitative colour differences in Port wines.

However, anthocyanins also partake in reactions that lead to the formation of modified anthocyanins. Several new and unusual anthocyanins (called vitisins) have been identified. They were found in red Port wines in trace amounts, and were formed in Port wines during maturation (Bakker & Timberlake, 1997; Bakker et al., 1997). One of these compounds, vitisin A, consists of malvidin 3-glucoside containing a $C_3H_2O_2$ link between carbon-4 and the 5-hydroxyl of the molecule. A tentative structure is proposed for vitisin B, which is decarboxy-vitisin A or malvidin 3-glucoside with a CH=CH structure linking carbon-4 and the 5-hydroxyl group. Vitisin A is less red than malvidin 3-glucoside and vitisin B is orange. Both vitisin A and B also occur in acylated forms, having the 6-position of the sugar acylated with acetic acid. A series of studies in model wines and Port wines with additions of pyruvic acid showed the formation of these vitisins, based on a reaction between malvidin 3-glucoside and pyruvic acid. These studies also confirmed the orange colour contribution these compounds make to Port wine, and vitisins could be formed even in the presence of acetaldehyde, when their formation is in competition with other polymerization mechanisms occurring in Port wines (Romero & Bakker, 2000a; 2000b; 2000c). Numerous other anthocyanins having undergone such a reaction have now been isolated and identified, belonging to the group of newly identified pyranoanthocyanins and described in more detail in Chapter 3.

These vitisin anthocyanins possess unusual chemical properties and have colour properties quite different from anthocyanins found in wines. Vitisin A is entirely protected from bleaching by sulfur dioxide, while vitisin B shows greater resistance than malvidin 3-glucoside. In wines low in acetaldehyde, some reversible bleaching of the anthocyanins can occur but in young Port wine this is unlikely. A more interesting point is the colour expression of vitisins; they are less sensitive to pH since they express more colour up to pH 7 than malvidin 3-glucoside, the dominant anthocyanin in most young wines. There is evidence of differences in the equilibrium concentrations of the various structures, with the dominant mixture containing the coloured flavylium structure and the quinonoidal base. The colour properties of the vitisins at wine pH are expected to influence the red wine colour, and may explain some of the observations made by many enologists regarding the colour changes in red Port wines during maturation, since at typical wine pH the vitisins are relatively more orange-brown than malvidin 3-glucoside, which has a much redder absorbance maximum than the vitisins. The greater colour intensity and the relatively more orange-brown colour of vitisins than the other monomeric anthocyanins in Port wines may mean that despite the low concentrations of the vitisins in Port wines, their contribution to the colour is relatively large.

All young red wines start as red Ruby-like wines; it is the length of maturation in old oak casks that slowly changes the deep red Ruby wine into a lighter orange-brown Tawny wine. One analytical method published tried to distinguish the two styles analytically and is based on measuring the colour hue of the wines through an increasingly thick layer of wine. Older Tawny style Port wines do get an apparently redder character while Ruby wines get an apparently browner character when the colour is measured through an increasingly thick layer of wine (Bakker & Timberlake, 1985c). The colour of Port wines can accurately be assessed by a trained panel and their qualitative and quantitative colour descriptions matched instrumental colour measurements of Port wines (Bakker & Arnold, 1993).

6.3.9 Volatile changes during maturation

During the long maturation period, the volatile components in Ports undergo considerable changes. Little information is available on the chemistry of these changes, although the data available on red table wines may be relevant (Chapters 3 and 4). The volatile components in Port wines are derived from the grapes, yeast fermentation and the fortifying spirit. However, the end character of the Port wine is to a considerable extent determined by the many processes taking place during the more oxidative maturation of these wines compared with table wines, such as oxidative changes, carbohydrate degradation, formation and hydrolysis of esters, formation of acetals and extraction from wood. As discussed above, the higher ethanol content compared with table wine is expected to influence the equilibria of esters and acetals and increase the extraction of compounds from the oak casks.

Of the more than 200 volatile components thus far detected in Ports, 141 have been wholly or partially identified (Simpson, 1980; Williams *et al.*, 1983; Maarse & Visscher, 1989). These volatile compounds consisted of 14 alcohols, one diol, two phenols, two alkoxy alcohols, two alkoxy phenols, five acids, five carbonyls, three hydroxy carbonyls, 81 esters, two lactones, nine dioxolanes, two oxygen heterocyclics, one sulfur-containing component, four nitrogen-containing components, six hydrocarbons and two halogen compounds.

A number of the identified volatile components are present in high enough contents to contribute to the overall Port aroma. The organoleptic importance of the various groups of volatile compounds did not wholly explain the sensory properties of Ruby or Tawny Ports (Williams *et al.*, 1983). Although there are some exceptions, it is not easy to discern the sensory contribution many of these volatile compounds make, since there is scant information on the typical content of the volatile compounds, nor on their typical threshold values in 20% v/v ethanol. Some of the volatile compounds in Port wines are the same as in red table wines, such as those formed during fermentation, but as a result of the different maturation processes there are qualitative and quantitative differences that are likely to contribute to the typical Port character. Table 6.6 shows some of the compounds formed in Ports and, where possible, their proposed origin is indicated.

Oxidation of alcohols during storage may account for the presence of aldehydes, in particular acetaldehyde is usually present in higher concentrations than in wine (see above). Cullere et al. (2007) determined the presence of oxidation related aldehydes in Port wines, many above their sensory threshold level. The branched aldehydes enhanced the dried fruit aroma and masked the more negative notes from the (e)-2-alkenals. An additive sensory effect was established for these compounds.

Oxidation of fusel and other alcohols in Port wines leads to carbonyls, some of which are thought to have relatively low odour thresholds, and to contribute a slightly rancid character to these fortified wines (Simpson, 1980). Although acetaldehyde will influence the maturation of the non-volatile components in the Ports its rôle as part of the overall Port aroma is not clear.

The maturation in the presence of higher ethanol and acetaldehyde concentrations as a result of the fortification, as well as the other alcohols, glycol and glycerol formed during the short fermentation can result in the formation of acetals. Acetals are generally formed in equilibrium with alcohols and aldehydes and their concentrations are thought to be higher in fortified wines than in table wines due to the higher alcohol concentration. However, they have little aroma themselves and are not thought to contribute significantly to the Port aroma. Aliphatic acetals usually contribute fruity or green aroma characteristics, whereas cyclic acetals have more pungent characteristics. Williams et al. (1983) found two acetals but their structure was not elucidated. Ferreira et al. (2002) determined acetal levels in Port wines stored under oxidative conditions to range from 9.4 to 175.3 mg L^{-1}. They found that the concentrations of 5-hydroxy-2-methyl-1,3-dioxane and 4-hydroxymethyl-2-methyl-1,3-dioxolane increased linearly with age. The flavour threshold determined in wine was 100 mg L^{-1}, hence these acetals are only expected to contribute to the 'old Port wine' aroma in wines older than 30 years. Interestingly, the authors demonstrated that the acetal formation reactions were blocked in the presence of sulfur dioxide, since it strongly binds with acetaldehyde (see Section 6.3.6).

Esters form the most abundant group of volatile compounds in table wines (Chapter 4) and also in Port wines, with ethyl esters and lower molecular weight acetates being most abundant in Ports (Williams et al., 1983). Most esters can be formed during the short yeast fermentation but the maturation will cause changes in equilibrium. The relatively large concentrations of succinates (approximately 100 µg g^{-1}) should contribute a wine-like and fruity aroma to the overall Port bouquet (Williams et al., 1983). Their formation is attributed to esterification and trans-esterification reactions during

Table 6.6 Volatile compounds typically found in Port wines, including their likely origin of formation.

Origin[a]	Chemical class	Compounds	Ref[b]
YF	Alcohols	2-Methyl-propan-1-ol (isobutanol)	1,2
YF		3-Methyl-butan-1-ol	1,2
YF		2-Methyl-butan-1-ol	1,2
YF		Hexan-1-ol	1,2
YF		(2R,3R)-butan-2,3-diol	1,2
YF		(2R,3S)-butan-2,3-diol	1,2
YF WE		2-Phenyl-ethan-1-ol	1,2
YF EM	Esters	Ethyl esters (C$_6$–C$_{14}$)	1,2
YF		3-Methylbutyl acetate	1,2
YF		Hexyl acetate	1,2
YF		2-Phenylethyl acetate	1,2
YF EM CD		Ethyl lactate	1,2
YF EM WE		Diethyl succinate	1,2
YF EM WE		Diethyl malate	1,2
EM		Methyl ethyl succinate	1,2
EM		Isopentyl succinate	1,2
OM		Ethyl 2-phenyl acetate	
EM		2-Methylpentanoate	3
EM		Ethyl 3-methylpentanoate	3
EM		Ethyl 4-methylpentanoate	3
EM		Ethyl cyclohexanoate	3
OM	Aldehyde	Benzaldehyde	
OA		Methional	5
OA		Phenylacetaldehyde	5
OA		(E)-2-Hexenal	5
OA		(E)-2-Heptenal	5
OA		(E)-2-Octenal	5
OA		Methylpropanal	5
OA		Methylbutanal	5
OA		2-Methylbutanal	5
WE	Lactones	Dihydro-5-butyl-4-methyl-3(H)-furan-2-one (oak lactone)	1.2
CD		3-Hydroxy-4,5-dimethyl-2(5H)-furanone (sotolone)	4
CD	Furans	2-Furaldehyde (furfural)	1,2
CD		2-Acetylfuran	1
CD		5-Methyl-2-furaldehyde	1,2
CD		Ethyl 4-oxo-pentanoate	1
CD		Ethyl furanoate	1,2
CD		5-Ethoxymethyl-2-furaldehyde	1
CD WE		5-Hydroxymethyl-2-furaldehyde	2
AF		5-Hydroxy-2-methyl-1,3-dioxane	6
AF		4-Hydroxymethyl-2-methyl-1,3-dioxolane	6

Table 6.6 *(Continued)*

Origin[a]	Chemical class	Compounds	Ref[b]
CarD	Norisoprenoid	β-Damascenone	7
CarD		β-Ionone	7
CarD		2,2,6-Trimethylcyclohexanone	7
CarD		2,2,6-Trimethyl-1, 2-dihydronaphtalene	7
CarD		Vitispirane	7

[a]YF: yeast fermentation, WE: wood extraction, EM: esterification during maturation, CD: carbohydrate degradation, OA: oxidation alcohols. AF: acetal formation. [b]CarD: carotenoid degradation. (1) Simpson (1980). (2) Williams *et al.* (1983). (3) Campo *et al.* (2007). (4) Ferreira *et al.* (2003). (5) Cullere *et al.* (2007). (6) Ferreira *et al.* (2002). (7) Ferreira & Pinho (2004).

Table 6.7 Concentration ranges in ng L^{-1} of ethyl 2-, 3- and 4-methylpentanoate, and ethyl cyclohexanoate in various commercial Port wines.

Wine type Threshold (ng L^{-1})	ethyl-2-mp 3	ethyl-3-mp 8	ethyl-4-mp 10	cyclohexanoate 1
Ruby	22	35	335	3.5
Tawny	36	43	442	14
White	53	39	442	50

Data adapted from Campo *et al.* (2007).

maturation, while diethyl succinate can also be derived from maturation in wood. Esters from 2-phenylethanol have fruity sweet aromas, contributing to the bouquet. Ethyl esters of medium-chain fatty acids have generally low flavour thresholds and are expected to contribute to the fruity aroma of Ports.

Recently, four new esters have been analysed and quantified in a range of wines, with low thresholds (Campo *et al.*, 2007). The authors suggested that ethyl 2-methylpentanoate, ethyl 3-methylpentanoate, ethyl 4-methylpentanoate and ethyl cyclohexanoate are formed by esterification reactions with ethanol and the corresponding acids formed by micro-organisms. Their concentrations are especially high in aged fortified wines and data for Port wines are shown in Table 6.7. The concentrations well above threshold level with OAV's between 3 and 50. These compounds are thought to contribute sweet fruity notes.

Oak lactone (β-methyl-γ-octalactone) is a typical wood extractive and has been reported in many wood-aged alcoholic beverages. It contributes to the nutty, oaky note of the aroma. Vitispirane has been isolated from Ports and other matured wines. Its contribution to flavour is described as camphorous and eucalyptus-like (Williams *et al.*, 1983), although it has also been described as flowery or fruity (see Jackson, 2008). Despite the prolonged wood ageing

Table 6.8 Concentration ranges (µg L^{-1}) of nor-isoprenoids compounds in Port wines.

Compound	Young wine	20 year old	40 year old
β-Damascenone	4.5–13.4	1.8–5.1	3.0–5.0
β-Ionone	0.3–1.4	0–0.4	0–0.7
2,2,6-trimethylcyclohexanone	0.2–0.8	0.7–1.6	1.1–2.7
Vitispirane	0.2–0.8	0.8–1.7	1.4–2.3
1,1,6-Trimethyl-1,2-dihydronaphtalene	0.1–1.9	1.3–3.9	1.5–7.7

Data adapted from Ferreira & Pinho (2004).

of, in particular, Tawny Ports, no phenolic wood-derived aldehydes are listed under Port wines (Maarse & Visscher, 1989). They are listed under red wines and are expected to be present in Tawny Ports.

Furan derivatives, which are readily derived from carbohydrate degradation, have generally sugary, oxidized aromas and probably contribute to the overall Port aroma. A number of these compounds have been identified (Table 6.6). Older Ports tend to have higher concentrations of volatile components, resulting from carbohydrate degradation and wood extraction, such as furfural and oak lactone. Presumably some changes occurring during bottle ageing of red wines also take place in Port wines, in particular in bottle-aged Vintage Ports.

Sotolon has been identified to give the typical 'nutty' and 'spicy' character to barrel-aged Port wines (Ferreira et al., 2003). Generally the levels were founds to increase linearly with oxidative ageing, typical for ageing of Port wines in oak vats. The concentrations ranged between 5 and 958 µg L^{-1} for wines between 1 and 60 years old. The flavour threshold in Port wines was 19 µg L^{-1}, hence the contribution of sotolon to flavour is significant after some ageing. Interestingly, additions of sotolon gave wines an increased sensory score of 'age'.

Breakdown of carotenoid compounds can be related to the formation of β-ionone and β-damascenone in old bottle aged Vintage style Ports, whilst vitispirane, 2,2,6-trimethylcyclohexanone and 1,1,6-trimethyl-1,2-dihydronaphtalene are more typical for old barrel aged Ports (Ferreira & Pinho, 2004, see Table 6.8). These compounds have low sensory threshold values. Since the fermentation Port wines is stopped by adding brandy approximately halfway, carotenoids from the grapes can still be found in young Port wines (see Ferreira & Pinho, 2004). This combined with considerably longer maturation times for some of these wines than for most table wines makes these compounds more typical for mature Port wines. Bottle-aged wines develop floral and violet notes, possibly related to these compounds. Ferreira & Pinho (2004) also determined some of the factors affecting their formation, such as oxygenation (only typical for barrel aged wines) and sulfur dioxide. Studies using supplementions in Port wines with β-carotene showed increases in the formation of β-ionone and β-cyclocitral, whilst lutein additions increased the formation of β-damascenone (Ferreira et al., 2008), confirming that these compounds are likely precursors.

6.4   Madeira

Madeira is a fortified wine, a wine style evolved on the island Madeira, approximately 600 km west of the Moroccan coast and almost 1000 km from Lisbon. Fortification of the wine was introduced and this helped to stabilize the wines during their voyage to the consumers. The process of Madeira production has been described in Chapter 1. The use of heat in wine maturation is fairly unique to Madeira wines, the so-called estufagem process. It is believed to be an accidental discovery; firstly the wines were fortified to help them survive the sea journey and secondly the high temperatures the wines were subjected to during their tropical sea journeys developed all the interesting flavours now valued in these wines.

The styles of Madeira wine range from very sweet to dry, some made from a single cultivar, others made from blends of grapes or wines and even solera type blending, as described for Sherries, can be used. The estufagem process is thought to be the main distinguishing feature of this fortified wine style. They are wines aged between 3–20 years, some can age even longer.

Until recently virtually no scientific papers were published on Madeira wine and most knowledge on these wines was recorded by wine writers. However, over the last decade or so scientific publications on aspects Madeira wines has started to emerge.

6.4.1   Madeira wine producers

Madeira is a small volcanic island, ideally situated on sea trade routes, which may well have helped to establish a wine industry. Typically, the grapes are grown in small vineyards, and each of the major wine-making companies buys the grapes directly from the growers. It is estimated that most vineyards are only a fraction of a hectare. Obviously, this requires an enormous organization, ensuring the grapes are all harvested at the correct time, in good condition, with varieties correctly identified, etc. to ensure adequate quality control.

Almost a decade ago the traditional bulk wine exports have all been stopped under the EU regulations and all wines exported have to be bottled in Madeira.

6.4.2   Commercial Madeira wine styles

There are a number of main styles, mostly depending on the age on the wine. Wines are generally sold without a vintage date on the label, since a solera system (as described for Sherry) is used to prepare the final blend. Part of the skill in this wine-making process is blending, ensuring that the desired styles of wines are maintained. Reserve wine is a blend where the youngest wine is five years old, Special Reserve wine has a minimum of ten years ageing, Extra Reserve has a minimum of 15 years maturity and Vintage is more than 20 years old. A Vintage Madeira is very rare and need not be announced until decades after the wine was made. Experts claim that Vintage Madeiras can last well over 100 years. A new style emerging is Colheita – a wine made from a single harvest; these wines are thought to be almost Vintage quality but released much earlier than Vintage wines.

The most highly regarded *Vitis vinifera* grapes are Sercial, Verdelho, Boal and Malvasia, however, most wine books claim that these main varieties only form a small part of the total grape production, up to 12.5% and the main red variety grown is Tinta Negra. Terms such as 'finest', 'selected' and 'choice' indicate wines made from blends of Tinta Negra grapes.

6.4.3   Wine writers' comments

The styles of wines are quite different, ranging from wines suitable as aperitif to digestif, depending on sweetness and intensity of flavours. The sweetness of the wines seems to be a key part of the description of its character, but as explained in wine-making, the amount of sugar left in the wine depends on when the grape spirit was added and is not a characteristic of the grape variety (see Chapter 1, Madeira). Johnson (1977) describes the wine as follows. Sercial is described as 'light, fragrant and slightly sharp, more substantial than Fino Sherry, but still a perfect aperitif'. Verdelho gives 'a peculiarly soft and sippable wine, the faint honey and distinct smoke make it a wine before or after meals'. Boal is 'lighter and less sweet than Malmsey, but still a dessert wine'. Malvasia is the 'sweetest of them all, probably the best, dark brown, very fragrant and rich, soft-textured and almost fatty, but with a tang of sharpness, a good after-dinner wine'.

6.4.4   Sensory properties

One of the first scientific papers on the sensory properties analysing 52 different Madeira wines described the wines in terms of colour, aroma, taste and 'global appreciation', using terms such as depth of colour, aroma quality, taste quality, body, finish, tipicity and sweetness (Nogueira & Nascimento, 1999). Surprisingly, two of the terms were strange aromas and strange tastes; there were no definitions of any of the sensory terms used. A detailed study on the sensory properties of four ten year old Madeira wines made from the four classic grape varieties Malvasia, Boal, Verdelho and Sercial and three younger wines showed that descriptors characterizing the wines were candy, nutty, maderized, toasty, laquer and dried fruit (Campo *et al.*, 2006).

6.4.5   Grapes and must

The main grapes used for Maderia wine production are white varieties of *Vitis vinifera*, and usually the grapes are vinified separately to produce their typical wine styles. Sercial generally is made into the driest wines, with about 25 g L^{-1} sugar and has a light golden colour, a firm acidity and this wine generally needs ten years maturation, although it is reported to last for a century. Wine books report that this grape is the same as the German Riesling grape, and hence a late ripener and not a good cropper. Producers on the island do not believe that Sercial is the same as Riesling. Verdelho is the softest of the four main grapes, giving the softest Madeira style wines and is usually medium dry (65 g L⁻¹ sugar). Boal is one of the most common grapes, giving rich, brown wines, with firm acidity, and sugar levels around 90 g L⁻¹. Malvasia, thought to originate from Greece, is also referred to as Malmsey and is often quoted as the original and best grape variety and was first planted in Madeira more than five centuries ago. It gives a dark brown wine, with a caramel richness, with a sugar content of 100 g L⁻¹ and good acidity.

Other grapes used are Terrantez (white), Bastado and Tinta Negra (both black), and Tinta Negra makes up currently the bulk of the commercial wine styles produced.

6.4.6 Base wines maturation

The character of these wines is determined by climate, the volcanic soil, grape varieties and to a considerable extent the unique production techniques used. Fermentation traditionally took place in concrete troughs or nowadays mostly in stainless steel fermentation tanks. Small producers also use wooden casks for fermentation. In order to retain natural sweetness in the base wines, the fermentation is stopped by the addition of neutral grape spirit (95% v/v). Nowadays all Madeira wines are fortified by adding neutral grape spirit (95% v/v ethanol) to 14-18% v/v ethanol. The wines are clarified and ready for the maturation with heat treatment.

The base wine is stored in casks or larger vats in a special lodge, referred to as the 'estufa', which is gradually heated to 45-50°C, the temperature is raised over a two week period. Usually the wines are kept at this temperature for at least three months. The temperature is allowed to drop slowly to ambient and the wines are matured in oak casks for a further two years, or more. More heat can be used during maturation by, for example, storing casks above the heating room, or in the sun.

Depending on the desired wine style, various maturation steps take place, the use of heat and plenty of time seem to be the common factors in the process. The wines are usually fined to remove the deposit formed during the heating and further aged as required. To ensure the required commercial blends can be made, special sweetening wines (surdo) are prepared by fortifying must just after fermentation has started, or by fortifying juice (abofado). Colouring wine is made by heat concentrating must to about a third of the volume, giving this colouring wine a dark brown colour and a caramel smell.

One of the first scientific papers on Madeira wines was published by Nogueira & Nascimento (1999), publishing data characterising Madeira wine. Small analytical changes were observed during the estufagem and further maturation over ten years; none of these were very significant or different from the expected changes. Of course, significant changes in volatile composition were observed, with three year old wines having a volatile acidity ranging from 0.34-0.50 g L⁻¹, whereas ten year old wines ranged from 0.88-1.0 g L⁻¹.

6.4.7 Volatile compounds

Despite Madeira wines being one of the classic fortified wines, until recently very little published scientific information was available but over the last decade some scientific data on the volatile compounds has become available. Presumably this can be attributed to the development of better, faster and cheaper analytical techniques for volatile compounds (Chapter 4). During the extensive ageing in wood, changes expected to be typical for these wines are wood derived compounds and sugar breakdown compounds, in addition to the changes in ester composition typical for maturing wines. One of the first papers including aspects of volatile composition by Nogueira & Nascimento (1999) indicated that the average 5-hydroxymethylfurfural content increased both with sweetness and with the age of the wine, as expected in these sweet wines, since this is a break down product of hexoses (Table 6.9).

Methanol increased, attributed to pectine break down and acetaldehyde also increased, attributed to the enzymic decarboxylation of pyruvic acid. Acetaldehyde can also be formed under oxidative conditions in wines by oxidation of ethanol (Wildenradt & Singleton, 1974), and has been reported to occur in Port maturation (Bakker & Timberlake, 1986). Whether these oxidative conditions prevail during maturation of Madeira wines is an interesting question but there is insufficient published information available to answer.

Alves et al. (2005) publishes a characterization of the aroma profile of Madeira wine of five grape varieties using sorptive extraction techniques for sample preparation. The general compositions of volatiles are shown in Table 6.10. The authors suggested that the C13 isoprenoids may affect the Madeira wines and they reported a good correlation between cis-oak lactone and ageing, and regarded it as an important compound to characterize older wines. Both trans-oak lactone (trans-5-butyl-4-methyl-4,5-dihydro-2(3H)-furanone) and cis-oak lactone (5-butyl-4-methyl-4,5-dihydro-2(3H)-funanone) were identified as significant aroma compounds derived from wood, in older Madeira wines. The cis-oak lactone has a sensory threshold of 87 µg L^{-1}, with aroma descriptors of woody, coconut, vanilla, berry and dark chocolate (quoted by Alves et al., 2005), although coconut seems to be the main term reported (see Table 4.21). Other trace aroma compounds these authors reported to be relevant to the Madeira wine aroma were 5-hydroxymethyl furfural and 2,3-dihydroxy-6-methyl-4H-pyran-4-one; both compounds are known to be formed by Maillard reactions. The aroma of 5-hydroxymethyl furfural is a combination of cinnamon and dried fruit.

Sotolon, previously identified in other sweet wines (see Table 4.21), has been identified in Madeira wines (Camara et al., 2004; Camara et al., 2006). The concentrations of sotolon linearly increased with the age of the wine, with none detected in wines being a year old, and up 100 µg L^{-1} in wines six years old to 1000 µg L^{-1} in 25-year-old wines. It is a strong odorant and using the odour threshold value 19 µg L^{-1} of Port wines (Ferreira et al., 2003) odour activity values greater than 20 were observed. The concentration averages quoted were 825 µg L^{-1} (Malvasia), 540 µg L^{-1} (Boal), 430 µg L^{-1} (Verdelho) and 258 µg L^{-1} (Sercial), ranging from sweet to dry wine. Furfural, 5-methylfurfural

Table 6.9 Ranges of some volatile compounds determined in dry to sweet Madeira wines, aged between 1–10 years.

Compound	Units	Range	Formation
5-Hydroxymethylfurfural	mg L^{-1}	171–575	Breakdown of hexose sugars, higher values in sweet wines
Methanol	mg/100ml	24–46	Hydrolysis pectins
Acetaldehyde	mg/100ml	34–65	Enzymic decarboxylation of pyruvic acid
Ethylacetate	mg/100ml	41–128	Yeast and acetic acid metabolism
Total higher alcohols	mg/100ml	81–150	Fermentation

Information adapted from Nogueira & Nascimento (1999).

Table 6.10 Percentages of volatile compounds detected in Madeira wines (Sercial, Verdelho, Boal, Malvasia), total 40 compounds.

Compounds	percentage	
	min	max
Esters	80.7	89.7
Carboxylic acids	1.6	4.2
Alcohols	3.5	8.2
Aldehydes	0.9	3.7
Pyrans	0.2	1.7
Lactones	<3	
Monoterpenes	0.1	1.4
Sesquiterpenes	0.1	0.8
C13 Norisoprenoids	1.7	6.5

Data adapted from Alves et al. (2005).

and 5-hydroxymethylfurfural were also formed in increasing amounts during the ageing of Madeira wines, but the odour thresholds of these compounds are high and hence their contribution to these wines is not thought to be very great. 5-Ethoxymethylfurfural can be formed from 5-hydroxymethylfurfural and gives spicy and curry notes.

Analysis of esters in Madeira wines made from Sercial, Verdelho, Boal, Malvasia and Tinta Negra, ranging from sweet to dry wines were analysed by Alves et al. (2005). The authors reported the dominating esters in order of abundance as ethyl octanoate, ethyl decanoate, ethyl decenoate and diethyl succinate (Table 6.11). Octanoate and decanoate fermentation esters were higher in dry and medium dry samples. C13 Norisoprenoids were also detected, below the reported threshold limits. However, although the quantification techniques have much improved over the last decade, the authors did point out that capacity limitations of the sampling techniques may not pick up very low level concentrations of sensory important aroma compounds accurately.

Table 6.11 Concentrations (μg L^{-1}) of esters found in five Madeira wines from single grape varieties (Sercial, Verdelho, Boal, Malvasia and Tinta Negro Mole) ranging from sweet to dry.

Compound	Concentration		Odour /flavour descriptors[b]	Threshold data[b]
	Min[a]	Max[a]		
Ethyl hexanoate	0.2	3.7	fruity, pineapple/ banana	1–36 ppb 850 in wine
Phenyl ethanol	0.0	8.1	fruity, honey-like	650 ppb
Diethyl succinate	0.9	65.6		
Ethyl octanoate	11.3	222.5	soapy/candlewax	8–12 ppb
Ethyl nonanoate	0.6	5.2	slightly fruity, oily, nutty	
Ethyl decenoate	0.1	112.8		
Ethyl decanoate	21.5	210.5	oily, fruity, floral	8–12 ppb
Ethyl dodacanoate	1.2	6.5	oily, fruity, floral	
Isoamyl octanoate	0.0	2.2		
Vitispirane	0.9	7.0	eucalyptus	800 μg L^{-1}
1,2 Dihydronaphtalene	0.7	12.5	petrol like	20 μg L^{-1}

[a]Alves et al. (2005). [b]See information in Table 4.10.

Some concentration differences were observed, depending on age or sweetness, as expected. For example octanoate and decanoate fermentation esters were higher in dryer wines, due to the longer fermentation times these wines had undergone. Many esters decreased in concentration during maturation. From available threshold data only some of the esters occur above the sensory threshold values, and are expected to contribute to the overall aroma of the wines. The aroma composition of Madeira wines made from the four grape varieties (Sercial, Verdelho, Boal and Malvasia) show a decrease in fatty acid esters (C6-C16) and increases in ethyl esters of diprotic acids, such as diethyl succinate (Camara et al., 2006).

Free terpenes in musts of Malvasia, Boal, Sercial and Verdelho were 132 μg L^{-1}, 77 μg L^{-1}, 72 μg L^{-1} and 63 μg L^{-1} respectively (Camara et al., 2004). Boal and Verdelho were characterized by neral, linalool, citronellol and β-ionone, Malvasia musts contained mostly neral and linalool, and Sercial musts contained mostly α-terpineol, β-damascenone and geraniol. β-Damascenone and β-ionone were present above threshold levels and give fruity aromas to these musts.

Terpenes and C13 norisoprenoids in Madeira wines have also been analysed (Camara et al., 2007), using a technique the authors previously verified (Camara et al., 2006). Wines were made from the four main grape varieties (Sercial, Verdelho, Boal and Malvasia) in three different vintages, the wines were sampled in all cases after eight months maturation. There were no statistically significant differences for any of the compounds in the wines made in the three years. Multivariate statistical analyses showed that the four varieties could be

Table 6.12 Concentrations in µg L^{-1} (minumum and maximum) of terpenes and C13 norisoprenoids identified in wines made from Sercial, Verdelho, Boal and Malvasia from three vintages after eight months maturation.

Compounds	Boal[a]	Malvasia[a]	Sercial[a]	Verdelho[a]	Sensory Threshold values
cis-Linalool oxide	0	0.5–0.7	0–0.04	0.3	7000 µg L^{-1b}
trans-Linalool oxide	0.2	1.3–1.7	0	0.2	65000 µg L^{-1b}
Linalool	8.7–9.9	8.0–9.4	13.2–14.6	5.4–5.9	50 µg L^{-1b}
α-Terpineol	2.9–3.6	12.5–14.1	4.0–5.4	8.7	400 µg L^{-1b}
Citronellol	12.9–19.6	7.9–9.2	2.7–3.9	0.9–1.0	18 µg L^{-1b}
Geraniol	2.3–2.4	3.7–4.5	5.7–6.3	1.0–1.1	130 µg L^{-1b}
Nerolidol	2.5–3.2	0	1.4–2.2	0	18 µg L^{-1b}
Farnesol	3.5–4.6	1.2–1.5	0.2–0.3	5.8–7.3	
Vitispirane (2 isomers)	1.8–2.8	0.4–0.8	0.5–1.2	0–0.8	
1,1,6-Trimethyl-1,2-dihydronaphtalene	0.9–1.5	0.2–0.3	0.3–0.4	0.6–0.7	
β-Damascenone	6.2–7.1	12.3–12.8	4.3–5.8	6.4–6.5	45 ng L^{-1c}

[a]Camara et al. (2007). [b]Information from Table 4.25. [c]Information from Table 4.14.

characterized typically by two or three compounds. Only citronellol in Boal wine is in the range of the sensory threshold level, so this compound may have a significant sensory impact on the wine. The other compound is β-damascenone, which has a very low sensory threshold, although there is some debate regarding the sensory role of β-damascenone in wine (see Section 4.3.3).

In another study of the 68 aroma compounds analysed in aged Madeira wines made from the four grape varieties (Sercial, Verdelho, Boal and Malvasia), 33 were found to be present in concentrations above their sensory detection level, although not all were formed during maturation (Campo et al., 2006). Many compounds were present in both young and old wines, and since the tipicity of the wines seem to depend on the long and heated maturation period, it seems that the typical aroma volatiles should be formed during this type of maturation. The authors stress that not all aroma volatiles had been identified, so some key Madeira odor compounds may still need to be identified. Compounds present in all four wines above their threshold levels are listed in Table 6.13. Sotolon, phenylacetaldehyde, (Z)-whiskylactone and some volatile phenols seem to be the main compounds identified so far typical for old Madeira. Acetals formed by acetalization reactions between acetaldehyde and glycerol leads to the formation of four hetrocyclic acetals, (cis-dioxane, trans-dioxane, cis-dioxolane and trans-dioolane), which were shown to be indicators of age in Maturing Madeira wines (Camara et al., 2003).

Young Tinta Negra Mole wines have been analysed before any ageing occurred (Perestrelo et al., 2006), as shown in Table 6.14. The most important

Table 6.13 Compounds present in four aged Madeira wines (Sercial, Verdelho, Boal and Malvasia), above sensory detection level.

Compound	Odour description
Phenylacetaldehyde	green, honey
(Z)-whiskylactone	coconut
2-Methoxy-4-vinylphenol	bitumen
Ethyl dihydroxycinnamate	flowery
Sotolon	spicy
2-Methoxyphenol	smoky
Ethyl Cinnamate	flowery

Information from Campo et al. (2006).

Table 6.14 Volatile compounds identified in Negra Mole wines, samples from young wines ranging from sweet to dry.

Compounds	Dry (%)	Medium dry (%)	Medium sweet (%)	Sweet (%)
Terpenes	0.05	0.12	0.14	0.14
Higher alcohols	71.47	54.14	44.58	46.09
Acetates	1.01	1.51	1.74	1.73
Ethyl esters	17.71	30.52	35.88	34.93
Fatty acids	7.53	10.07	12.21	12.29
Carbonyl compounds	1.54	1.19	3.46	2.33
Lactones	0.15	0.11	0.10	0.07
Acetals	0.01	0.03	0.05	0.05
Furan compounds	0.51	1.21	1.59	1.49
Sulfur compounds	0.30	0.58	0.68	0.68
Volatile phenols	0.22	1.73	1.15	1.68

Date from Perestrelo et al. (2006).

flavour compounds of these young wines were were ethyl octanoate, phenylethanal, ethylhexanoate, octanoic acid and 2-phenylethyl acetate. It would be interesting to see how these wines changed during maturation.

Acetals have been found in maturing Madeira wines, at levels higher than typical for wines matured under less oxidative conditions (Camara et al., 2003). Oxidative conditions led to higher concentrations of acetaldehyde and increased formation of heterocyclic acetals (1,3-dioxanes and 1,3-dioxolanes), although there was no measured effects of the heating conditions on their formation. There were linear increases with age for both cis-5-hydroxy-3-methyl-1,3-dioxane and cis-4-hydroxymethyl-2-methyl-1,3-dioxolane. Absolute quantities were not reported, so their contribution to Madeira flavour still need to be determined, however, information on Port wines indicates their significance only in very old Port, in part due to the high threshold levels for these compounds (see Section 6.3.9).

Bibliography

General texts

Bakker, J. (1995) Flavour interactions with the food matrix and their effects on perception. In: *Ingredients Interactions, Effects on Food Quality* (ed. A.G. Gaonkar), pp. 411–439. Marcel Dekker Inc.

Broadbent, M. (1979) *Wine Tasting*. Christies' Wine Publications, London.

Clarke, O. (1996) *New Essential Wine Book*. Webster/Mitchell Beazley, London.

Clarke, O. & Mead, M. (2001) *Grapes and Wines*. loc. cit. Ch. 2.

Jackson, R.S. (1994, 2000) *Wine Science, Principles, Practice, Perception*. 2nd edn. Academic Press, San Diego.

Maarse, H. & Visscher, C.A. (1989) *Volatile Compounds in Food, Qualitative and Quantitative Data Volume II*, 6th edn. TNO-CIVO Food Analysis Institute, Zeist (The Netherlands).

Ribéreau-Gayon, P., Glories, Y., Maujean, A. & Dubourdieu, D. (2006) *Handbook of Enology, Vol I, Wine and Wine Making*, 2nd Edn. John Wiley & Sons, Ltd, Chichester.

Ribéreau-Gayon, P., Glories, Y., Maujean, A. & Dubourdieu, D. (2006) *Handbook of Enology, Vol II, The Chemistry of Wine, Stabilization and Treatments*, 2nd Edn. John Wiley & Sons, Ltd, Chichester.

Robinson, J. (1993, 1995) *Jancis Robinson's Wine Course*. BBC Books, London.

References: Sherry

Arctander, S. (1967) Perfume and Flavour Chemicals. loc. cit. Ch. 4.

Amerine, M.A., Berg, H.W., Kunkee, R.F. *et al.* (1980) *The Technology of Wine Making*. Avi Publising, Westport, USA.

Bakker, J. (1993) Sherry, Chemical composition and analysis. In: *Encyclopaedia of Food Science, Food Technology and Nutrition* (eds. R. Macrae, R.K. Robinson & M.Y. Sadler), pp. 4122–4126.

Begoña Cortes, M., Moreno, J.J., Zea, L., Moyano, L. & Medina, M. (1999) Response to the aroma fraction in Sherry wines subjected to accelerated biological ageing. *Journal of Agriculture and Food Chemistry*, **47**, 3297–3302.

Benitez, P., Castro, R., Natera, R. & Garcia-Barosso, C. (2005) Effects of grape destemming on the polyphenolc and volatile content of fine Sherry wine during alcoholic fermentation. *Food Science and Technology International*, **11**, 233–242.

Broadbent, M. (1979) *Wine Tasting*. Christies' Wine Publications, London.

Brock, M.L., Kepner, R.E. & Webb, A.D. (1984) Comparison of volatiles in Palomino wine and submerged culture *flor* Sherry. *American Journal of Enology and Viticulture*, **35**, 151–155.

Campo, E., Cacho, J. & Ferreira, V. (2007) Solid phase extraction, multidimensional gas chromatography determination of four novel aroma powerful ethyl esters. Assessment of their occurrence and importance in wine and other alcoholic beverages. *Journal of Chromatography A*, **1140**, 180–188.

Campo, E., Cacho, J. & Ferreira, V. (2008) The chemical characterisation of the aroma of dessert and sparkling white wine (Pedro Ximenex, Fino, Sauternes and Cava) by gas chromatography-olfactometry and chemical quantitative analysis. *Journal of Agricultural and Food Chemistry*, **56**, 2477-2484.

Clarke, O. (1996) *New Essential Wine Book*. Webster/Mitchell Beazley, London.

Criddle, W.J., Goswell, R.W. & Williams, M.A. (1983) The chemistry of Sherry maturation II. An investigation of the volatile components present in standard Sherry base wines. *American Journal of Enology and Viticulture*, **34**, 61-71.

Dubois, P., Rigaud, J. & DeKimpe, J. (1976) Identification de la dimethyl-4,5-tetrahydrofuranedione-2,3 [Sotolon] dans le vin jaune du Jura. *Lebensmittel – Wissenshaft & Technologie*, **9**, 366-368.

Esteve-Zarzoso, B., Peris-Toran, M.J., Garcia-Maiques, E., Uruburu, F. & Querol, A. (2001) Yeast population dynamics during the fermentation and biological aging of Sherry wines. *Applied and Environmental Microbiology*, **67**, 2056-2061.

Estrella, M.I., Hernandez, M.T. & Olano, A. (1986) Changes in polyalcohol and phenol compound contents in the aging of Sherry wines. *Food Chemistry*, **20**, 137-152.

Estrella, M.I., Alonso, E. & Revilla, E. (1987) Presence of flavonol aglycones in Sherry wines and changes in their content during ageing. *Zeitschrift für Lebensmitteln Forschung*, **184**, 27-29.

Fagan, G.L., Kepner, R.E. & Webb, A.D. (1981) Production of linalool, cis and trans nerolidol, and trans-trans farnesol by *Saccharomyces fermentati* growing as a film simulated wine. *Vitis*, **20**, 36-42.

Gonzalez Gordon, M. (1972) The Sherry viniculture. In: *Sherry, the Noble Wine*, pp. 110-136. Cassell & Co. Ltd., London.

Goswell, R.W. (1986) Microbiology of fortified wines. In: *Developments in Food Microbiology 2* (ed. R.K. Robinson), pp. 1-19. Elsevier Applied Science Publishers, London.

Goswell, R.W. & Kunkee, R.E. (1977) Fortified Wines. In: *Economic Microbiology volume 1, Alcoholic Beverages* (ed. A.H. Rose), pp. 477-535. Academic Press, London.

Kunkee, R.E. & Bisson, L.F. (1993) Wine-making yeasts. In: *The Yeasts*, Vol. 5 (eds. A.H. Rose & J.S. Harrison), 2nd edn., pp. 69-127. Academic Press, London.

Maga, J.A. (1989) The contribution of wood to the flavour of alcoholic beverages. *Food Reviews International*, **5**, 39-99.

Martin, B., Etievant, P.X. & Henry, R.N. (1990) The chemistry of sotolon: a key parameter for the study of a key component of *flor* Sherry wines. In: *Flavour Science and Technology* (eds. Y. Bessiere & A.F. Thomas), pp. 53-56. John Wiley & Sons, Ltd, Chichester.

Martinez de la Ossa, E., Perez, L. & Caro, I. (1987a) Variations of the major volatiles through the aging of Sherry. *American Journal of Enology and Viticulture*, **38**, 293-297.

Martinez de la Ossa, E., Caro, I., Bonat, M., Perez, L. & Domecq, B. (1987b) Dry extract in Sherry and its evolution in the aging process. *American Journal of Enology and Viticulture*, **38**, 321-325.

Mesa, J.J., Infante, J.J., Rebordinos, L., Sanchez, J.A. & Cantoral, J.M. (2000) Influence of the yeast genotypes on the enological characteristics of Sherry wines. *American Journal of Enology and Viticulture*, **51**, 15-21.

Montedoro, G. & Bertuccioli, M. (1986) The flavours of wines, vermouth and fortified wines. loc.cit., Ch. 4.

Moyano, L., Zea, L., Villafuerte, L. & Medina, M. (2009) Comparison of odor-active compounds in Sherry wines processed from ecologically and conventionally grown Pedro Ximenez grapes. *Journal of Agricultural and Food Chemistry*, **57**, 968–973.

Muller, C.J., Kepner, R.E. & Webb, A.D. (1978) 1,3-Dioxanes and 1,3-dioxolanes as constituents of the acetal fraction of Spanish Fino Sherry. *American Journal of Enology and Viticulture*, **29**, 207–212.

Nykänen, L. (1986) Formation and occurrence of flavor compounds in wine and distilled alcoholic beverages. *American Journal of Enology and Viticulture*, **37**, 84–96.

Rapp, A. & Mandery, H. (1986) Wine aroma. *Experientia*, **42**, 873–884.

Reader, H.P. & Dominguez, M. (1994) Fortified wines: Sherry, Port and Madeira. In: *Fermented Beverage Production* (eds. A.G.H. Lea & J.R. Piggott), pp. 159–207. Blackie, Glasgow.

Webb, A.D. & Noble, A.C. (1976) Aroma of Sherry wines. *Biotechnology and Bioengineering*, Vol. XVIII, 939–1052.

Wurz, R.E.M., Kepner, R.E. & Webb, A.D. (1988) The biosynthesis of certain gamma-lactones from glutamic acid by film yeast activity on the surface of *flor* Sherry. *American Journal of Enology and Viticulture*, **39**, 234–238.

Zea., L., Moyano., L. & Medina, M. (2008) Odourant active compounds in Amontillado wines obtained by combination of two consecutive ageing processes. European Food Research and Technology, **227**, 1687–1692.

References: Port wine

Bakker, J. (1986) HPLC of anthocyanins in Port wines; determination of ageing rates. *Vitis*, **25**, 203–214.

Bakker, J. (1993) Port, Chemical composition and analysis. In: *Encyclopaedia of Food Science, Food Technology and Nutrition* (eds. R. Macrae, R.K. Robinson & M.Y. Sadler), pp. 3658–3662. Academic Press, London.

Bakker, J., Bridle, P., Timberlake, C.F. & Arnold, G.M. (1986a) The colour, pigment and phenol content of young Port wines: effects of cultivar, season and site. *Vitis*, **25**, 40–52.

Bakker, J. & Arnold, G.M. (1993) Analysis of sensory and chemical data for a range of red Port wines. *American Journal of Enology and Viticulture*, **44**, 27–34.

Bakker, J., Bellworthy, S., Hogg, T.A., Kirby, R.M., Reader, H.P., Rogerson, F.S.S., Watkins, S.J. & Barnett, J.A. (1996) Two methods of Port vinification: a comparison of changes during fermentation and of characteristics of the wines. *American Journal of Enology and Viticulture*, **47**, 37–41.

Bakker, J., Bellworthy, S., Reader, H.P. & Watkins, S.J. (1999) The effect of enzymes during vinification on colour and sensory properties of Port wines. *American Journal of Enology and Viticulture*, **50**, 271–276.

Bakker, J., Bridle, P., Honda, T., Kuwano, H., Saito, N., Terahara, N. & Timberlake, C.F.(1997) Isolation and identification of a new anthocyanin occurring in some red wines. *Phytochemistry*, **44**, 1375–1382.

Bakker, J. & Timberlake, C.F. (1985a) The distribution of anthocyanins in grape skin extracts of Port wine cultivars as determined by high performance liquid chromatography. *Journal of the Science of Food and Agriculture*, **36**, 1315–1324.

Bakker, J. & Timberlake, C.F. (1985b) The distribution and content of anthocyanins in young Port wines as determined by high performance liquid chromatography. *Journal of the Science of Food and Agriculture*, **36**, 1325–1333.

Bakker, J. & Timberlake, C.F. (1985c) An analytical method for defining a Tawny Port wine. *American Journal of Enology and Viticulture*, **36**, 252–253.

Bakker, J. & Timberlake, C.F. (1986) The mechanism of colour changes in ageing Port wine. *American Journal of Enology and Viticulture*, **37**, 288-292.

Bakker, J., Preston, N.W. & Timberlake, C.F. (1986b) Ageing of anthocyanins in red wines: comparison of HPLC and spectral methods. *American Journal of Enology and Viticulture*, **37**, 121-126.

Bakker, J., Picinelli, A. & Bridle, P. (1993) Model wine solutions: colour and composition changes during ageing. *Vitis*, **32**, 111-118.

Bakker, J. & Timberlake, C.F. (1997) The isolation and characterisation of new color stable anthocyanins occurring in some red wines. *Journal of Agricultural and Food Chemistry*, **45**, 35-43.

Broadbent, M. (1979) *Wine Tasting*. Christies' Wine Publications, London.

Clarke, O. (1996) *New Essential Wine Book*. Webster/Mitchell Beazley, London.

Cullere, L., Cacho, J. & Ferreira, V. (2007) An assessment of the role played by some oxidation related aldehydes in wine aroma. *Journal of Agricultural and Food Chemistry*, **55**, 876-881.

Ferreira, A.C.S., Barbe, J.C. & Bertand, A. (2002) Heterocyclic acetals from glycerol and acetaldehyde in Port wines: Evolution with ageing. *Journal of Agricultural and Food Chemistry*, **50**, 2560-2564.

Ferreira, A.C.S., Barbe, J.C. & Bertand, A.T. (2003) 3-Hydroxy-4,5-dimethyl-2(5H)-furanone: A key odorant of the typical aroma of oxidative aged Port wine. *Journal of Agricultural and Food Chemistry*, **51**, 4356-4363.

Ferreira, A.C.S. & de Pinho, P.G. (2004) Nor-isoprenoids profile during Port wine ageing – influence of some technological parameters. *Analytica Chimica*, **513**, 169-176.

Ferreira, A.C.S., Monteiro, J., Oliveira, C.M. & de Pinho, P.G. (2008) Study of major aromatic compounds in Port wines from carotenoid degradation. *Food Chemistry*, **110**, 83-87.

Goswell, R.W. (1986) Microbiology of fortified wines. In: *Developments in Food Microbiology 2* (ed. R.K. Robinson), pp. 1-19. Elsevier Applied Science Publishers, London.

Goswell, R.W. & Kunkee, R.E. (1977) Fortified wines. In: *Economic Microbiology Volume 1, Alcoholic Beverages* (ed. A.H. Rose), pp. 477-535. Academic Press, London.

Liu, S.Q. & Pilone, G.J. (2000) An overview of formation and roles of acetaldehyde in wine making with emphasis on microbiological implications. *International Journal of Food Science and Technology*, **35**, 49-61.

Nykänen, L. (1986) Formation and occurrence of flavor compounds in wine and distilled alcoholic beverages. *American Journal of Enology and Viticulture*, **37**, 84-96.

Picinelli, A., Bakker, J. & Bridle, P. (1994) Model wine solutions: effect of sulphur dioxide on colour and composition during ageing. *Vitis*, **33**, 31-35.

Romero, C. & Bakker, J. (2000a) Effect of acetaldehyde and several acids on the formation of vitisin A in model wine anthocyanin and colour evolution. *International Journal of Food Science and Technology*, **35**, 129-140.

Romero, C. & Bakker, J. (2000b) Effect of storage temperature and pyruvate on kinetics of anthocyanin degradation, Vitisin A derivative formation, and color characteristics of model solutions. *Journal of Agricultural and Food Chemistry*, **48**, 2135-2141.

Romero, C. & Bakker, J. (2000c) Anthocyanin and colour evolution during maturation of four Port wines: effect of pyruvic acid addition. *Journal of the Science of Food and Agriculture*, **81**, 252-260.

Simpson, R.F. (1980) Volatile aroma components of Australian Port wines. *Journal of the Science of Food and Agriculture*, **31**, 214-222.

Singleton, V.L. (1992) Tannins and the qualities of wines. In: *Plant Polyphenols* (eds. R.W. Hemingway & P.E. Laks), pp. 859-880. Plenum Press, New York.

Wildenradt, H.L. & Singleton, V.L. (1974) The production of aldehydes as a result of oxidation of polyphenolic compounds and its relation to wine ageing. *American Journal of Enology and Viticulture*, **25**, 119-126.

Williams, A.A., Lewis, M.J. & May, H.V. (1983) The volatile components of commercial Port wines. *Journal of the Science of Food and Agriculture*, **34**, 311-319.

References: Madeira

Alves, R.F., Nascimento, A.M.D. & Nogueira, J.M.F. (2005) Characterization of the aroma profile of Madeira wine by sorptive extraction techniques. *Analytica Chemica Acta*, **546**, 11-21.

Camara, J.S., Herbert, P., Marques, J.C. & Alves, M.A. (2004) Varietal flavour compounds of four grape varieties producing Madeira wines. *Analytica Chimica Acta*, **513**, 203-207.

Camara, J.S., Marques, J.C., Alves, M.A. & Ferreira, A.C.S. (2003) Heterocyclic acetals in Madeira wines. *Analytical and Bioanalytical Chemistry*, **375**, 1221-1224.

Camara, J.S., Marques, J.C., Alves, M.A. & Ferreira, A.C.S. (2004) 3-Hydroxy-4,5-dimethyl-2(5H)-furanone levels in fortified Madeira wines: relationship to sugar content. *Journal of Agricultural Food Chemistry*, **52**, 6765-6769.

Camara, J.S., Alves, M.A. & Marques, J.C. (2006) Review. Development of headspace solid-phase micro-extraction-gas chromatography-mass spectrometry methodology for analysis of terpenoids in Madeira wines. *Analytica Chemica Acta*, **555**, 191-200.

Camara, J.S., Alves, M.A. & Marques, J.C. (2006) Changes in volatile composition of Madeira wines during their oxidative ageing. *Analytica Chimica Acta*, **563**, 188-197.

Campo, E., Fereira, V., Escudero, A., Marques, J.C. & Cacho, J. (2006) Quantitative gas chromatography – olfactometry and chemical quantitative study of the aroma of four Madeira wines. *Analytica Chimica Acta*, **563**, 180-187.

Camara, J.S., Alves, M.A. & Marques, J.C. (2007) Classification of Boal, Malvazia, Sercial, and verdelho wines based on terpenoid patterns. *Food Chemistry*, **101**, 475-484.

Campo, E., Cacho, J. & Ferreira, V. (2007) Solid phase extraction, multidimensional gas chromatography determination of four novel aroma powerful ethyl esters. Assessment of their occurrence and importance in wine and other alcoholic beverages. *Journal of Chromatography A*, **1140**, 180-188.

Elliott, T. (2010) *The Wines of Madeira*. Trevor Elliott Publishing, Gosport.

Ferreira, A.C.S., Barbe, J.C. & Bertand, A.T. (2003) 3-Hydroxy-4,5-dimethyl-2(5H)-furanone: A key odorant of the typical aroma of oxidative aged Port wine. *Journal of Agricultural and Food Chemistry*, **51**, 4356-4363.

Johnson, H. (1977) *The World Atlas of Wine*. Mitchell Beazley, London.

Nogueira, J.M.F. & Nascimento, A.M.D. (1999) Analytical characterisation of Madeira wine. *Journal of Agricultural Food Chemistry*, **47**, 566-575.

Perestrelo, R., Fernandes, A., Albuquerque, F.F., Marques, J.C. & Camara, J.S. (2006) Analytical characterisation of the aroma of Tinta Negra Mole red wine: Identification of the main odorant compounds. *Analytica Chimica Acta*, **563**, 154-164.

Chapter 7
Formation Pathways in Vinification

7.1 Introduction

Flavour components in fruit and vegetables, and in their derived beverages, are produced biogenetically and/or by thermal processing. The developing flavours and aromas towards maturity result from biogenesis by the action of endogenous enzymes on non-volatile precursors in intact tissue. Biogenesis of new flavours in fruit and vegetables occurs by exogenous enzymes entering damaged tissues and their juices, as in fermentation to produce alcoholic beverages (notably wine) from the grape. These enzymes are mostly present in the micro-organisms (yeasts, moulds and bacteria) that can enter, some causing 'decayed' or 'spoilage' flavours. The original flavours from the fruit may or may not persist through a fermentation.

Other beverages, such as coffee brews, derive their flavour/aroma largely from roasting or thermal processes on the raw material, which make use of the well known Maillard reaction between amino acids (from proteins) and reducing sugars (like glucose and fructose). These non-enzymatic processes produce both new non-volatile substances, responsible for taste and new volatile substances that are responsible for aroma. The final beverage will also contain unchanged non-volatile and volatile substances from the original raw material, such as green coffee beans. Other chemical reactions may also occur.

The Maillard reaction is generally considered only to occur with any significant rate above 50°C. The contribution of this reaction to the flavour of most table wines, produced largely by fermentation, is therefore very small. In some wines (especially Madeira and thermo-vinified wine), there will be evidence of Maillard-type reaction products. Maturation (ageing) of wines taking substantial periods of time will also give additional flavour changes by chemical reactions, such as the production of furans and furanones from wooden barrels that can have been 'charred'. Yeasts will then no longer play a rôle (except when matured *sur lie*), but airborne bacteria may be active and

Wine Flavour Chemistry, Second Edition. Jokie Bakker and Ronald J. Clarke.
© 2012 Blackwell Publishing Ltd. Published 2012 by Blackwell Publishing Ltd.

others introduced. Both Jackson (2008) and Ribéreau-Gayon *et al.* (2006) have provided very detailed accounts of all stages in vinification and should be consulted for further information.

In considering the formation pathways for flavour components in wine, it is convenient to examine first the process variables in the fermentation stage of vinification. These are (1) the raw materials, i.e. the grapes; (2) strain of yeast used; (3) temperature; (4) maceration; (5) clarification procedures; and (6) the nutrient medium. These descriptions will not cover all the flavour effects found in practice, so that it is also convenient to consider briefly the source of each group of flavours separately. The effects of various process variables have been closely studied over the decades and continue as a normal exercise in any craft but such studies have also been carried out in a rigorous academic manner in research institutes.

Over the last decade more studies have been done on the role of yeasts and their interaction with the nutrients in the must during fermentation and the effect on the aroma content and overall wine quality and adds to our current understanding (for reviews see Swiegers & Pretorius, 2005; Swiegers *et al.*, 2005; Ugliano *et al.*, 2007; Fleet, 2008; Ugliano & Henschke, 2009). Some of this information is discussed below. The reviews give detailed information regarding the formation biochemical pathways, enzymes involved and even the genetic information coding for the enzymes on the yeast, although this information is not yet available for all compounds.

The biochemical pathways of the flavour formation in grapes is also becoming a more prominent subject of study and the grapevine genome has now been sequenced (see Palaskova *et al.*, 2008). So far only limited information has been published regarding grape terpene biosynthesis and carotenoid breakdown leading to the formation of β-ionone.

Once the volatile composition can be matched to sensory perception and it is known how grapes, yeast and fermentation conditions affect the formed aroma profile of the wine, wine makers may have the knowledge to plan in advance the properties they want their wine to have and make the wine accordingly!

7.2   Process variables in vinification

7.2.1   Grapes

As already described, a wide range of grape varieties/cultivars is available and used for wine-making. There are known differences of composition, especially of varietal aroma (see Chapter 4), although other differences in precursors for flavour are not generally well delineated. Considerable information is available on the chemical composition related to grape maturity and harvesting conditions. The content and proportions of free amino acids (present in grapes/musts, $1-4\,g\,L^{-1}$), like proline and others, can differ markedly from one variety to another and will therefore be responsible for some differences found in volatile compound content, from their breakdown during fermentation.

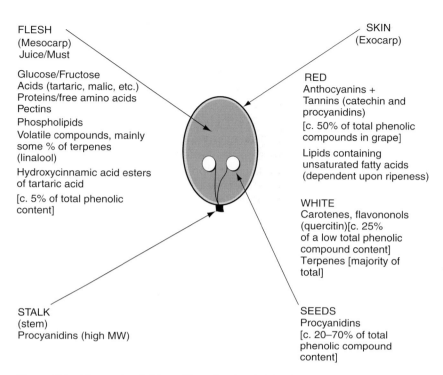

FLESH
(Mesocarp)
Juice/Must

Glucose/Fructose
Acids (tartaric, malic, etc.)
Proteins/free amino acids
Pectins
Phospholipids
Volatile compounds, mainly
some % of terpenes
(linalool)
Hydroxycinnamic acid esters
of tartaric acid
[c. 5% of total phenolic
content]

SKIN
(Exocarp)

RED
Anthocyanins +
Tannins (catechin and
procyanidins)
[c. 50% of total phenolic
compounds in grape]

Lipids containing
unsaturated fatty acids
(dependent upon ripeness)

WHITE
Carotenes, flavononols
(quercitin)[c. 25%
of a low total phenolic
compound content]
Terpenes [majority of
total]

STALK
(stem)
Procyanidins (high MW)

SEEDS
Procyanidins
[c. 20–70% of total
phenolic compound
content]

Figure 7.1 (Approximate) Location of extractable chemical compounds in wine grapes (red and white) during vinification.

Note: See Chapters 3, 4 and 5 for detailed information on contents, including trace elements and minerals.

In a Doctoral thesis by Milley (1988) quoted by Ribéreau-Gayon *et al.* (2006), the content of free amino acids in three different grape varieties used for Champagne (Chardonnay, Pinot Noir and Pinot Meunier) is given at four different stages of ripening (September/October). Chardonnay is particularly characterized by a high proline content. Sulfur-containing amino acids, such as cysteine and methionine, are of general potential interest, since they are precursors of highly odoriferous sulfur compounds in wines. Numerous examples of amino acids as precursors will be noted in Section 7.4.

The location of the different chemical compounds in the actual wine grape (both red and white) is important in vinification, determining the type of many of the process stages used. A guide to their location is illustrated in Figure 7.1.

The maturity of the grape at harvest will determine the composition and as a result the potential aroma and quality of the wine. Traditionally the sugar–acid ratio has served as an indicator of ripeness. However, there is more interest in optimum maturity to gain the best wine quality of the grapes. A study in the volatile composition of Cabernet Sauvignon grapes during ripening showed the evolution of these compounds; terpenes characterized early berry development, whereas at later stages, aldehydes followed by alcohols tended to dominate (Kalua & Boss, 2009). The authors suggested that alcohol to aldehyde ratio may possibly be used to predict the picking date for optimum grape and wine aroma. Another study to predict the flavour potential of grapes used acid and enzymic hydrolysis, the former method showing some potential (Loscos *et al.*, 2009).

7.2.2 Yeast strain

Yeast is the primary agent of change in grape 'must' to make wine, as already described briefly in Chapter 1. Several kinds of yeast may be involved in vinification but the most important is *Saccharomyces* (genus) *cerevisiae* (species). Some confusion on relevant species still arises, which is not entirely clarified by taxonomic evidence. The *Saccharomyces* genus has been divided according to cell shape, usually spherical to ellipsoidal but may be cylindrical or even apiculate, so that wine yeasts were once characterized as being *Saccharomyces ellipisoideus*.

Some oxygen is required for yeast to synthesize sterols and long-chain fatty acids, needed for the cell membranes of the yeast. Normal grape handling during de-stemming and crushing usually allows the must to absorb sufficient oxygen (see Chapter 3) for yeast growth. Fermentation by the yeasts does not require oxygen. Most yeasts grow, as is happening in a grape must, by vegetative reproduction; and wine yeasts grow fairly well under anaerobic conditions but with a lower yield. *Saccharomyces ellipsoideus* in vinification is now classified as *Saccharomyces* with *cerevisiae* as the correct species name; but there will be (not clearly delineated) varieties or strains of this species according to source/method of manufacture. There are also several other known species such as *S. bayanus, S. chevalieri, S. italicus* and *S. heterogenicus* but experimental data in a Doctoral thesis (Soufleros, 1978), quoted by Ribéreau-Gayon (1978), has indicated that an identical amount of the total higher alcohol and total fatty acid esters is formed under identical fermentation conditions from these species. However, other species such as *S. baillii* were shown in the same data to give substantially lower amounts. *S. cerevisiae* is also used in the brewing of beer, particularly ales, which are bottom fermented, but *S. uranum* is deliberately used in the manufacture of lagers, since it liberates significantly larger amounts of octanoic and decanoic acids during fermentation, providing a distinctive flavour. *S. apiculatus* (lemon-shaped) has been reported to give a higher quantity of esters/amyl alcohol, and *S. oviformis* to give aldehydes and ethyl propanoate in particular. The effects of yeast strain on important volatile compounds for flavour have been discussed by a number of workers, especially by Soles *et al.* (1982) in relation to the differences in ester production.

Much research has been focused on the effect of yeast on aroma production and its impact on wine quality, comprehensively reviewed by Bisson & Karpel (2010) and numerous reviews therein (Ugliano & Henschke, 2009; Swiegers & Pretorius, 2005). Of course, it has been known for some time that many volatile compounds are produced by yeasts during fermentation, one of the reasons wine has such a different aroma from grape juice. The flavours formed during the yeast fermentation depend on numerous factors (cultivar,

Table 7.1 Volatile compounds formed during *Saccharomyces* fermentation.

Compound class	Compounds
Alcohol	ethanol, propanol, butanol, 2-methylpropanol, 2-methylbutanol, 3-methylbutanol, hexanol, phenylethanol
Acid	acetate, propanoate, butyrate, 2-methylpropanoate, 2-methylbutanoate, 3- methylbutanoate, hexanoate, octanoate, decanoate
Ethyl ester	ethylbutyrate, ethyl hydroxybutyrate, ethyl hexanoate, ethyl octanoate, ethyl decanoate, ethyl propanoate, ethyl 2-methylpropanoate, ethyl 2-methylbutyrate, ethyl 3-methylbutyrate,
Acetate ester	ethyl acetate, hexyl acetate, 2-methylpropylacetate, 2-methylbutylacetate, 3-methylbutylacetate, phenethylacetate
Carbonyl	acetaldehyde, diacetyl, acetoin
Sulfur compounds	hydrogen sulfide, methanethiol, ethanethiol, dimethylsulfide, dimethyldisulfide, dimethyltrisulfide, diethylsulfide, diethyldisulfide, methionol, methional
Thiol	4-mercapto-4-methylpetan 2-one, 3-mercaptohexan-1-ol, 3-mercaptohexylacetate, furfurylthiol

Information from Bisson & Karpel (2010).

ripeness, fermentation temperature and many processing parameters), including the yeast population. With the availability of more accurate methods of aroma analyses as well as techniques to determine the genetic differences between yeasts, this has become a fruitful area of research. Bisson & Karpel (2010) listed a total of eight alcohols, nine acids, 15 esters, three carbonyls, ten sulfur and four thiol compounds being formed during the fermentation (Table 7.1).

Fusel alcohols, esters and sulfur-containing volatiles are commonly present above their threshold values and will influence wine flavour. The sulfur compounds can easily have a negative impact on wine flavour (see Table 4.28). In many cases the precursors as well as the genetic codes for the enzymes involved in the formation have been identified, although this work is by no means complete. For example a recent study on yeast involved in the release of 4-mercapto-4-methyl pentan-2-one (4-MMP) elucidated the four genes and enzymes involved in the release of this volatile thiol during the fermentation (Howell et al., 2005).

For the choice of yeast to have a dramatic impact on the resulting wine quality, impact aroma compounds would need to be affected, rather than the general fruity and floral compounds contributing generally to wine flavour. Also the precursors need to be present in sufficient concentration in order for the yeast to synthesize, liberate, or convert them into aroma compounds, so processing should be adapted to control the flavour formation. There appears to be a great diversity in the flavour impact genes of *Saccharomyces*, which may in future be used to select the yeast able to enhance the desired flavours in a particular wine. New recombinant yeast strains may well give more options to the wine maker than currently available.

Wine fermentation has been traditionally associated with 'natural' yeasts present on the surface of grapes and indeed present generally in winery premises, which in the case of red wine production are adequately present, though said to be often less so for white wines. Whilst *Saccharomyces ellipsoideus* will, however, be the predominant species, there will be others, including 'wild' strains and such yeasts as *Kloeckera* spp., *Torulopsis* and *Hansenula* present on the surface of grapes. However, as fermentation proceeds, *S. cerevisiae* will be the dominant active fermenting yeast. *Kloeckera* spp., for example, are inhibited at alcohol levels above 3–4%, and most other wild yeasts species are either slow growing or are inhibited by sulfur dioxide and by alcohol or the absence of oxygen. Other yeasts and parasitic fungi, such as *Candida* and *Botrytis*, can occur on diseased and damaged grapes, which again are overwhelmed by the growth of the fermenting yeasts. Acetic acid bacteria will also be found in damaged and diseased fruit. *Hansenula* and *Pichia* are associated with surfaces of wineries in addition to other yeasts, though hygienic operation in modern wineries lowers their significance.

There is a modern tendency associated with a definite emphasis on hygiene generally throughout a winery, for purposeful inoculation of the grape must with outside manufactured active dry *S. cerevisiae* based on laboratory strains (which, however, may not be totally pure), with white wines and to a lesser extent with red wines. It is certainly required after the vinification, such as for the production of sparkling wines, and is essential when pasteurized must is used. This subject causes considerable discussion and controversy, drawing parallels with the slightly different and often preferred taste of cheeses like Brie, when a choice can be made from non-pasteurized or compulsorily pasteurized milk (Barr, 1988). Potentially bland and uniform wines are thought to be produced when only cultured yeasts are used. Those in favour of using inoculum like the rapid onset of fermentation, reducing the risk of the growth of less desirable organisms, potentially giving off-flavours.

There are documented differences between wines depending whether cultured or wild yeasts were used for fermentation. A comparison of Chardonnay wines allowed to ferment with indigenous wild yeast and Chardonnay wines made by fermenting with an inoculated strain of *S. cerevisiae* showed a greater variability in flavour compounds in the wines fermented with wild yeasts; and higher concentrations of 2-methylpropanol, 2-methylbutanoic acid, ethyl 2-methylpropanoate, ethyl decanoate, and ethyl dodecanoate were considered the most significant differences contributing to the flavour of wines (Varela *et al.*, 2009).

Different yeasts (species and strains) will contain the same enzymes but their relative proportions and catalytic activity may well vary, so providing some of the differences noted for the same environment. During fermentation on a sugar substrate, yeasts are growing and reproducing, where they can do so asexually by budding and sexually by the formation of spores, though the latter seldom occurs in must. The chemical composition of yeasts is of relevance. The yeast cell is surrounded by a strong and mechanically refractory cell wall, which may account for 20–30% of the cell solids, consisting of glucans and glycoproteins. The cell wall surrounds the plasma membrane, which regulates the transport of chemical substances into and out of the cell. *S. cerevisiae* contains some 45% estimated protein on a dry basis and a number of vitamin substances in small amounts.

Spoilage yeasts such as *Brettanomyces*, if not controlled, can lead to the undesirable flavours of ethyl phenols (Section 7.4.9).

7.2.3 Malo-lactic organisms

Another type of fermentation to be considered in any overall vinification is the malo-lactic acid conversion. The organism for this type of fermentation is *Leucanostoc oenos*, which grows after alcoholic fermentation, as also discussed in Chapter 1. Malo-lactic fermentation of model systems with grape flavour precursors using *Oenococcus oeni*, *Lactobacillus brevis* and *Lactobacillus casei* showed that these organisms released small amounts of terpenes, norisoprenoids, volatile phenols and vanillins into the model system (Hernandez-Orte *et al.*, 2009). Quantitative differences were tentatively attributed to the nature of the enzymes of the malo-lactic organisms. The enzymes were able to affect volatile changes even when the organisms failed to carry out the malo-lactic fermentation.

7.2.4 Temperature

The temperature at which the fermentation should be conducted for optimal effect has been studied by numerous investigators, past and present in different locations, both in respect of alcohol production and for volatile compound formation (esters and higher alcohols, in particular) and also for the extraction of phenolic compounds from the grape skins.

Standard operating temperature

A controlled fermentation temperature is regarded as highly important for white wines and it is generally assumed to be kept on the cool side, i.e. 15–17°C. For red wines a higher temperature is generally selected, ca. 25°C. In red wine production, a shorter hotter fermentation extracts more phenolic compounds (from the skins), whilst a longer cooler one produces more aromatic substances (esters). Barr (1988) comments that a red wine intended to be drunk young, i.e. without containing too many tannin substances, is fermented at 25°C, whilst Peynaud (1986) reports that, for a fine wine, the ideal fermentation temperature should be no more than 30°C, irrespective of grape variety. Ribéreau-Gayon (1978), on the other hand, comments that in the production of red wines the fermentation should be at considerably higher than 30°C. At some Châteaux it is claimed higher temperatures are actually used, e.g. 32–34°C.

Ribéreau-Gayon (1978) quotes some interesting data from a Doctoral thesis of a pupil (Soufleros, 1978). For example, at 20°C, and a normal pH of 3.4, 10.8 mg L^{-1} of total fatty esters were generated compared with 7.8 mg L^{-1} at 30°C. With a pH of 2.9, even lower ester contents were recorded at the same temperature.

Robinson (1995) and Jackson (2008) comment that wine makers in the New World vinify white wine at the lower end of the 12–17°C range, so that the tropical fruit flavours (like pineapple) are especially generated. Conversely, Europeans tend to ferment Chardonnay at 18–20°C. The effect of temperature has also been studied by Ough *et al.* (1979, 1986), in particular in respect of the generation of isoamyl (pentyl) acetate (fruity/pears, bananas) and of 2-phenyl ethyl acetate (fruity/apricot). These authors show a higher content for isoamyl acetate at 13°C than at 30°C; the formation of 2-phenyl ethyl acetate peaks at 20°C. Other esters, such as ethyl octanoate and decanoate (the oily/soapy flavours) are produced optimally at 13–15°C. As described under maturation in Chapter 4, the final ester composition of a wine, particularly white, will be determined by the period and type of ageing practiced, due to changes in the ester composition during ageing. White wines drunk very young will have a dominant ester flavour resulting from low temperature fermentation.

Fermentation temperature has been shown also to influence the content of higher alcohols (fusel oils) in the work of Soufleros (1978) already quoted. Higher temperatures, in conjunction with lower pH values, gives a markedly lower fusel oil content.

Fermentation temperature is clearly important for red wines (and also white) since it influences the required extraction of phenolic substances, though this will also be largely determined by grape skin contact and variety of grape chosen. Most popular wine books (Clarke & Rand, 2001; Robinson, 1995) indicate fermentation temperatures for red wines to be 25–30°C; however, there is a trend to the use of 'cool' fermentation for red wines, in Australia, to gain a 'fruity' red wine to be drunk young.

Thermovinification

Thermovinification is a process that can be used for red wines to increase the contents of some phenolic compounds and of some volatile compounds during subsequent conventional fermentation. It consists of taking de-stemmed grapes and heating them up to temperatures in the range 60–87°C to disrupt the cell structure and facilitate subsequent extraction, especially of anthocyanins from the skins. The holding time of the heated pulp varies between two minutes at 87°C to several hours at lower temperatures; lower temperatures can only be used for sound, non-mouldy fruit, since the enzyme laccase remains active even at 60°C. The pulp is then pressed and once the juice has cooled to 25°C the yeasts are added and fermented without skins and seeds.

The process has been investigated in detail by Fischer *et al.* (2000) in Germany and the results statistically compared with those of standard tank vinification procedures conducted at the same time. The standard procedure had three sub-variants of operation in respect of handling/submerging the grape skins during fermentation; and three different red wine grape varieties, Pinot Noir, and two cultivars (Domfelder and Portugieser) specific to Germany. There were a number of interesting conclusions drawn; thus, the use of thermovinification very markedly increases the amount of certain esters, in particular 3-methyl-1-butyl (isoamyl) acetate (see above) and hexyl acetate, though the results and their interpretation were not always clear. There were important claimed differences in the extraction of certain phenolics between the different in-tank operations – thus quercetin (glycoside) content was some 2–8 times higher in wines vinified in 'pump over' tank than 'punch down'. Differences in higher alcohol content and alkanoic acids were also shown. There were no sensory data, though based on the detailed chemical analysis the authors suggested that thermovinification would provide a more fruity style of red wine. The authors also said that thermovinification is mostly applied in cool climate viticulture to grape varieties of low anthocyanin concentration, yielding not only fruity red wine with strong colour but low tannin concentration. It is the method of choice for big wineries and co-operatives.

7.2.5 Clarification procedures

Ribéreau-Gayon (1978), quoting his own work of 1975, comments on the desirability of allowing the fermentation for white wines to take place in clarified juice or must. Clarification, for example, by racking, increased the production of total fatty acid esters significantly and lowered that of the higher alcohols. Must clarification by use of pectolytic enzymes can influence type of phenol content, as described in Section 7.4.9. Numerous clarification methods are available, all described in depth by Ribéreau Gayon et al. (2006).

7.2.6 Nutrient medium in fermentation

A correct composition of the nutrients in the medium for the growing yeast cells is important, especially in relation to nitrogenous components. A lack of nitrogen can lead to a 'stuck' fermentation. This may happen, for example, with clarified musts, where protein content has been lowered. An excess of nitrogen can, however, give much cell multiplication but a reduced conversion of sugar into alcohol.

The nitrogen in the must will depend on vineyard management, such as any nitrogen additions in the vineyard and of course the wine maker can make an addition of nitrogen to the must. It has become more apparent that the nitrogen available for yeast fermentation can significantly influence the non-volatile and volatile flavour components in wine, and for more detailed information the reader can refer to a number of reviews (Bell & Henschke, 2005; Ugliano et al., 2007; Ugliano & Henschke, 2009).

Nitrogen addition in the vineyard influenced berry composition and therefore may be expected to influence fermentation and the resulting wine.

Generally, increases in esters are reported with use of nitrogen in the vineyard, however, there is insufficient information in literature to understand how wine composition and quality may be affected. Nitrogen application in the vineyard does give an increase in total nitrogen compounds in grapes, such as amino acids. The amino acid content may well have more influence on the aroma profile of a wine than perviously assumed but there is insufficient information available.

Yeasts assimilate nitrogen and the types and concentration of nitrogen available in must regulates the growth and metabolism of the yeast, including the sugar, nitrogen and sulfur pathway, all influencing the aroma formation and hence having a possible impact on the sensory properties of the resulting wine. In many wine regions, the available nitrogen in the must is low, possible limiting yeast growth, fermentation rate and increasing the risk of a stuck fermentation. A shortage of nitrogen can give high concentrations of fusel alcohols and there is an increased risk of sulfur off-flavour formation. However, a recent study threw doubt on the benefit of adding nitrogen to to reduce the formation of sulfur compounds, since the final Shiraz wines thus made all had higher levels of volatile sulfur compounds, irrespective of the fermenting yeast used (Ugliano et al., 2009). Shiraz is a variety particularly prone to form high levels of volatile sulfur compounds and a reductive aroma character. Wine makers oxygenate their fermentations to help to counter this. An excess of nitrogen can give an estery off-flavour, and is a limiting composition of must and the formation of aroma compounds, affecting the aroma and quality of wine. Other factors of importance are yeast strain and fermentation conditions.

Vilanova et al. (2007) studied aroma compound formation in chemically defined wine like media. There were differences in formation of aroma compounds, attributed to the differences in pathway of formation. They concluded that the addition of nitrogen in the form of diammonium phosphate shows a peak of higher alcohols formed at about 200-300 mg L^{-1} addition, after which the amounts formed tailed off considerably. The production of fatty acids ethyl esters and acetate esters, compounds responsible for the fruity characteristics of wine, generally increased with greater additions of nitrogen. The concentration of branched esters, now thought to contribute red berry flavour to wine, tended to decrease with increasing nitrogen additions. These preliminary trials suggest that nitrogen additions to white wine juice may influence the aroma and sensory quality of the resulting white wine. Since red wines are made differently from white wines (see Chapter 1), with more of the flavour generally believed to be derived from maturation, there is much less information on the effect of nitrogen on red wines.

The sulfur metabolism of yeasts give a number of sulfur containing compounds, many unpleasant and both yeast strain and nutrient composition determine their formation. Vitamins, such as biotin and pantothenic acid will also influence volatile formation in must but the influence of other factors, such as metal ions and phosphate, is not documented.

7.2.7 Maceration

An important process variable in vinification is the contact time and type of the grape skins with the grape juice and flesh either in the course of the fermentation or before, the so-called maceration.

Maceration primarily affects the extraction of phenolic substances, affecting bitterness, astringency and colour, as already described in Chapters 1 and 3, and in the section on thermovinification in this chapter. The extraction of volatile compounds can also be affected by skin contact, in particular that of terpenes. Different techniques will be used on account of the differential location of these compounds in the juice, flesh, skins, and indeed in the seeds and stems, though both the last mentioned are often avoided for use and often removed at crushing. There is the effect of crushing and draining and, particularly in red wine, vinification, the various ways of maintaining contact of the skins with the fermenting juice. Terpenes tend to reside more in the skin of grapes than in their flesh or juice but their distribution varies markedly between varieties, including Muscadet, and the particular terpenoid composition also varies.

7.3 Production of ethyl alcohol

The major function of the yeast, *Saccharomyces cervisiae*, in wine fermentation is, of course, the production of ethyl alcohol (ethanol, C_2H_5OH) from the sugars (glucose and fructose) present in the juice. This yeast will also ferment any added sucrose but not other sugars like arabinose, rhamnose and xylose, which may also be present in small quantity in the must.

The particular enzyme present in the yeast has the general name zymase but in fact, yeast contains several enzymes, including invertase, which is necessary to split the sucrose into its component sugars (glucose and fructose).

The mechanism of the metabolic pathway from glucose and fructose to ethyl alcohol has been well established; the conversion proceeds primarily via the so-called Embden–Meyerhof glycolytic pathway oxidation to pyruvate, then to acetaldehyde and ethyl alcohol. For growth and reproduction, yeast cells require a steady supply of ATP (adenosine triphosphate) together with the reducing power of NADH (nicotinamide adenine dinucleotide). There are metabolic intermediates, which result in the noted formation of succinates, glycerol, acetoin, diacetyl (buta-2,3-dione), acetic and succinic acids (Amerine & Joslyn, 1970). This complex pathway can be understood in detail by referring to Jackson (2008) and, of course, to numerous other texts on biochemistry.

Notably, the production of alcohol during fermentation assists the physical extraction of numerous compounds (e.g. terpenes) from grape cells, which appear in the fermented wine.

7.4   Production of individual groups of compounds

7.4.1   Esters

The majority of esters are formed during fermentation, with the important variables of temperature and clarification affecting content and type of compound in white wine, as already noted. These syntheses are controlled enzymatically. The synthesis of acetate esters is fairly well documented and is linked to the lipid and acetyl-CoA metabolism. The esters formed vary with different yeast strains and the esterase content, a diverse group of enzymes, able to cleave and form ester bonds. The ester concentration in the young wine depends on its rate of formation during the fermentation but also on the amount released from the yeast into the wine. Ethyl acetate is usually present in the highest concentration and is formed in small amounts by yeast during fermentation, but also by acetic acid bacteria during ageing in oak.

Ethyl esters occur in lower concentrations and the formation of ethyl fatty acid esters is not well understood. Ethyl esters are secreted slower and less efficiently from the yeast into the fermentation than acetate esters.

There are numerous influences on the amounts of esters formed. An increase in grape maturity tends to give a lower formation of acetate esters during fermentation and a higher formation of ethyl esters. Grape solids are a source of lipids and affect yeast growth and a small quantity of solids during fermentation stimulates ester production. Factors affecting the ethyl ester formation during fermentation are temperature, with higher fermentation temperature resulting in ethyl octanoate and ethyl decanoate (Saerens *et al.*, 2008). These authors suggested that genetic expression of the genes and enzymes involved in the biosynthesis of ethyl esters was not a limiting factor in their formation, the limiting factor is the availability of precursors in the must. An addition of medium chain fatty acids to the fermentation resulted in an increased formation of ethyl esters.

The effect of fermentation temperature has long been exploited, especially in the fermentation of white wines, which have fruity characters produced during low temperature fermentation. Lower fermentation temperature (15°C) tends to favour ethyl esters.

The ester concentration tends to increase with an increase in available nitrogen, although the there is no clear established link between individual esters and amino acids. Addition of ammonium salt before the fermentation enhances the ester production.

However, certain subsequent reactions occur non-enzymatically; thus in young wines the esters of low molecular weight acids (e.g. ethyl acetate) are generally in excess of their equilibrium constant. Hydrolysis back to the constituent alcohol and acid can then occur, favoured by high temperature and low pH. Other esters of higher molecular weight acids are present at low equilibrium constant levels so that, as noted, they tend to increase in amount on ageing, as for example with succinic esters. Low SO_2 levels are stated to favour ester synthesis and retention, whilst intercellular grape fermentation (e.g. carbonic maceration) and the absence of oxygen during yeast fermentation further enhance ester formation. Acetic acid bacteria can synthesize ethyl acetate but acetic acid can also react non-enzymatically with ethanol to form ethyl acetate.

Another example of ester formation during maturation are the four esters (ethyl 2-, 3- and 4-methylpentanoate and ethyl cyclohexanoate) recently reported in a range of wines by Campo *et al.* (2007). They are formed during maturation by esterification reactions with ethanol and the corresponding acids formed by micro-organisms. Their concentrations vary, and are especially high in aged wines.

7.4.2 Aldehydes

Acetaldehyde, the main aldehydic constitutent, is an early metabolic product of fermentation but can then fall in amount; another source is the oxidation of ethyl alcohol by *o*-diphenols present (Appendix I) as the must is fermented to 'dryness'. During fermentation acetaldehyde is an intermediate in the conversion from pyruvate to acetaldehyde and then ethanol. The acetaldehyde level generally drops to a low concentration at the end of the fermentation. Fermentation conditions influence the concentration of acetaldehyde, such as a higher fermentation temperature leads to a higher concentration.

Certain C_6 aldehydes have been described as important such as hexanal and the hexenals, which are thought to result from enzymatic cleavage of oxidized linoleic and linolenic acids, respectively, as first explained by Drawert (1974). This occurs during crushing, when auto-oxidation of the small quantity of lipids (oils) in the grapes can occur. The C_6 aldehydes are formed by enzymic oxidation of C_{18} unsaturated fatty acids. The corresponding alcohols may be also formed at the same time but subsequent fermentation will also entirely convert these aldehydes into alcohols (Appendix I.1.3). Saturated higher aldehydes are derived from the biosynthesis of fatty acids from the acetyl-CoA, and usually produced in trace amounts.

The amount of lipid substances in grapes is not precisely known though some data on component linoleic and linolenic acids is available from Roufat *et al.* (1987), referenced by Jackson (2008), showing a decline in ripening. This would support the contention of a 'grassy' flavour arising only from unripe grapes. Though some 10 mg of oxygen per litre of grapes (0.31 millimole) are believed to be taken up in crushing, it would be difficult to translate this figure into an expected unsaturated aldehyde content through stoichiometric equations and, of course, the growing microflora on crushed grapes would also be expected to take up oxygen.

The formation of several other aldehydes during ageing in-cask has already been discussed in Chapter 4. Extracted compounds during fermentation can also be changed by yeast enzymes, also referred to in Chapter 4.

7.4.3 Ketones

Ketones are generally produced during fermentation, including diacetyl, though the latter may be more likely produced through the action of spoilage lactic acid bacteria, or by lactic acid bacteria carrying out the malo-lactic fermentation. Diacetyl formation in wine depends on a great many factors, reviewed by Bartowsky & Henschke (2004). At the completion of the yeast fermentation, the concentration is typically less than $1 mg L^{-1}$, and will have little impact on the wine aroma (see Chapter 4).

The malo-lactic fermentation forms the main quantity of diacetyl, depending on strain. Generally, *Pediocuccus* and *Lactobacillus* produce high concentrations of diacetyl, but *Oenocuccus oeni* produces lower concentrations and the range is wide ($0.15-4 mg L^{-1}$). However, there are numerous facors influencing diacetyl formation during this process. A low inoculation concentration, some oxygen supply during malo-lactic fermentation, a higher citric acid concentration, a low temperature (18°C rather than 25°C) and a low pH value all seem to stimulate the concentration of diacetyl formed. Sulfur dioxide also plays a role in this process, inhibiting the growth of yeast and bacteria and stabilizing the diacetyl concentration.

The ketones (ketonic terpenes) β-damascenone, α-, and β-ionone with flavour significance are believed to exist in grapes or at least musts. These so-called isoprenoid ketones are postulated to arise from the enzymatic oxidation and cleavage of β-carotene (and other carotenoid substances) during the crushing of the grapes (Ribéreau-Gayon *et al.*, 2006). There may also be increase in amount as the result of 'in-bottle' ageing. Oak ageing may also release some α- and β-ionone (Ribéreau-Gayon *et al.*, 2006).

7.4.4 Acetals

Acetals (7.I) are found in wines, particularly in fortified wines such as Sherries, Port wines and Madeira, and are formed almost entirely by conventional chemical reaction between components subsequent to their production in the vinification process (fermentation). Oxidative storage conditions, such as for Madeira and Port wines, enhances the levels of aldehydes and subsequently the formation of acetals, see Chapter 6 for more information.

Acetalization is the reversible reaction between an alcohol (or polyol) and an aldehyde, as shown for (7.I):

$$RCHO + 2R^1OH \rightleftharpoons RHC\begin{matrix} OR^1 \\ \\ OR^1 \end{matrix} + H_2O$$

7.I

Formation is enhanced by excess alcohol and low pH. When R is CH_3, as in ethanal, and R^1 is C_2H_5, as in ethanol, we may get the simplest acetal, 1,1-diethoxyethane, also known as diethyl acetal. There is an isomer 1,2-diethoxyethane.

Acetaldehyde will similarly react with other high alcohols present, e.g. propanol. Other aldehydes present, such as pentanal and 2-phenylacetal-dehyde, will form acetals that are known to be present.

Reaction with ethanal may also occur with butan-1,2,-diol (glycol) or glycerol present as shown below.

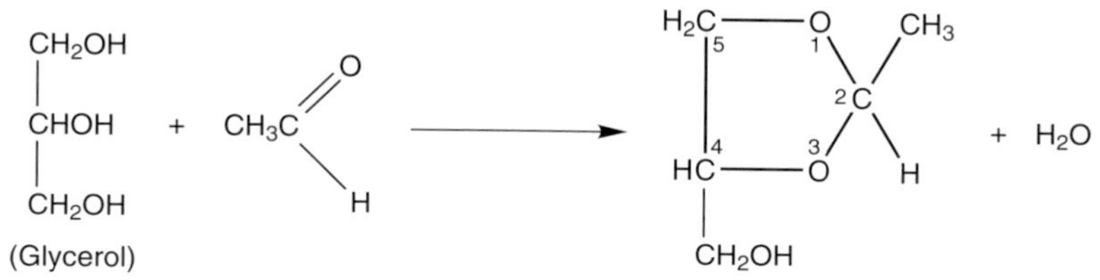

7.II

The glycerol acetal formed is *cis*-4 hydroxymethyl-2-methyl-1,3-dioxalane (7.II). There has, however, to be sufficient free aldehyde present for reaction, which is less likely in table wines compared with Sherry, Port wines or Madeira.

7.4.5 Higher alcohols

As already mentioned, straight-chain higher alcohols or fusel oils are largely the products of fermentation and closely parallel the generation of ethyl alcohol. Sponholz (1988) quoted by Jackson (2008) says that the formation of higher alcohols is markedly influenced by winery practices, so that synthesis is favoured by the presence of oxygen, high fermentation temperatures and the presence of suspended material into fermenting mash. However, Ribéreau-Gayon et al. (2006) suggest the opposite to be true in Section 7.2.4. Overall, factors stimulating the fermentation generally stimulate the formation of higher alcohols.

Chaptalization and pressure tank fermentation also tend to enhance synthesis. Chen (1978), also quoted by Jackson (2008), says that certain higher alcohols may originate from grape-derived aldehydes (as we have seen through crushing in Section 7.4.2) and by the reductive denitrification of amino acids (see Section 7.4.1) or via synthesis from sugars. It seems that a higher amino acid content in the must leads to a greater higher alcohol content, since amino acids are the precursor for a number of higher alcohols. Additional higher alcohols are formed from metabolic action of spoilage yeasts and bacteria. Pleasant smelling lighter alcohols may be produced by particular micro-organisms, e.g. oct-1-en-3-ol as in grapes with 'Noble Rot' (see Section 7.5).

According to the early twentieth century work of Ehrlich, the production of fusel oils is due to the need for the living yeast cell to provide itself with protein by disruption of amino acids in the fermenting mash. Many higher alcohols are formed by decarboxylation followed by reduction of α-keto acids, produced as intermediate compounds in the amino acid synthesis and catabolism. Leucines give rise to isoamyl alcohol (3-ethyl-butan-1-ol), isoleucine to rotatory amyl alcohol (2-methyl-butan-1-ol), and valine to isobutyl alcohol (2-methylpropan-1-ol). These branched chain alcohols are formed via the so-called Ehrlich pathway, involving amino transferases, decarboxylases and dehydrogenases. The addition of leucine, etc. will increase the overall content of fusel oil but by the addition of other nitrogen compounds, which are more easily assimilable by the yeast, the formation of higher alcohols during fermentation may be avoided. Early on, it was also found that the active amyl alcohol content in fusel oils prepared from potatoes, corn and molasses, differ markedly with higher amounts of the latter. Subsequent scientific opinion, however, suggests that the protein requirements of the yeast are largely met by synthesis of inorganic nitrogen present. The aromatic alcohols 2-phenyl-ethanol, tyrosol and tryptophol have phenyl alanine, tyrosine and tryptophan as their precursors, whilst methionol is formed from methionine.

Glycerol is is formed as part of the conversion of sugar to alcohol (see Ugliano & Henschke, 2009) and its content in wine ranges from 5–14 g L^{-1}, with red wines tending to have higher levels than white wines. Higher levels are found in wines made from grapes affected by *Botrytis* (see Section 7.5). The formation of glycerol during fermentation can be affected by the available nitrogen for yeast fermentation, as reviewed by Ugliano *et al.* (2007). Glycerol itself is sweet, however, its contribution to the sweetness or mouthfeel of a wine is still not clear. The threshold concentration for sweetness is high (5.2 g L^{-1}, Noble & Bursick, 1984, quoted by Ugliano & Henschke, 2009) and the high acidity in wines is expected to interact with residual sweetness, so the extent to which glycerol contributes to sweetness or mouthfeel aspects is still unknown.

7.4.6 Furanones and lactones

These compounds can be present in wines via a number of pathways. The simpler lactones like γ-butyrolactone (dihydro-3(*H*)-furan-2-one) can arise in the fermentation, by the lactonization of γ-hydroxybutanoic acid, a somewhat unstable substance, itself being formed by the deamination and decarboxylation of free glutamic acid or from protein present. Lactones can be present in the grapes, thus in Riesling and Merlot (see Chapter 4) and Jura wine, Port wines and Madeira (sotolon). Sotolon, however, is more characteristic of botrytized wines (see Section 7.5), but can be produced by a condensation reaction between γ-ketobutyric acid and ethanol. They can be formed from amino or organic acids, such as glutamic and succinic acid.

These compounds can be formed from saccharide fragmentation and the Maillard reaction, as discussed by Grosch (2001). Camara *et al.* (2004) determined up to 1 mg L^{-1} sotolon being formed in sweet Madeira wines and they observed that its formation was very highly correlated with the formation of furfural, 5-methylfurfural and ethyl 5-hydroxymethylfurfural. Therefore they concluded that sotolon synthesis in Madeira is probably related to sugar degradation as a result of Maillard reactions.

Other lactones present arise during the ageing processes, particularly of course the oak lactones (see Chapter 4) during 'in-cask' ageing in oak barrels.

These substances appear to be already present in the untreated oak, and the favourable *cis*-variant increases in amount after 'toasting' of the oak barrels. Solarone is reputed to develop during 'in-bottle' ageing, although its concentrations remain generally low.

7.4.7 Acids

Of the few volatile acids found in wines, some acetic acid is produced by the yeast fermentation but any excess is most likely to have been produced by the acetobacter micro-organism in unhygienic conditions (see Chapter 1). Anaerobic acetic acid bacteria metabolize ethanol into acetic acid, causing spoilage, often as a consequence of excess oxygenation. The *Saccharomyces* strain in wine-making is also known to produce acetic acid, the amount produced depending on the strain. Aetic acid is formed also by oxidation of acetaldehyde. Similarly, the presence of propanoic and butanoic acids is due also to micro-organisms. The higher aliphatic acids C_6 (hexanoic), C_8 (octanoic) and C_{10} (decanoic), barely volatile, but reportedly present in wines (together with their esters) result from the yeast fermentation itself, formed as part of the saturated fatty acid metabolism. Branched chain fatty acids are formed by oxidation of aldehydes formed from α-keto acids during the amino acid metabolism.

The main non-volatile acids (tartaric, malic and citric) are already present in the grape must. Succinic acid will be present either/both from the lactic acid bacteria (undesirable) or the malo-lactic fermentation process. The sugar acids, gluconic, glucuronic and galacturonic, formed by oxidation of glucose, are associated primarily with botrytized grapes.

7.4.8 Amines

As free amino acids are present in wine musts, not surprisingly, by decarboxylation, small amounts of amines can be formed, or otherwise by deamination of aldehydes, though bacteria will be needed for this purpose. Small amounts of amines can be formed by yeasts, especially during the early part of the fermentation. Yeast autolysis can also release amines in wine. Their contribution to flavour development is uncertain and only some amines are volatile themselves, usually at higher pH values.

However, certain of these amines are also reportedly converted into amides by *Saccharomyces cerevisiae*, whilst yeasts generally convert amines into alcohols. Amides do not appear to affect wine flavour.

7.4.9 Phenols (volatile)

Phenols not originally present in the must are formed either in the fermentation or generally released in ageing. Two types of phenols are regarded as important to 'phenol' taint (see Chapter 4), vinyl- and ethylphenols. Vinylphenols are formed by enzymatic decarboxylation by the yeast during fermentation, selectively, from two cinnamic acids (p-coumaric and ferulic) present. The effect can be noted in white wines but is lower in red wines, where inhibition by other phenols is occurring. Phenolic acid decarboxylases, usually derived from spoilage organisms, decarboxylate phenolic acids into volatile phenols. *Saccharomyces* brewing strains can also produce phenolic acid off-flavours in beer but in most strains of *S. cerevisiae* the phenolic acid decarboxylase activity is very low, hence little or no volatile phenols are formed during fermentation.

There are differences in content of cinnamic acids for red and white wines, depending upon grape variety and ripening conditions, thus concentrations are higher in ripe grapes and those grown in hot climates. An example of varietal difference is that Sémillon or French Colombard often have a higher p-coumaric acid and ferulic acid content than Sauvignon Blanc. The use of certain pectolytic enzymes in must clarification, with contaminant cinnamic esters and the yeast strains used are also influential in determining content of cinnamic acid. It is desirable to keep the cinnamic acid content as low as possible. This subject is very fully reported by Ribéreau-Gayon *et al.* (2006) and comprehensively reviewed by Suarez *et al.* (2007).

The presence of ethylphenols, relatively infrequent, arises not during fermentation but rather during ageing, due to the effect of a particular yeast, *Brettanomyces/Dekkera* (Chatonnet *et al.*, 1992/1993; quoted by Ribéreau-Gayon *et al.*, 2006, and reviewed by Suarez *et al.*, 2007). They claimed that the occurrence of ethyl phenols has become more prominent in aged red wines, in particular due to the current trend requiring ageing in wooden barrels, as it appears *Brettanomyces* species can thrive in old wood. Wines are aged in wood without any physical clarification processes, which is thought to enhance the quality of the wine leaving the aroma more intact, but this also poses an increased risk of spoilage by slow growing yeasts such as *Brettanomyces bruxellensis*, *Brettanomyces anomalus* and *Saccharomyces baiili*, as well as some lactic acid bacteria, all capable of producing undesirable off-flavours. These organisms can grow in the presence of ethanol and require minimal nutrients. In particular, ageing wines at higher pH values in oak barrels increases the risk of spoilage. With the trend in picking grapes very ripe, at which stage the acidity is lower, this may increasingly be the case.

Numerous factors influence the formation of undesirable volatile phenols, in particular the concentration of precursors and the size of the *Brettanomyces/Dekkera* population but also alcohol concentration of wine, see Suarez *et al.* (2007). Prevention seems to be difficult, although good housekeeping helps and detailed information is given regarding the methods for cleaning wooden barrels in order to reduce the problem. It is crucial that barrels are cleaned well, pH of the wine is low, adjusted if necessary, storage is at low temperature and sulfur dioxide is added to protect the wine, since these yeasts are controlled by the sulfur dioxide used. Fining of wine with various fining agents or membrane filtration prior to maturation also reduces the risk.

Interestingly, Suarez *et al.* (2007) discussed that the formation of highly stable pigments based on malvidin 3-glucoside are reported to play a role in the prevention of ethylphenol formation. Benito *et al.* (2009) tested Saccharomyces strains with different hydroxycinnamate decarboxylase activities in fermenting red wine musts to enhance the formation of stable anthocyanins and minimize the production of 4-ethylphenol. They were able to increase the formation of vinylphenolic pyroanthocyanins (see Chapter 3) and reduce the concentration of 4-ethylphenol and 4-ethylguaicol, generated by vinylreductase of *Dekkera bruxellensis*. They concluded that generating vinyl phenols early in the wine-making process allowed them to bind with anthocyanins forming stable coloured compounds and may reduce the risk of off-flavour formation at later stages during barrel ageing.

The use of oak barrels after toasting, during ageing is the main factor in determining the presence of the other phenols identified in the wine; in particular, (iso-)eugenol in large amounts and the cresols in very small amounts. Data has been presented (by Ribéreau-Gayon *et al.*, 2006) relating the degree of toasting (none, light, heavy) to the extractability of the various phenols. Syringol, for example, is only really available until after heavy toasting of the barrels before use for wine maturation.

7.4.10 Terpenes

Terpenes in plants are secondary constituents and its biosynthesis starts with acetyl CoA, as described by Maicas & Mateo (2005). Although micro-organisms can synthesize terpenes, their formation by yeasts has not been observed. Synthesis of terpenes is thought to take place in grape berries. Part of the genome of Pinot Noir has been shown to have the capability to produce terpenes, although this variety does not normally produce significant levels (see Ebeler & Thorngate, 2009).

It is now established that many terpenes (the alcohols and polyols) are present in either free or bound form. The latter are glycosides based on glucose, arabinose, rhamnose and apiose, with diglycosides most frequent. The bound forms are, of course, non-volatile and do not contribute to wine aroma, unless they lose their sugar attachment. The proportion in the free and bound form, in ripe grapes, vary with variety, both within the Muscat aromatic group where the content of both is high and other neutral varieties where the content is low. Data from Günata (1984) quoted by Ribéreau-Gayon *et al.* (2006) shows higher amounts of monoterpenes, both in the free and bound form in the skins of white Muscat grapes, followed by that in the juice and pulp. Such grape varieties are often given some maceration on the skins during wine-making to extract the terpenes.

The free terpenes pass substantially unchanged during fermentation. Though grapes do contain β-glycosidases that can release more free volatile, odoriferous-free terpenes, they are unable to do so under normal vinification conditions. Various fungal β-glycosidases have been used in experiments to enhance the terpene flavour, although precursors for potential off-flavours can also be released, so care has to be taken in selecting an enzyme. For detailed information on the hydrolysis of terpenes refer to the review by Maicas & Mateo (2005). Fleet (2008) also mentions the potential of non-*Saccharomyces* yeast species being possible producers of more effective glycosidases to remove the sugars of glycosilated terpenes.

The final terpene composition of a wine will also depend upon the type and time of maturation subsequently prescribed, as already described in Chapter 4. The odoriferous compound, called Ho-trienol for simplicity, is derived from a terpen-diol, by acid hydrolysis during the fermentation. This terpen-diol and other terpene polyols present in grapes are not themselves odoriferous.

7.4.11 Pyrazines

Methoxypyrazines are nitrogen-containing heterocyclic compounds biogenically produced in many fruits and vegetables, including grapes, by the metabolism of amino acids. A large part of the 2-methoxy-3-isobutylpyrazine in Cabernet Sauvignon grapes is located in their skins from which they are extracted in red wines.

The content depends strongly upon climatic and soil conditions, so that unripe grapes have the highest amounts, declining with maturity. Direct sunlight on the grape bunches on the vine reduces its 2-methoxy-3-isobutylpyrazine content. However, one study seems to come to the opposite conclusion (Sala *et al.*, 2004), suggesting the lack of photodegradation of 2-methoxy-3-isobutylpyrazine in sunlight did not occur in shaded grapes. They concluded that during maceration 2-methoxy-3-isobutylpyrazine was also released from the solid parts of the grapes, whilst malo-lactic fermentation did not influence the concentration.

A recent model study on the formation of possible flavour compounds in wine via Maillard and Strecker degradation using temperatures and pH values akin to wine storage conditions and carbonyl compounds present in wines (glyoxal, methylglyoxal, diacetyl, pentan-2,3-dione, acetoin and acetol) showed the possibility of some of these compounds forming under storage conditions (Pripis-Nicolau *et al.*, 2000). In the presence of sulfur amino acids, in particular cysteine, the authors reported the formation of hetrocyclic compounds, in particular pyrazines, methyl pyrazines, methyl thiazoles, acetylthiazolines, acetylthiazolidines, trimethyloxazole and dimethylethyloxazoles. These model observations will need to be confirmed in further studies.

7.4.12 Sulfur compounds

There are five different groups of sulfur compounds in wine: thiols, sulfides, polysulfides, thioesters and heterocyclic compounds. Only a few contribute positively to the aroma of wine, as discussed in Chapter 4, most cause off-flavours, often referred to as reduction defects and are produced during alcoholic fermentation. There is a range of origins of sulfur compounds in wine, but wine makers would like to be able to avoid their formation. Juice settling, for white wines, use of sulfur dioxide and selected yeast strain helps to prevent their formation.

Table 7.2 Concentrations of sulfur compounds formed during wine-making.

Compound category	Concentration range detected (μg L^{-1})
Sulfides	0–480
Mercaptans	0–16
Thioacetates	0–56
Thioalcohols	0–4500
Tiazoles	0–14
Others	0–5

Data adapted from Ugliano & Henschke (2009).

During normal wine-making low concentrations of hydrogen sulfide (3–4 μg L^{-1}) and methanthiol (<1 μg L^{-1}) are produced by yeasts. The latter is formed from methionine by deamination, then decarboxylation to methional, followed by reduction to methionol. Hydrogen sulfide is formed by the sulfate reduction sequence pathway. A shortage of nitrogen during the fermentation is one of the main causes of hydrogen sulfide accumulation. Autolysis at the end of fermentation can also release hydrogen sulfide into the wine. Numerous other sulfur-containing (off) odours are formed as a result of reactions of the reactive hydrogen sulfide, for example reaction with ethanol gives ethanethiol, a mercaptan. Some sulfide compounds can be eliminated from wine by aeration. Ugliano & Henschke (2009) reviewed the sulfur metabolism in *Saccharomyces cerevisiae*, although not all the genetic information is available. Only very small concentrations are formed, but because these compounds have generally very low odour threshold values, their sensory impact can be significant. Table 7.2 shows typical concentrations in wine.

Recently, five somewhat complex volatile thiols positively contributing to the wine aroma have been identified, especially in Sauvignon Blanc wines and are listed in Chapter 4. Three of these (e.g. 4-mercapto-4-methylpentan-2-one or 4MMP, 4-mercapto-4-methylpentan-2-ol or 4MMPOH and 3-mercaptohexanol or 3MH) have also been found in the must of Sauvignon Blanc as their non-odorous conjugate form with S-cysteine, which releases the free volatile form during fermentation by the presence of a specific β-lyase in the yeast. Only a small concentration of thiols is released, proportional to the concentration of precursors. Most of the 4MMP and 4-MMPOH precursors are found in the flesh. The strain of *Saccharomyces cerevisiae* used is important, to ensure optimum activity of this enzyme to release the volatile thiol compounds from cysteinylated precursors.

Furfurylthiol is formed by yeasts, using furfural extracted from oak during fermentation as a precursor but the mechanism is not yet known.

A model study on the formation of possible flavour compounds in wine via Maillaird and Strecker degradation by Pripis-Nicolau *et al.* (2000) showed that in the presence of sulfur amino acids, in particular cysteine, some sulfur compounds were formed and they suggested that just after the fermentation sufficient carbonyl compounds, such as diacetyl, may be present for reactions with cysteine to occur. Numerous off-flavours, such as hydrogen sulfide, methanthiol, dimethyldisulfide and carbon disulfide, were also formed. It will be interesting to see whether these reactions observed in models can also occur during wine storage, this may give future opportunities to allow formation of selected flavour compounds by ensuring the correct storage conditions and fermenting with yeasts that give the precursors required for such reactions.

7.5 Noble Rot

Following infection, which may take place early in grape growth and which requires release of nutrients through the grape skin, the micro-organism *Botrytis cinerea* (a mould) then injects active enzymes and cause changes in the infected grapes. These changes are in addition to a drying effect by removal of water from within the grape and cause concentration of the sugars. This drying effect limits the extent of secondary invasion by other bacteria and fungi. The chemical changes that occur within the grape are considered (Jackson, 2008; Ribéreau-Gayon *et al.*, 2006) to be:

(1) Pectolytic enzyme action. The pectins of the cell wall are degraded, causing collapse and death of tissues, leading to moisture loss.
(2) Formation of gluconic acid. This is more likely, however, to result from invasion by acetic acid bacteria, which will also produce acetic acid and ethyl acetate.
(3) Increase of citric acid content. The increase will be roughly proportional to the weight loss by drying.
(4) Terpene conversion. Linalool, geraniol and nerol are metabolized to the less volatile compounds, β-pinene, α-terpineol and various pyran and furan oxides, so that the Muscat variety of grape will be particularly affected. Over 20 terpene derivatives have been isolated from infected grapes.
(5) Ester degradation. Non-specific.
(6) New syntheses. In particular, sotolon, 4,5-dimethyl-3-hydroxy-2(5*H*)-furanone, which is reported to have a 'toasty/spicy' fragrance, and 1-octen-3-ol ('the mushroom' alcohol), are formed.
(7) Thiamine deficiency. This factor increases the synthesis of sulfur-binding compounds during fermentation, so that the free SO_2 content in wine may be markedly reduced unless much higher additions of total SO_2 are used to achieve the required free SO_2 content. Acetic acid bacteria also produce sulfur-binding compounds, like 2- and 5-oxoglutaric acids (2-oxo-pentan-1,3-dioic acid).
(8) Chemical changes by the activity of laccases. These enzymes, present in *Botrytis cinerea*, are most likely to be influential in the must of the grape during fermentation. They will oxidize 1,2-, 1,3- and 1,4-diphenols and the

anthocyanins, and are particularly active at the pH of wine, so that they need to be inhibited by higher concentrations of total SO_2 (~50 mg L^{-1}) at pH 3.4 in wine. In must, much higher levels of SO_2 are needed to inhibit laccase.

Botrytis cinerea has several other effects, significant in the vinification process:

(1) Synthesis of high molecular weight polysaccharides, polymers of mannose and galactose, increasing the acetic and glycerol content in fermentation, high polymers of glucose (glucans), which form strand-like colloids in the presence of alcohol, which can clog filters in filtration.
(2) Calcium salt instability. *B. cinerea* produces an enzyme, which oxidizes galacturonic acid, a breakdown product of pectin, into galactaric acid, the calcium salt of which becomes insoluble, leading sometimes to sediments in bottles of botrytized wine.

A number of sulfur compounds have been identified in wines made from *Botrytis* infected grapes, determined in Sauternes wines after fermentation and are thought to contribute to the aroma (Sarrazin *et al.*, 2007; Bailly *et al.*, 2006). The most significant compounds are listed in Table 7.3.

Sensory evaluations showed the additive effects between these sulfur compounds. Most of the polyfunctional thiol compounds disappeared after two years of bottle ageing (Bailly *et al.*, 2009), only 3-sulfanylhexan-1-ol remained present above threshold concentration. However, other key odorants determined in young Noble Rot wines could still be found after about six years bottle ageing, such as α-terpineol, sotolon, fermentation alcohols, esters, and oak related maturation compounds (guiacol, vanillin, eugenol, β-damascenone, *trans*-non-2-enal, β-methyl-γ-lactone, γ-nonalactone and furaneol). A new compound, abnexon, was isolated and identified. It is formed at about threshold level in these aged wines (about 7 µg L^{-1}), and contributes a honey, spicy aroma to aged Sauternes wines made from *Botrytis* affected grapes.

Table 7.3 Sulfur compounds and their sensory property in wines made from *Botrytis* infected grapes.

Compound	Sensory property	Ref.
3-Sulfanylpentan-1-ol	citrus	[a]
3-Sulfanylheptan-1-ol	citrus	[a]
2-Methyl-3-sulfanylbutanol	raw onion	[a]
3-Sulfanylhexan-1-ol	fruity	[a]
3-Sulfanyl-3-methylbutanal	bacon, petroleum	[b]
2-Methylfuran-3-thiol	bacon, petroleum	[b]

[a]Sarrazin *et al.* (2007). [a]Bailly *et al.* (2006).

Bibliography

General texts

Jackson, R.S. (2008) *Wine Science, Principles, Practice, Perception*, 3nd edn. Academic Press, San Diego.

Ribéreau-Gayon, P., Glories, Y., Maujean, A. & Dubordieu, D. (2006) *Handbook of Enology, Vol. 2 The Chemistry of Wine, Stabilization and Treatments*. John Wiley & Sons, Ltd, Chichester. loc. cit.

References

Amerine, M.A. & Joslyn, M.A. (1970) Table Wines: The Technology of their Production, quoted by Jackson, R.S. *Wine Science* (1994) loc. cit. pp. 236–241.

Bailly, S., Jerkovic, V., Marchand-Brynaert, J. & Collin, S. (2006) Aroma extraction dilution analysis of Sauternes wines. Key role of polyfunctional thiols. *Journal of Agriculture and Food Chemistry*, **54**, 7227–7234.

Bailly, S., Jerkovic, V., Meuree, A., Timmermans, A. & Collin, S. (2009) Fate of key odourants in Sauternes wines through aging. *Journal of Agriculture and Food Chemistry*, **57**, 8557–8563.

Barr, A. (1988) *Wine Snobbery*, p. 247. Faber and Faber, London.

Bartowsky, E.J. & Henschke, P.A. (2004) The 'buttery' attribute of wine -diacetyl-desirability, spoilage and beyond. *International Journal of Food Microbiology*, **96**, 235–252.

Bell, S.J. & Henschke, P.A. (2005) Implication of nitrogen nutrition for grapes, fermentation and wine. *Australian Journal of Grape and Wine Research*, **11**, 242–295.

Benito, S., Palomero, F., Morata, A., Uthurry, C. & Suarez-Lepe, J.A. (2009) Minimisation of ethylphenol precursors in red wines via the formation of pyranoanthocyanins by selected yeasts. *International Journal of Food Microbiology*, **132**, 145–152.

Bisson, L.F. & Karpel, J.E. (2010) Genetics of yeasts impacting on wine quality. *Annual Review of Food Science and Technology*, **1**, 139–162.

Camara, J.S., Marques, J.C., Alves, M.A. & Ferreira, A.C.S. (2004) 3-Hydroxy-4,-5-dimethyl-2(5H)-furanone levels in fortified Madeira wines: relationship to sugar content. *Journal of Agricultural Food Chemistry*, **52**, 6765–6769.

Campo, E., Cacho, J. & Ferreira, V. (2007) Solid phase extraction, multidimensional gas chromatography determination of four novel aroma powerful ethyl esters. Assessment of their occurrence and importance in wine and other alcoholic beverages. *Journal of Chromatography A*, **1140**, 180–188.

Chatonnet, P., Boidron, J.N., Dubordieu, D. & Pons, M. (1992) quoted by Ribéreau-Gayon *et al.* (2000) *Handbook of Enology*, loc. cit. pp. 218–225.

Chen, E.C-H (1978). The relative importance of Ehrlich and biosynthetic pathways to the fermentation of fusel alcohols, quoted by Jackson, R.S. (1994) loc. cit. p. 185.

Drawert, F. (1974) Wine making as a biotechnological sequence. In: *Chemistry of Wine Making*. Advances in Chemistry Series, 157, pp. 1–10. American Chemical Society, Washington.

Ebeler, S.E. & Thorngate, J.H. (2009) Wine chemistry and flavour: Looking into the crystal glass. *Journal of Agricultural and Food Chemistry*, **57**, 8090–8108.

Fischer, U., Strasser, M. & Gutzler, K. (2000) Impact of fermentation technology on the phenolic and volatile composition of German red wines. *International Journal of Food Science and Technology*, **35**(1), 81–94.

Fleet, G. (2008) Wine yeasts for the future. *Federation of European Microbiological Sciences Yeast Research*, **8**, 978–995.

Grosch, W. (2001) *Volatile Compounds in Coffee: Recent Developments* (eds. R.J. Clarke & O.G. Vilzthum), pp. 68–89. Blackwell Publishing Ltd., Oxford.

Günate, Z. (1984) *Terpenes*, quoted by Ribéreau-Gayon, P. *et al.* (2006) *Handbook of Enology II*, loc. cit. pp. 206–211.

Hernandez-Orte, P., Cersosimo, M., Loscos, N., Cacho, J., Garcia-Moruno, E. & Ferreira, V. (2009) Aroma development from non-floral grape precursors by wine lactic acid bacteria. *Food Research International*, **42**, 773–781.

Howell, K.S., Klein, M., Swiegers, J.H., Hayasaka, Y., Elsey, G.M., Fleet, G.H., Hoy, P.B., Pretorius, I.S. & de Barros Lopes, A. (2005) Genetic determinants of volatile-thiol release by *Saccharomyces cerevisiae* during wine fermentation. *Applied and Environmental Microbiology*, **71**, 5420–5426.

Kalua, C.M. & Boss, P.K. (2009) Evolution of volatile compounds during the development of Cabernet Sauvignon grapes (*Vitis vinifera L.*) *Journal of Agriculture and Food Chemistry*, **57**, 3818–3830.

Loscos, N., Hernandez-Orte, P., Cacho, J. & Ferreira, V. (2009) Comparison of suitability of different hydrolytic strategies to predict aroma potential of different grape varieties. *Journal of Agriculture and Food Chemistry*, **57**, 2468–2480.

Pripis-Nicolau, L., de Revel, G., Bertand, A. & Maujean, A. (2000) Formation of flavor components by the reaction of amino acid and carbonyl compounds in mild conditions. *Journal of Agriculture and Food Chemistry*, **48**, 3761–3766.

Ribéreau-Gayon, P. (1978) Wine flavour. In: *Flavour of Food and Beverages* (eds. G. Charalambous & G.E. Inglett), p. 370. Academic Press, New York.

Ribéreau-Gayon, P., Glories, Y., Maujean, A. & Dubordieu, D. (2000) *Handbook of Enology*, Vol. I, loc. cit. pp. 193–161 (carotenoid breakdown), pp. 383–389 (barrel toasting).

Robinson, J. (1995) *Jancis Robinson's Wine Course*, pp. 81–82. BBC Books, London.

Saerens, S.M.G., Delvauz, F., Verstrepen, K.J., Van Dijck, P., Thevelein, J.M. & Delvaux, F.R. (2008) parameters affecting ethyl ester production by *Saccharomyces cerevisiae* during fermentation. *Applied and Environmental Microbiology*, **74**, 454–461.

Sala, C., Busto, O., Guasch, J. & Zamora, F. (2004) Influence of vine training and sunlight exposure on the 3-alkyl-2-methoxypyrazines content in musts and wines from the *Vitis vinifera* variety Cabernet sauvignon. *Journal of Agriculture and Food Chemistry*, **52**, 3492–3497.

Sarrazin, E., Shinkaruk, S., Tominaga, T., Bennetau, B., Frerot, E. & Dubourdieu, D. (2007) Odourous impact of volatile thiols on the aroma of young botrytized sweet wines: Identification and quantification of new sulphanyl alcohols. *Journal of Agriculture and Food Chemistry*, **55**, 1437–1444.

Soles, R.M., Ough, C.S. & Kunkee, A. (1982) Ester concentration differences in wine fermented by various species and strains of yeast. *American Journal of Enology and Viticulture*, **33**, 94–98.

Soufleros, A. (1978) University thesis quoted by Ribéreau-Gayon, P. (1978). Wine Flavour. In: *Flavour of Foods and Beverages* (eds. G. Charalambous & G. Inglett), p. 369 (different yeasts), p. 370 (ester formation). Academic Press, London.

Sponholz, W.R. (1988) Alcohols derived from sugars and other sources, quoted by Jackson, R.S. (2000), loc. cit. p. 185.

Suarez, R., Suarez-Lepe, J.A., Morata, A. & Calderon, F. (2007) The production of ethylphenols in wine by yeasts of the genera *Brettanomyces* and *Dekkera*: A review. *Food Chemistry*, **102**, 10-21.

Swiegers, J.H., Bartowsky, E.J., Henschke, P.A. & Pretorius, I.S. (2005) Yeast and bacterial modulation of wine aroma and flavour. *Australian Journal of Grape and Wine Research*, **11**, 139-173.

Swiegers, J.H. & Pretorius, I.S. (2005) Yeast modulation of wine flavour. *Advances in Applied Microbiology*, **37**, 131-175.

Ugliano, M., Henschke, P.A., Herderich, M.J. & Pretorius, I.S. (2007) Nitrogen management is critical for wine flavour and style. *Wine Industry Journal*, **22**(6), 24-30.

Ugliano, M. & Henschke, P.A. (2009) Yeasts and Wine Flavour. In: *Wine Chemistry and Biochemistry* (eds. M.V. Moreno-Arribas & M.C. Polo), pp. 313-392. Springer ScienceBusiness Media, LLC.

Ugliano, M., Fedrizzi, B., Siebert, T., Travis, B., Magno, F., Versini, G. & Henschke, P.A. (2009) Effect of nitrogen supplementaion and *Saccharomyces cerevisiae* on hydrogen sulfide and other volatile sulphur compounds in Shiraz fermentaion and wine. *Journal of Agriculture and Food Chemistry*, **57**, 4948-4955.

Varela, C., Siebert, T., Cozzolino, D., Rose, L., McClean, H. & Henschke, P.A. (2009) Discovering a chemical basis for differentiating wines made by fermentation with 'wild' indigenous and inoculated yeasts: role of yeast volatile compounds. *Australian Journal of Grape and Wine Research*, **15**, 238-248.

Vilanova, M., Ugliano, M., Varela, C., Siebert, T., Pretorius, I.S. & Henschke, P.A. (2007) Assimilable nitrogen utilisation and production of volatile and non-volatile compounds in chemically defined medium by *Saccharomyces cerevisiae* wine yeast. *Applied Microbial Biotechnology*, **77**, 145-157.

Appendix I

I.1 Chemical formulae nomenclature

IUPAC recommended, unless otherwise stated; other chemical and trivial names are also given.

I.1.1 Nomenclature for a homologous series of compounds (Greek number/word system) (see Table I.1)

Table I.1 Nomenclature.

Greek hydrocarbon (alkanes)			
Number name	**Compound name C_nH_{2n+2}**	**Radical name**	**Formula name C_nH_{2n+1}**
methu = wine	C_1 methane	methyl	methan-
aitho = burn	C_2 ethane	ethyl	ethan-
propulon = entrance	C_3 propane	propyl	propan-
bouturon = butter	C_4 butane	butyl	butan-
5. Pente	C_5 pentane	pentyl	pentan-
6. Hex	C_6 hexane	hexyl	hexan-
7. Hepta	C_7 heptane	heptyl	heptan-
8. Okta	C_8 octane	octyl	octan-
9. Ennea	C_9 nonane	nonyl	nonan-
10. Dekak	C_{10} decane	decyl	decan-
11. Endeka	C_{11} undecane	undecyl	undecan-
12. Dodeka	C_{12} dodecane	dodecyl	dodecan-
13. Treikaideka	C_{13} trtridecane	trtridecyl	trtridecan-
14. Tesseres	C_{14} tetradecane	tetradecyl	tetradecan-
15. Pentakaideka	C_{15} pentadecane	pentadecyl	pentadecan-
16. Hexaideka	C_{16} hexadecane	hexadecyl	hexadecan-

(Continued)

Wine Flavour Chemistry, Second Edition. Jokie Bakker and Ronald J. Clarke.
© 2012 Blackwell Publishing Ltd. Published 2012 by Blackwell Publishing Ltd.

Table I.1 *(Continued)*

Greek hydrocarbon (alkanes)			
Number name	**Compound name C_nH_{2n+2}**	**Radical name**	**Formula name C_nH_{2n+1}**
17. Heptaideka	C_{17} heptadecane	heptadecyl	heptadecan-
18. Octaideka	C_{18} octadecane	octadecyl	octadecan-
19. Nonaideka	C_{19} nonadecane	nonadecyl	nonadecan-
20. Eicosi	C_{20} eicosadecane	eicodecyl	eicosadecan-

Unsaturated hydrocarbons (C_nH_{2n+1}) are more correctly named (methene), ethene, propene, etc.; their radicals, (methenyl), ethenyl, propenyl, etc. and their names in formulae, (methen-), ethen-, prop-1-en- (or –2-), etc.

I.1.2 System for substituent groups (derivatives) (see Table I.2)

Table I.2 Substituent groups.

Alcohol	**Aldehyde**	**Acid**	**Ketone**
–C–OH	–CHO	–COOH	–C=O
-ol	-al	-oic	-one (suffixes)

e.g. C_4, butanol, butanal, butanoic, butanone. N.B. The same total carbon numbering, as in the initiating hydrocarbon, is retained in derivative alcohols, aldehydes, acids and ketones.

I.1.3 System for substituting in long-chain compounds

4 3 2 1
CH₃.CH₂CH₂CH₂.OH   Butan-1-ol (normal or primary butyl alcohol)

CH₃.CH₂.CHOHCH₃   Butan-2-ol (secondary butyl alcohol)
                       (equivalent to Butan-3-ol)
                       or 1-methyl-propan-1-ol

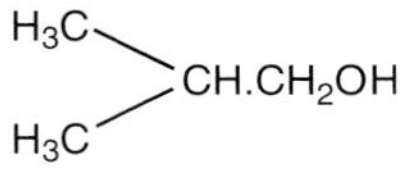

H₃C
    ⟩CH.CH₂OH   2-methyl-propan-1-ol (also a primary alcohol)
H₃C   (isobutyl alcohol or fermentation butyl alcohol)
                       (iso- means equivalent to primary)

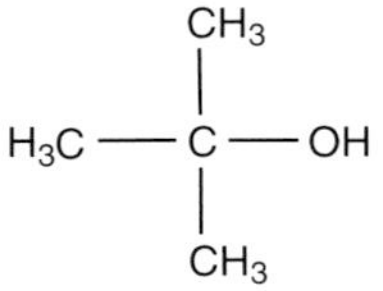

            CH₃
             |
H₃C — C — OH   1,1-dimethyl-ethan-1-ol (tertiary butyl alcohol)
             |
            CH₃

I.1.4 System for characterizing esters

A carboxylic acid (suffix, -oic) has its ending changed to -oate, with the name preceded by the name of the radical group of the esterifying alcohol, e.g.:

butanol (butyl alcohol) + butanoic acid = butyl butanoate
                                        (butanoic acid, butyl ester;
                                        or butanoic ester of butanol)

Alternatively, esters can be named from the acylating group (i.e. from the acid), R.COO, by use of the suffix -oyl; thus caffeoyl quinic acid, or cinnamoyl tartaric acid (from the hydroxyl group).

I.1.5 System for characterizing unsaturated compounds

$$7 \quad 6 \quad 5 \quad 4 \quad 3 \quad 2 \quad 1$$

e.g. $CH_3.CH:CH.CH:CH.CH_2.COOH$

hept-3, 5-dien-1-oic acid

I.1.6 Systems for esters, thiols and thio-compounds

Esters: e.g. CH_3OCH_3 methoxy methane; then, diethoxy-, etc. (other names are dimethyl ether or oxide, or methyl oxymethane).

Thiols: e.g. CH_3SH methan-thiol (other names are methyl thio-alcohol, methyl mercaptan, mercapto-methane).

Thio ether: e.g. CH_3SCH_3 dimethyl sulfane, dimethyl sulfide (other names are dimethyl thioether, thiobismethane).

I.1.7 Miscellaneous IUPAC recommendations

Stereochemistry (1) *Z* or *E* for *cis/trans* isomers. (2) Chiral orientation, use of *R* and *S* instead of D and L. (3) Optical activity, represented by (+) or (−) instead of d and el.

Linear hydrocarbon chains; numbering system, ω from end carbon atom.

I.1.8 Alternative chemical names (see Table I.3)

Table I.3 Alternative names.

Chemical group	Example name	Other chemical name
Alcohols	Butan-1-ol	1-Hydroxybutane
Aldehydes	Butan-1-al	Oxa-1-butane
Ketones (NB 1-one not possible)	Butan-2-one	Keto-2-butane (oxo-2-butane, methyl ethyl ketone)
Acids (carbon numbering system retained)	Decan-1-oic acid	1-Decyclic acid
Acids	e.g. butan-1,4-dioic acid. Ethane dicarboxylic acid, dicarboxyethane.	
Alcohols	Methanol, carbinol, then methyl carbinol for ethanol, and so on.	

Carbon-numbering system not retained in alternative names, except in ring systems cyclohexane.

I.1.9   Numbering systems for ring compounds (see Table I.4)

Table I.4   Ring compounds.

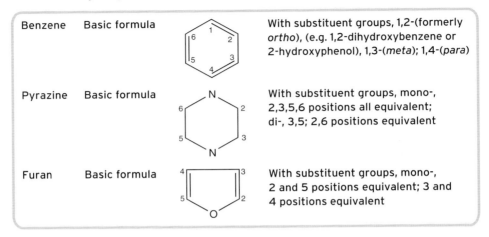

Benzene	Basic formula		With substituent groups, 1,2-(formerly *ortho*), (e.g. 1,2-dihydroxybenzene or 2-hydroxyphenol), 1,3-(*meta*); 1,4-(*para*)
Pyrazine	Basic formula		With substituent groups, mono-, 2,3,5,6 positions all equivalent; di-, 3,5; 2,6 positions equivalent
Furan	Basic formula		With substituent groups, mono-, 2 and 5 positions equivalent; 3 and 4 positions equivalent

I.1.10   Trivial and common names for derivative alkanes and other compounds

The names of organic volatile and other compounds in the tables in the text and Appendix II are given in both their IUPAC or chemical, and their traditional or trivial forms, as far as possible. Note, however:

(1) Saturated compounds (aldehydes, alcohols and acids). Common or trivial names accepted by IUPAC up to C_4, e.g. butyric (butan-), C_3 propionic (propan-), C_2 acetic, acetaldehyde, (ethan-), C_1 formic (methan-), but C_5 valeric is pentan-, and so on up to C_{20}. Isopentyl alcohol is acceptable for 3-methyl-butan-1-ol, but not isopentanol or isoamyl alcohol, etc. Oxo-, but not keto- is used for ketones. Radical names also used for all alcohols, e.g. propyl and hexyl alcohols.

(2) Unsaturated hydrocarbon radicals, e.g. C_2 vinyl (acceptable for ethenyl), C_3 allyl or acrylic (for prop-2-enyl).

(3) Trivial or traditional names recommended for complex hydroxy aliphatic acids, as in wine chemistry, such as tartaric, malic, citric, lactic, succinic and pyruvic acids.

(4) Use of the term to mean 'acyl' (–CO–) from the corresponding acid name (old nomenclature) is not generally recommended:
  ● Monoketones, e.g. acetyl methane, CH_3COCH_3, for propanone or dimethyl ketone; propionyl methane, for butan-2-one or methyl ethyl ketone.
  ● Diketones, e.g. diacetyl, $CH_3CO.CO.CH_3$, for butan-2,3-dione or biacetyl; acetylacetone, for pentan-2,3-dione or ethyl methyl ketone.

(5) Phenol is recommended for hydroxybenzene. For phenolic compounds, names such as catechin are generally retained. Cinnamic acid names are based upon propenoic acid derivatives.

(6) Terpenes, common or trivial names, such as limonene and α-terpineol, and furanones, such as furaneol (see information for particular cases but full chemical names generally recommended).

(7) Use of α, β, etc. for substituents in linear chains has been replaced by numbering system as in I.1.3 and I.1.5 above. The same applies to ring structures.

I.1.11 General

Reformed systems for nomenclature of organic compounds were first considered and devised (e.g. the suffix -ol for alcohols) by an International Commission in Geneva in 1898, and taken up by the International Union of Pure and Applied Chemistry (IUPAC). Their recommendations are published from time to time (e.g. 1993). Register names (e.g. *Chemical Abstracts* and *Macmillan Dictionary of Organic Compounds* may have an inverted order for chemical names – substituent groups last).

I.2 Stereochemistry

Many pairs of chemical compounds, called isomers, each having the same molecular weight and linear representation of their constituent atoms/groups of atoms in their molecules, will show differences in actual molecular structure on account of different spatial distributions of attached component atoms. These compounds fall into one of two subdivisions, called (1) enantiomers (Gk. *enantios* – opposite) and (2) geometric isomers, in ways to be described. Some pairs of compounds may fall into both subdivisions. Both subdivisions may reflect colour/flavour and other physically measurable differences within the pairs. The second group may reflect some chemical reaction differences between the compounds, due to steric hindrance.

A further group of isomers, which have some important significance in wine chemistry, is known as tautomers (Gk. *tautos* or *to auto* – the same thing), and are characterized by a slight difference in linear representation of their molecules. Tautomerism refers to the ability of certain compounds to react in two different ways, dependent on their molecular structure, thus giving rise to two separate series of derivatives by which a compound can be identified.

I.2.1 Enantiomers

These compounds generally require the presence of at least one asymmetric carbon in the molecule. An asymmetric atom is an atom that is linked to four different monovalent atoms or groups. In this way, for a single asymmetric carbon atom, two spatially different forms of a given molecule are possible. These structural forms are not superimposable but are, in fact, mirror images. The two compounds will have identical chemical and physical properties, but they will differ markedly in their capacity to rotate the plane of polarized light, either to the left or to the right, to which they are exposed in suitable equipment. Numerous such pairs of compounds are familiar in wine chemistry, thus

2-hydroxypropanoic acid (lactic acid), $CH_3.C^*HOH.COOH$, with one asymmetric carbon atom (indicated by an asterisk) can exist in two different structures (a pair of enantiomers) and the molecule can be drawn as shown in I.I.

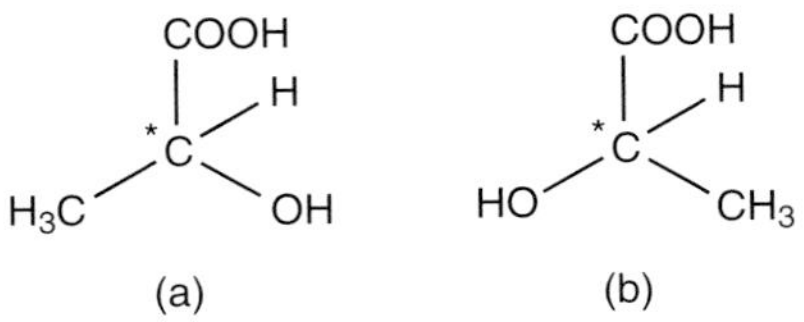

(a)   (b)

I.I

By an older convention, the structure (a) was accorded the designation D, now a small capital letter (D), and (b) L, now L. These designations refer solely to the arbitrarily stated orientation of the OH group, thus OH in (a) is shown on the right (Dextro-), and OH in (b) on the left (Laevo-). This type of designation is still used for the molecules of well known natural substances, for example, also, so-called 'active' amyl alcohol (2-methyl-butan-1-ol), $CH_3.CH_2.C^*H(CH_3)$. CH_2OH and especially for the aldo- and keto-hexoses, e.g. D-glucose, $C_6H_{12}O_6$.

These long-chain sugar molecules will have several asymmetric carbon atoms (four in aldohexoses and three in ketohexoses) when looked at in their straight-chain form, and will have 16 and eight different spatial structures (eight and four pairs of enantiomers). In fact, the compounds are all known and have individual names. From these, dextro-glucose (based upon the configuration shown in Table 3.4) is of the most interest in wine chemistry. However, in its ring molecular structure, as glucose is believed to exist in aqueous solution, the molecule shows an additional asymmetric atom (No. 1, as shown also in Table 3.4), which gives the substance a characteristic optical rotation value. To the D- designation is added the symbol α for the main cyclic form, and β for the other. The molecular configuration represented by the D and L structures is again conventionally designated by the right-hand positioning of a particular hydroxyl group (i.e. situate at the No. 4 carbon atom in the chain) and similarly for α and β.

Optical activity

Optical activity is measured in $[\alpha]_D^{T°}$ specific rotation units from experimental determinations, where D is the sodium line at 589 nm and temperature $T°C$; its value is calculated from $\alpha \times 100/l \times c$, where l is the length in decimetres of liquid traversed, c is the number of grams substance in 100 ml solution, and α the angle of rotation observed, to the right or to the left. Its numerical value differs greatly for different compounds, dependent upon the groups attached (e.g. chromophores). It is not clearly related to the structural configurations described, of 'handedness'. If the enantiomer rotates the plane of polarized light to the right, so-called dextrorotatory, it is designated as the + form (formerly lower case d); if to the left, laevorotatory (formerly lower case l). However, the molecular configuration (arbitrarily) is usually also quoted, as either D or L, in a straight-chain compound. In the simplest cases, D-structures will be associated with (+) optical dextro-rotation, and L- with (−). Whilst the main glucose is D-α, it is dextrorotatory, as is the D-β glucose but at a lower numerical value; fructose D-β is laevorotatory.

Chirality

This subject of enantiomerism is now more generally referred to as chirality (Gk. *cheiros*, meaning 'hand'), a term first mentioned by Lord Kelvin in 1904. As is now increasingly recognized, the term has special significance to odour perception for volatile compounds. In non-volatile compounds, physical property differences can occur as a consequence of certain types of chiral difference of otherwise identical compounds and flavour/taste differences within pairs of enantiomers.

A detailed and clear description of chirality is given in the textbook, *Organic Chemistry* (Clayden *et al.*, 2001). These authors start their chapter with the representation of the lactic acid molecule, drawn as either (a) or (b) shown in I.II.

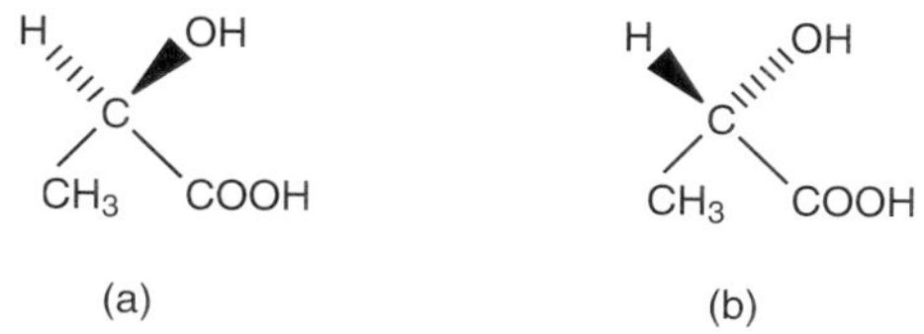

(a)   (b)

I.II

The plane of symmetry is conceived by drawing the carbon skeleton lying in the plane of the paper, with the other two groups H and OH either pointing back symbolized by (||||) or pointing forward (◀). Configuration (a) is called the *R* form (from *R. rectus* – right) and (b), the *S* form (*S. sinister* – left), which are the equivalents of the old D and L designations.

When there are two asymmetric carbon atoms (stereogenic or chiral centres), the structural definitions necessary become more complicated, thus for the most prevalent wine acid (2,3-dihydroxy-butan-dioic acid) tartaric acid, HOOC.C*HOH.C*HOH.COOH.

The following four structures, stereoisomers ($2^2 = 4$) are possible, and two so-called diastereoisomers ($2^{2-1} = 2$) each with an expected pair of enantiomers, as follows: (H groups are omitted from the diagrams for convenience).

(1) With the OH group in the *syn* position (two enantiomers, one diastereo-isomer) (I.III)

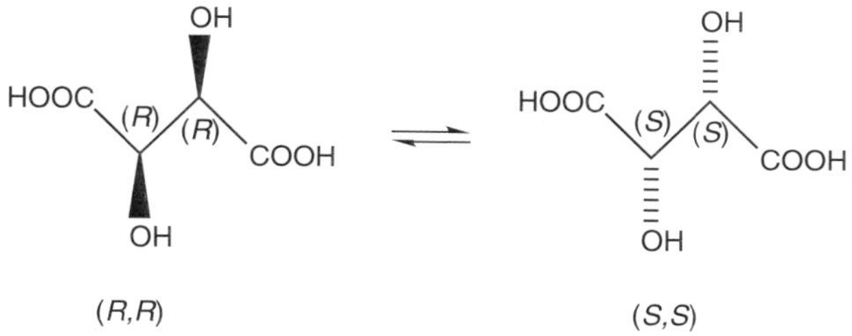

(R,R)   (S,S)

I.III

(2) With the OH groups in the *anti*-position (another diastereoisomer, no enantiomers) (I.IV).

(R,S)   (S,R)

I.IV

The designations *R* and *S* for multi-chiral centres in general are now determined by a somewhat complex set of rules to cover all cases, the so-called C-I-P rules of 1952, described in the book *Organic Chemistry*, but not necessary to examine here. However, the structures described above as (*R,R*) and (*S,S*), in Section (1), represent true enantiomers, that is, they are mirror images, but careful examination of the structures (*R,S*) and (*S,R*) in (2) will reveal that they are not true enantiomers, as they are superimposable. They have, in fact, the same molecular structure and are so-called achiral. The latter compound is known and is called meso-tartaric acid (Gk. *mesos*, middle). In addition, it has zero optical activity (internal compensation), compared with the *RR* compound, which has (+) and the *SS* (−). In practice, there will also be a so-called racemic mixture, which is an equi-mixture of the two optically active forms, resulting in zero optical activity. The meso compound has a lower melting point and a lower water solubility than either of the two enantiomers, which have equal numerical values of these physical properties. Pasteur, in his studies on vinification around 1848, was especially interested in the optical activity and other physical properties of tartaric acid. He also devised a method of separating the enantiomers in a racemic mixture. He made use of the selective destructive effect of certain yeasts and moulds on a d(+) ammonium tartrate over its l(−) enantiomer. He also found that a racemic mixture of sodium ammonium tartrate could be separated after recrystallization from aqueous solutions, since the crystals of the two enantiomers had a slightly different shape.

Recently it has been found that, though pairs of enantiomers in general have identical physical and chemical properties and behave identically, this is no longer true when they are placed in a chiral environment. This latter phenomenon, in fact, enables the separation of enantiomeric volatile compounds in GC equipment by use of a stationary phase, which is made chiral by bonding with enantiomerically pure compounds. Pasteur's separations of tartaric acid salts were, probably, also chirally determined. More importantly, many pure compounds in an enantiomeric pair possess different odour/aroma characteristics in many instances, presumably related to the chirality of the olfactory organ receptors at the back of the nose. The effectiveness of compounds used as drugs in modern medicine also depends, in many instances, on their chiral structure. This discovery has given great impetus to developing methods of maximizing separation of enantiomers, and of so-called asymmetric synthesis of organic compounds. The slight differences of sweetness and acidity found within enantiomers of sugars and organic acids, perceived by the tongue, may well have a chiral basis.

Several volatile compounds in wine have been reported as having different odour characteristics in their R and S structures, though both enantiomers may not actually be present in particular wines. A compound with a single asymmetric carbon atom, such as oct-1-en-3-ol, $CH_3(CH_2)_4$. *CHOH.CH:CH$_2$, which is reportedly present in botrytized grapes, in its S form is said to have a 'vegetable, mouldy' flavour, whilst the R form has a 'green mushroomy, meaty' flavour, with a mushroom odour (quoted, Flament, 2001), though the differences in this instance may not be great. Chiral centres can be at a distance from each other in the molecule, but the lack of symmetry may be established. Limonene and carvone are clear examples, where (+) and (−) forms are reported to give compounds of markedly different odours and thresholds. Other examples will be found in the main text in Chapter 4, particularly amongst the flavanones and terpenes, though the presence of which enantiomer or both may not as yet have been clearly established. There does not appear to be much information in the literature on the different odours of the corresponding meso compounds.

Occurrence of enantiomers

In nature, biogenetic pathways appear to produce one form only; for example, through one important amino-acid precursor, alanine (and others) present in the laevorotatory, S form in plants, though R in bacterial cell walls. Malic acid (2-hydroxybutan-dioic acid), the 'apple' acid or 'ordinary' malic acid is laevo(−) rotatory. Tartaric acid, the 'ordinary' tartaric acid already found in grapes, is dextro(+)rotatory, and its formula in full, once d-tartaric acid, is now given as, {R,(R*,R*)}-2,3-dihydroxy-butandioic acid; though better, (2R,3R)(+)-2,3-with R* indicating the existence of the chiral centres as already described, or numerals before the R indicating position of the chiral centres.

Fermentation processes appear to produce the racemic form (as with the lactic acid bacteria to form dl-lactic acid). The optically active amyl alcohol (2-methyl-butan-1-ol) formed by the fermentation of sugars by *Saccharomyces cerevisiae* by conversion from 1-leucine is said to be laevorotatory, in contrast to the so-called fermentation amyl alcohol (3-methyl-butan-1-ol), which is inactive, non-chiral, deriving from isoleucine.

I.2.2 Geometrical (stereo-) isomers

So-called geometrical isomerism is found mainly in compounds with two carbon atoms linked by a double bond. When each of the two carbon atoms contain the same two but different groups, e.g. hydrogen/hydroxyl, they can be arranged and fixed in two separate ways, giving rise to so-called 'cis' and 'trans' forms, e.g. in general terms.

The terms cis and trans refer to the side contiguity of the atoms or groups, a and b, with cis meaning on the same side for each. According to the new IUPAC nomenclature, these are now called 'Z' (Ger. *zusammen* – together) and 'E '(Ger. *entgegen* – opposite) respectively (I.V).

cis(Z)   trans(E)

I.V

The compounds within a pair will have markedly different physical and chemical properties, and importantly, different flavour characteristics in many cases. Chemical reactivity differences can occur due to steric hindrance with some molecular structures. Examples have been described in the main text, such as for the unsaturated aldehydes/alcohols, hydroxycinnamic acids and 2-furanones; for example, the *cis* form of 2-nonenal has a known lower threshold value than the *trans*, and has a different odour, but both these features vary markedly with concentration. The 'oak lactone' has both geometric isomers but also optical activity variants in each, giving rise to four compounds with different flavour characteristics.

Biogenetic pathways usually produce only one form, e.g. linoleic and linolenic acids in vegetable oils are always in the *cis* form at each of their double bonds. Chemical transformations of these compounds, as in conjugation and hydrogenation, produce a quantity of *trans* forms. Unsaturated aliphatic aldehydes, like 2-nonenal and 4-heptenal, produced by the enzymatic oxidation of linoleic acid seem to be formed in both *cis* and *trans* forms. There seems, however, to be a tendency for similar groups to take up spatial positions, as far away as possible from each other, thus favouring the *trans* forms.

I.2.3 Tautomerism

Tautomerism has also been described as dynamic isomerism of two possible different molecular structures for the same molecule, though these will not be too dissimilar. Numerous examples are now known, falling into different chemical classes but the triad system, with three carbon atoms and in particular, the arrangement in the keto-enol system, C=C-OH, is of the greatest interest in wine chemistry. Compounds with this latter grouping and capacity for change, exhibit the properties of both ketones and hydroxyl compounds, thus for resorcinol (1,3-dihydroxybenzene, or *m*-hydroxy-phenol) we have the situation shown in I.VI.

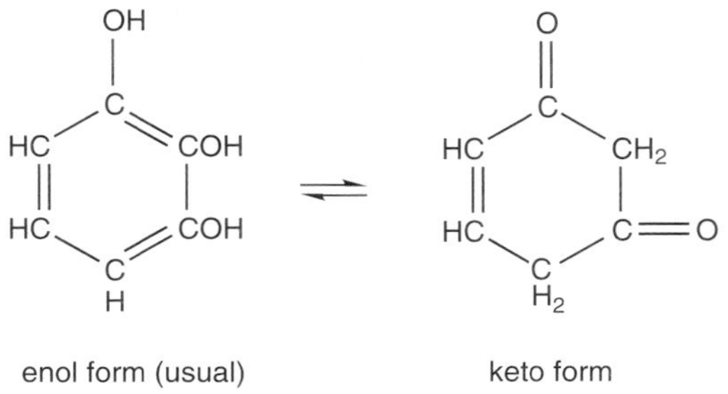

enol form (usual)   keto form

I.VI

Similarly catechol (1,2-), hydroquinone (1,4-) and trihydroxybenzenes (1,3,5-, phloroglucinol) exhibit tautomerism, which is relevant to oxidative changes in wines.

A three-carbon system, as in $-C=C-C$, again with a mobile carbon atom, is again of interest to the wine industry, as in the following examples:

(1) Eugenol(2-methoxy-4(prop-2-enyl)phenol)↔iso-Eugenol[-(prop-1-enyl)].
(2) Linoleic and linolenic acids, to their conjugated forms.

$$-CH=CH.CH_2.CH=CH- \leftrightarrow -CH_2.CH=CH.CH=CH-$$

I.3   Chemistry of the oxidation of organic compounds

Oxygen (Gk. *oxus* – sharp) is an element in period 2, group 16 of the periodic table, with an atomic number of 8. Its atom has an electron distribution around the nucleus, described as 2,6,0, i.e. total of eight electrons, in the different energy shells, s, p and d. These quantum shells or orbitals are represented by the notation, $1s^2$, $2p^2$, $2d^4$. The electrons in the No. 2 levels have a total of six valency electrons, thus the oxygen atom normally only forms two paired bonds, :O:, i.e. a normal valency of two in compounds.

Molecular oxygen (O_2, dioxygen) necessarily has shared orbitals from the two identical atoms but now with twelve electrons in the valence orbits. The molecular orbitals have a distribution designated as $1\sigma^2_g$, $2\sigma_u$, $3\sigma^2_g$, $1\pi^4_u$, $2\pi^2_g$ related to the corresponding atomic orbitals. The suffixes, g and u, refer to bonding capability, according to the orbital, σ or π. In the oxygen molecule, with the half-filled orbitals of $2\pi_g$, the outermost electrons have parallel spins in different orbitals. With filled orbitals, paired electrons have opposite spin. The numbers in superscript are again the number of electrons in each orbital. Molecular structure is a highly complex topic but the reader is referred to *Inorganic Chemistry* (Shriver *et al.*, 1994) for detailed information.

The oxygen molecule is regarded as particularly interesting and different from others, in that the lowest energy configuration has two unpaired electrons in different orbitals, which makes it also diamagnetic (i.e. tending to move into magnetic fields). Further electrons present as in O_{2-} and O_2^- can be accommodated in the $2\pi_g$ orbitals; the g orbitals are regarded as anti-bonding. These orbital configurations described do not, however, fully characterize energy levels. The oxygen molecule in the ground state is also characterized by the term $O_2{}^3\Sigma$, so-called triplet oxygen.

In this condition it has a radical character, enabling participation in chain reactions, as in some oxidative processes occurring in wines, which have been described in the text and are further discussed in the following sections. The oxygen molecule can also occur in two so-called singlet states. The first singlet state, $O_2{}^1\Sigma$ has paired electrons in the same two orbitals as in the ground state (but with higher energy); and another, represented as $^1\Delta$, with electrons paired only in one orbital (with intermediate energy). The latter can survive long enough to participate in chemical reactions. It can be generated in solution by energy transfer from photo-excited molecules.

The other allotrope of oxygen is ozone (O_3), which is a highly reactive substance in oxidation processes but of lesser interest in wine chemistry.

I.3.1 Auto- and enzymatic oxidation of lipids

The auto-oxidation of lipids, in particular the unsaturated acids, oleic, linoleic and linolenic and their glycerol esters, is a complex phenomenon but has been studied in detail by many investigators over recent decades. One proposed reaction mechanism, based upon a free radical chain process and applicable to commercial oils and fats, has been described by Belitz & Grosch (1987). It consists of three steps as follows:

(1) Initiation (by light/heat):

RH (unsaturated acid) → R$^{\bullet}$ (fatty acid free radical)

(2) Propagation (i.e. a chain reaction):

R$^{\bullet}$ + O$_2$ (ground state) → ROO$^{\bullet}$ (peroxide free radical)
ROO$^{\bullet}$ + RHROOH (fatty acid peroxide) + R$^{\bullet}$ (fatty acid free radical)

(3) Termination (in the absence of oxygen):

2ROO$^{\bullet}$ → inactive dimer
2R$^{\bullet}$ → inactive dimer
ROO$^{\bullet}$ + R → inactive dimer

RH will be a typical alkyl compound, with a labile hydrogen atom, such as that of a methylene group (–CH$_2$–), adjacent to two carbon atoms linked by a double bond. This is a three-carbon tautomeric system, one unit as in the oleic, two in the linoleic and three in the linolenic acid molecules. The mobility of the hydrogen atoms can be explained in terms of electron concentration.

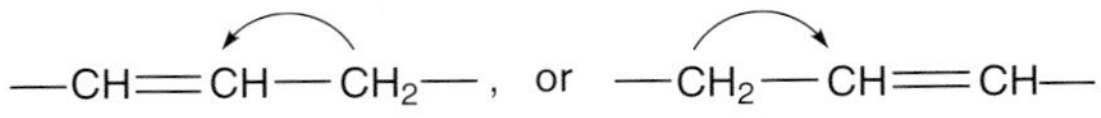

$$—CH{=}CH—CH_2—, \quad or \quad —CH_2—CH{=}CH—$$

Apart from an inherent chain reaction, the other important feature of this process is the formation of relatively stable peroxides. However, they usually decompose with the formation of secondary reaction products, under the same influences of light/heat, metal catalysts and other initiating agents, as in Step 1 above. Several chemical groups of secondary products are possible, but of the greatest interest for wine are those arising from peroxide formation and chain scission, e.g. formation of unsaturated aldehydes (Fig. I.1).

The triad (C-C-C) in the initial radical form of RH will be subject to double-bond shifting in a dienoic or trienoic acid, so that we can find conjugation of the double bonds in the final reaction products. The usual *Z* (*cis*) double bond system affected by the methylene group is transformed into an *E* (*trans*) configuration. The other unaffected double bonds will, however, retain their original configuration (i.e. usually *cis*, in natural fatty acids).

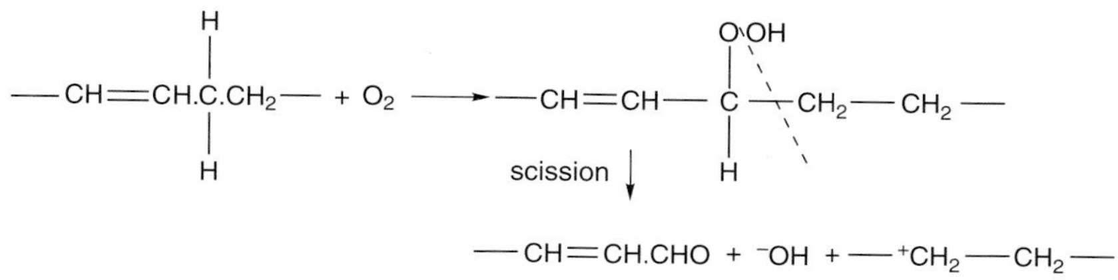

Figure I.1 Oxygen attack at a methylene group adjacent to a double bond and subsequent reaction.

Retardation, though still in the presence of oxygen, can be effected by certain antioxidant compounds, particularly with edible fats containing oleic acid as the only fatty acid present; thus, for example, by the use of tocopherol, which has a phenolic hydrogen atom.

Peroxy free-radicals react preferentially, leading to a fat hydroperoxide, as is the case in a normal uninhibited oxidation reaction. However, a tocopherol free radical, which is produced simultaneously, has much less affinity to react with the oxygen molecules present, so the reaction chain is stopped. Some phenolic compounds found in grapes/wine, such as quercetin, are reported to be good antioxidants for oils and fats.

The decomposition reaction sequence described above was first used by Drawert (1974) to explain the formation of hexanal from linoleic acid, and of *cis*-hex-3-en-1-al and *trans*-hex-2-en-1-al from linolenic acid. The initial formation of the required hydroperoxide within plant tissues as in grapes was, however, considered to be oxygen activated by a lipoxygenase enzyme, and the scission by an 'aldehydase'. Reduction to the corresponding saturated alcohols (taking place largely in the fermentation) to the unsaturated alcohols, was also described as being due to an 'alcohol oxido-reductase' (probably a dehydrogenase). However, with subsequent oxidation during maturation, there may be some reversion to the original aldehydes. The precise methylene group to be peroxidized will differ in different examples. Semmelroch & Grosch (1996) have provided similar schemes to show the production of 2-nonenal and 4-heptanal.

I.3.2 Oxidation–reduction of alkyl alcohols and aldehydes

Some explanatory mechanisms have been published for these reactions, which occur frequently in wines. Ethanol can be oxidized to ethanal by loss of hydrogen, and in reduction, vice versa, by gain,

$$CH_3.CH_2.OH \rightarrow CH_3.CHO + H_2$$

Typically, this reaction is said to take place in the presence of a dehydrogenase, together with a co-enzyme, nicotinamide adenine dinucleotide, NAD, thus

$$NAD + \text{reduced substrate} \rightarrow NADH + \text{oxidized substrate}$$

This co-enzyme is also used in the Krebs cycle for converting citric acid to α-keto-glutaric acid (2-oxobutandioic acid).

It is probable that these enzymes are no longer active after fermentation and clarifications. The proposed mechanism for the observed formation of ethanal from alcohol during maturation has been ascribed to the presence of phenolic substances and of ferric/cupric ions (Wildenradt & Singleton, 1974). In contrast, Ribéreau-Gayon *et al.* (2000) say that it is the procyanidins, polymerized during their oxidation that can then oxidize other components such as ethanol.

I.3.3 Oxidation of phenolic compounds

Many explanatory mechanisms have been published for the observed oxidation phenomena of phenolic compounds, which are particularly important in wines, as described in the main text of Chapter 3. It is generally considered that, of all the numerous compounds in wines, proteins, carbohydrates, etc., the phenolic compounds present, such as the tannins and anthocyanins contained in the grapes/must, are oxidized first, on exposure to oxygen, during vinification and subsequently. Although during the fermentation the fermenting yeasts will take up all the available oxygen, presumably the conditions in the fermenting must are non-oxidative. Oxygen-generated free radicals from other sources can be mopped up by phenolic substances, before undesirable chain reactions set in. Some of these substances, therefore, have well-publicized antioxidant roles, playing a part in different medical conditions. These mechanisms can be highly complex and indeed controversial in many cases.

The chemical reaction properties of the simple mono-, di-, and tri-hydroxybenzenes (phenols), which are the basic building blocks of the more complex polyphenolic compounds, mostly present in wines, need to be understood first of all.

(1) Oxidative. Dihydroxybenzenes, of which there are three: (a) hydroquinine (*p*-hydroxyphenol) is very easily oxidized to *p*-quinone (1,4-diketobenzene), but is of more interest as a photograph developer than in wine; (b) catechol (*o*-hydroxyphenol) of greater significance in wines, also oxidizes to a quinone (*ortho*-) (1,2-diketobenzene), easily reverting back, however, in acid solutions, but in the course of oxidation, hydrogen peroxide is produced, which is available for oxidation of other substances; and (c) resorcinol (*m*-hydroxyphenol) does not form a quinone on oxidation. There are three possible trihydroxybenzenes: (d) pyrogallol (pyrogallic acid, 2,3-dihydroxyphenol), an easily oxidizable substance, e.g. turning alkaline solutions brown, explaining its use in early oxygen gas analyses; (e) phloroglucinol (3,5-dihydroxyphenol), which is also a tautomeric substance, reacting also in a non-benzenoid tri-keto form; and (f) hydroxy-hydroquinone (2,4-dihydroxyphenol), of lesser interest. The oxidation characteristics of all these substances will be different when present in combined form.

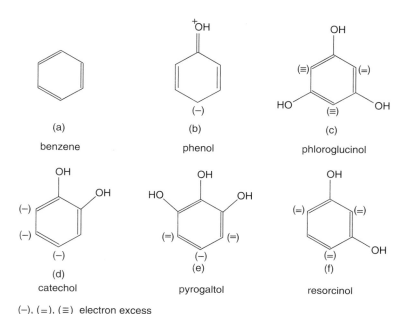

(a)
benzene

(b)
phenol

(c)
phloroglucinol

(d)
catechol

(e)
pyrogaltol

(f)
resorcinol

(−), (=), (≡) electron excess

Figure I.2   Mesomeric effects in phenols.

(2) Colouring reactions. These phenols are generally colourless in aqueous solution, but catechol is noted for the formation of red compounds with alkaline ferric chloride solutions. Cyanidins and other polyphenolics use colouring reagents in their traditional analysis (see main text).

(3) Substitution reactions. The ease of substitution of hydrogen atoms in the benzene ring of phenolic compounds is commonly associated with the type of electron redistribution in the ring, as a result of the effect of the substituent hydroxyl groups. It is well known that, in a simple phenol (monohydroxybenzene), there is a so-called mesomeric effect (Gk. *mesos*, middle) in a carbon chain to give an electron excess at the o- (1,2) and p- (1,4) positions or nodes, together with a net positive charge at the oxygen atom of the hydroxyl group (Fig. I.2). Thus, substitution reactions are easier on phenol than benzene and in particular at the 2, 4 and 6 positions. This effect does not occur, however, at the meta-position (3). Ribéreau-Gayon *et al.* (2000) describe this effect, particularly with the trihydroxy-benzene (phloroglucinol), with pyrogallol, and with the dihydroxybenzene (catechol), when the electron excess becomes additive, at the unsub-stituted adjacent ortho-positions. In this way, electrophilic reagents (not necessarily compounds with a net positive charge) are facilitated in their substitutive attack at these nodes.

In actual polyphenolic compounds with pyran rings, as in wines, there will be positions with mobile or labile hydrogen atoms in their molecules, susceptible to oxidative attack, in a similar fashion as lipids discussed in Section I.3.2. Use of these concepts can explain the results of oxidation of a flavan-3-ol or monomeric catechin in producing a wide range of intermediate compounds but ultimately additional polymeric substances. One such mechanism was well described in a Doctoral thesis of De Freitas (1995), quoted by Ribéreau-Gayon et al. (2000). These polymeric substances are increasingly brown with increasing size, and eventually they form precipitates.

Flavan-3-ol (I.VII) has a molecular formula, $C_{15}H_{14}O_6$, with a molecular weight of 290.

Structural formula of flavan-3-ol, showing $C_2{}^*$ and $C_3{}^*$ atom in the C ring; $C_{4'}$ atom in the B ring

I.VII

The proposed scheme includes the following sequences:

(1) Mobile hydrogen at C2 in the C-ring is removed by an O_2 molecule to form first a flavan radical and a -OOH radical. The flavan radical then either reacts:
   (a) with further oxygen to form a peroxy flavan radical which converts to a hydroperoxide in a chain reaction, which transforms into polymeric forms; or,
   (b) with a peroxy radical, removes a hydrogen atom at C4 in the B benzene ring (catechol). This latter radical then converts into an o-quinone (keto-C4), and then again transforms into polymeric forms.
(2) Homolytic breakdown occurs at the hydroxyl group at C4 in the B ring producing flavan and hydrogen radicals. This flavan radical will give rise to a range of compounds, quinones and compounds with linked units, all transformable again into polymeric substances.

The mechanism of the final polymerizing stages is not described, though di-ortho-quinones seem to be important immediate precursors. There is some difference in speed of oxidative reaction, dependent upon the exact molecular and chiral structure of flavan-3-ol; thus (−)-epicatechin (R at both C2 and C3) is said to be more rapidly oxidizable than (+)-catechin (R at C2 and S at C3), possibly due to steric hindrance.

Oxidation of procyanidins

Red wine contains substantial quantities of so-called procyanidins that are already polymers of catechins (C) or epicatechins (eC) in several different structures (types, eight B dimers, C trimers and higher polymers), as first described in Chapter 3. The interflavan linkage is of two types, firstly a C4–C8 bond with the following combination of units as shown below.

Procyanidin	B_1	B_2	B_3	B_4
Units (upper/lower)	eC–C	eC–eC	C–C	C–eC

Secondly, a C4–C6 bond, with procyanidins named from B_5 to B_8 and a similar sequence of units. The ease of oxidation of procyanidin dimers with a C4–C8 interflavan bond (types B_1 to B_4) is stated (Ribéreau-Gayon et al., 2000) to depend upon the upper structural unit, whilst those with a C4–C6 linkage (types B_5 to B_8) depend upon the lower unit. Catechin (+) in B_3, B_4 (upper unit) and in B_5, B_7 (lower unit) molecules make them more oxidizable than those containing (–)-epicatechin in B_1, B_2 and in B_8, though the reason is not entirely clear. These procyanidins in acid solution are, however, also unstable even in the absence of oxygen to form further polymeric forms, usually higher; thus a B_2 dimer will produce a C_1 trimer. The B_1 form is the most prevalent dimer in grapes. Acid hydrolysis can involve tautomeric enol-keto change and then the formation of a reactive cationic form. It is also stated that: (1) for the same basic unit (C or eC) oxidizability increases with the degree of polymerization; thus, C_1 trimer > B_2 > monomer, for eC-; (2) oxidizability increases with increased substitution in the B ring, thus gallocatechin > epicatechin. However, oxidizability is not defined, and some of the information appears to conflict with values of relative redox potential determined in model solutions for some of these substances, described in Section 3.8.3 (Chapter 3).

Even higher polymers are possible due to interaction of procyanidins with ethanal, itself formed by their presence during oxidation. Certain of the higher polymers may not have actual linkages, but are so-called 'disorganized' polymers. Anthocyanins can also form complexes with procyanidins or tannins. All these possible changes are especially related to changes in the basic taste characteristics of bitterness and astringency, particularly during maturation, as discussed in Chapter 3.

Anthocyanins are also oxidizable in acid solutions, though malvidin 3-glucoside with two methoxy groups instead of hydroxyl, and only one hydroxyl group in the B-benzene ring, is stated more resistant to oxidation than cyanidin 3-glucoside (with two hydroxyl groups). The mesomeric effect will be lowered in the former compound. However, over time, all monomeric anthocyanins in matured wines disappear (Chapter 6).

Oxidation of non-flavanoid phenolic compounds

Caffeoyl-tartaric acid or caftaric acid [caffeic acid is 3-(3,4-dihydroxyphenyl)-prop-2-enoic acid or 3,4-dihydroxycinnamic acid], and p-coumaryl-tartaric acid or coutaric acid [coumaric acid is 3-(4-monohydroxyphenyl)-prop-2-enoic acid, or *p*-hydroxy-cinnamic acid] are both regarded as highly oxidizable, with the caffeic compound with two hydroxy groups, the most. Ester type linkage is through one of the enantio-hydroxy groups (said to be C2) of the tartaric acid (butan-2,3-diol-1,4-dioic acid). The mechanism of oxidation has been studied by Cheynier *et al.* (1989) quoted by Ribéreau-Gayon *et al.* (2000), also (1991) referenced by Jackson (2008) to experimental work in model solutions using polyphenol oxidase.

General

A review on phenolic compounds and polyphenol oxidase in relation to the browning of grapes and wines by Macheix *et al.* (1991) gives an excellent insight in the chemically complicated reactions involved. They conclude, after considering the available literature, that the following five points can be made:

(1) Coupled oxidation reactions occur, involving compounds not oxidized by polyphenol oxidase (grape enzyme triggering reactions which cause browning).
(2) Caftaric acid plays a special role, by starting coupled oxidation reactions.
(3) Glutathione limits browning by rapidly reacting with the caftaric acid quinone. The ratio between levels of caftaric acid and glutathione is thought to be one of the important parameters.
(4) Flavan-3-ols, in particular their monomers and dimers, are determining the extent of browning in wines.
(5) The mechanisms involved in oxidation are similar in enzymatically controlled and purely chemical reactions. Enzymatic catalysis occurs only during the first stage of oxidation, and reactions continue purely chemically, in particular the polymerization leading to brown end products. Enzymatic oxidation causes the rapid trigger to browning reactions.

The substrates most rapidly oxidized by polyphenol oxidase are (+)-catechin, (−)-epicatechin, caffeic acid, catechol and 4-methylcatechol. The caftaric acid quinone reacts most easily with glutathione to form 2,5-glutathionyl caftaric acid, which is not a substrate for polyphenol oxidase (see Macheix *et al.*, 1991).

I.3.4 Oxidation–reduction (redox) potentials

Oxidation is always accompanied by reduction and vice versa, which can be expressed in the equation,

$$a\text{Ox.}_A + b\text{Red.}_B \rightarrow a'\text{Red.}_A + b'\text{Ox.}_B$$

where Ox._A is the oxidized state of substance A, and Red._B is the reduced state of substance B, and so on; a, a' are the number of unit molecules involved for substance A in the oxidized and reduced state respectively, and similarly for b and b' for substance B. A so-called reaction quotient (Q) expresses the relationship between the molar concentrations [X] of these substances in their different states (as in Equation I.1).

$$Q = \frac{[Red._A]\, a'\, [Ox._B]b}{[Ox._A]\, a\, [Red._B]b'}$$

(I.1)

The Nernst equation expresses the relationship of this ratio Q with the electric potentials of the system, determinable from measurements in a suitable galvanic cell with appropriate electrodes (Equation I.2).

$$E = E_0 - \frac{RT}{nF}\, \ln Q$$

(I.2)

The reaction described above will proceed spontaneously if $E > 0$. E_0 is the so-called standard or normal potential, which is the value of E at equilibrium; as shown in Equation I.3, where K is the known equilibrium constant. R, T, nF, are described in Chapter 3. Whilst K in practice can range from 10^{34} to 10^{-34}, the corresponding E_0 will only range $+2$ to -2.

$$E_0 = \frac{RT}{nF}\, \ln K$$

(I.3)

Redox potentials are conveniently thought of as half-reactions of so-called couples ($Ox._A/Red._A$), thus $Ox._A + e^- \to Red._A$, which is an expression in the so-called reduction mode, since a substance in the oxidized form has to gain one or more electrons in order to be oxidized. The standard reduction potential for this half reaction can be stated, say E_0^A. Similarly, $Ox._B + D \to Red._B$, with a standard potential of E_0^B. For the moment, we can ignore the number of electrons which may be involved. By substracting, we have,

$$Ox._A - Ox._B \to Red._A - Red._B$$

so the overall standard potential for the total reaction is given by, $E_0^t = E_0^A - E_0^B$ (Equation I.4).

$$
\begin{aligned}
\ln Q &= \frac{\ln[Red._A] + \ln[Ox._B]}{\ln[Ox._A] + \ln[Red._B]} \\
&= \ln\left[Red._A\right] + \ln\left[Ox._B\right] - \ln\left[Ox._A\right] - \ln\left[Red._B\right] \\
&= \left(\ln\left[Red._A\right] - \ln\left[Ox._A\right]\right) + \left(\ln\left[Ox._B\right] - \ln\left[Red._B\right]\right) \\
&= \ln\left[Red._A / Ox._A\right] - \ln\left[Ox._B / Red._B\right] \\
&= E_0^A - E_0^B
\end{aligned}
$$

(I.4)

The potential of each couple is therefore given by Equation I.5.

$$\frac{RT}{nF}\, \ln\, \left[\text{Reduced forms}\right] / \left[\text{Oxidized form}\right]$$

(I.5)

Numerical values for R, T, and F have been given in the main text: ln is the natural $\log_e$. This theoretical treatment of the subject follows that given by

Shriver *et al.* (1994). Applying this treatment to actual oxidation reactions, such as the oxidation by dissolved oxygen of solutes such as Fe^{2+}, we have two half-reactions,

$$O_2(g) + 4H^+ + 4e^- \rightarrow 2H_2O, \quad E_0 = +1.23 \text{ V}$$
$$Fe^{3+} + e^- \rightarrow Fe^{2+}, \quad E_0 = +0.76 \text{ V}$$

both in the reduction mode, with published standard potential values which, by difference, give the overall reaction as follows,

$$O_2(g) + 4H^+ + 4Fe^{2+} \rightarrow 4Fe^{3+} + 2H_2O, \quad E_0 = +0.46 \text{ V}$$

The oxidation of ferrous ions to ferric ions would be spontaneous, but the standard potential is not large and hence, in practice, the reaction would be slow. The E values at any given concentration of reactants can now be determined from the Nernst equation.

In the first half-reaction, above, Ribéreau-Gayon *et al.* (2000) state that

$$E_H(\text{volts}) = 1.23 + \log[H^+]^4[O_2]/[H_2O]^2$$

so that $= 1.23 - 0.059 \times 7 + 0.14\log_{10}[O_2]$

In distilled water, $[O_2]$ is a concentration in mole units (quantity, $g/32$). By multiplying through by 1000, E will be in millivolts and the oxygen concentration more conveniently in millimoles, for dissolved oxygen in water and wines. The figure 7 is the pH for distilled water only. The origin of the other two constants in the equation is not clear. The ratio used is oxidized/reduced with the positive sign instead of reduced/oxidized as in the original Nernst equation but the result is mathematically the same.

I.4 Estimation of partition coefficients of volatile compounds in air/water

Estimates may be calculated, provided we first know the vapour pressure of the compound of interest at the required temperature. Vapour pressure data is widely available in tables. If necessary, available data can be graphically presented by plotting the logarithm of vapour pressure (P_j^s, in mmHg or kPa) against the reciprocal of the absolute temperature ($1/T_{oK}$) over a range of values. A typical straight-line plot is shown in Figure I.3 for alkyl unsaturated aldehydes. Notably, the lines for each compound are not exactly parallel; equations are available to express relationships, such as $\log P = a.1/T + b$, where a and b are constants. Vapour pressure at any other temperature may be found by interpolation or extrapolation from the plot. It will be noted also that the vapour pressures of the higher homologues at room temperature can be very low. Also needed are the vapour pressures of water over the same range of temperatures, but these are very accurately known.

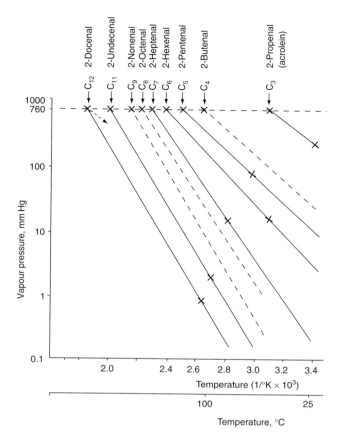

Figure I.3 Logarithm of vapour pressure (P_j^s, in mmHg or kPa) versus reciprocal of the absolute temperature (1/T°K) for some alkyl unsaturated aldehydes.

The so-called activity coefficient ($\gamma_{j\text{-}w}^\infty$) of the particular component at infinite dilution in water also needs to be known, but it can be estimated in either of two ways:

(1) Activity coefficient reflects the inherent characteristic of a compound not easily mixing with water and not forming a perfect solution, obeying Raoult's Law. For this reason, the saturation water solubility ($S_{j,w}^t$) of the volatile compound (j) in water (w) expressed in molar units at a temperature of T°C, as a reciprocal value is equal to the dimensionless activity coefficient, i.e. $1/S = \gamma^\infty$. Data on the actual saturation solubilities of many volatile compounds in water is scarce, though generally available for room temperatures for well known compounds. The technique of obtaining reasonable accuracy in measurement at inevitably low solubilities needs care (Buttery, 1969).

(2) The Pierotti correlations (1959) showed that Equation I.6 applies to a wide range of compounds, where n_1 is the total number of carbon atoms in the

molecule of compound. A, B and C are temperature dependent variables, values of which are available from Pierotti's published tables.

$$\log \gamma^{\infty}_{j,w} = A + Bn_1 + C/n_1 \tag{I.6}$$

For normal alkanoate esters, $A = -0.950$, $B = 0.64$ and $C = 0.260$ at 20°C. However, account has also to be taken of secondary linkages between two radical groups, so that $1/n_1$ is replaced by $(1/n_1^1 + 1/n_1^{11})$. For ethyl butanoate, with the molecular formula of $CH_3.CH_2.CH_2.CO.OC_2H_5$, though n_1 is $4 + 2 = 6$, n_1^1 is 4, but n_1^{11} in fact is 3, to include also the carbon atom to the polar group, and is not 2. Once $\gamma^{\infty}_{j\text{-}w}$ is known by either method, the value of $K_{j,a\text{-}w}$ is determined from Equation I.7, where P_j^s is the vapour pressure of the pure substance at the same temperature. The value of the relative volatility (α^{∞}) can be obtained from $k_{j,a\text{-}w}/0.97 \times 10^{-6} \times P_w^s$, where P_w^s is the vapour pressure of pure water at the given temperature.

$$K_{j,a\text{-}w} = \gamma^{\infty}_{j\text{-}w} \times 0.97 \times 10^{-6} \times P_j^s \tag{I.7}$$

These calculations will now be illustrated for ethyl acetate (ethanoate), though there are some directly or indirectly determined values of $K_{j,a\text{-}w}$ for use. Estimated values may be usefully compared.

(1) The saturation solubility is reported by Lide (2001, loc. cit.) at 8.1% w/ws (8.1 parts per 100 parts solution) at 25°C, and the molecular weight is 88. The mole fraction solubility is therefore $(8.1/88)/[(8.1/88) + (91.9/18)]$, i.e. 0.0176 so that $1/S = 56.8$, say 57. The vapour pressure of ethyl acetate is 93 also at 25°C, so that $K_{j,a\text{-}w}$ at 25°C then calculates as $57 \times 0.97 \times 10^{-6} \times 93 = 0.50 \times 10^{-2}$.

(2) Pierotti correlations are only given for esters at a temperature of 20°C. For ethyl acetate, $n_1 = 4$; $n_1^1 = 2$ and $n_1^{11} = 3$ (Equation I.8). Therefore, $\gamma_j = 70$, $P = 70$, and $K = 70 \times 70 \times 0.97 \times 10^{-6} = 0.47 \times 10^{-2}$.

$$\log \gamma^{\infty}_{j\text{-}w} = -0.930 + (0.64 \times 4) + 0.260\left(\frac{1}{2} + \frac{1}{3}\right) = 1.864 \tag{I.8}$$

These figures may be compared with the data in Table I.5, for directly determined values at higher temperatures (Gretsch et al., 1991) and extrapolated at others from a straight line relationship of $\log K_j = $ Constant/T. At 25°C, the values at $\times 10^2$, 0.52 extrapolated from direct, and 0.50 estimated, from (1) above, at 20°C these values are 0.40 (direct) and 0.47 (from (2) above) respectively.

It is apparent that the solubility of ethyl acetate in water increases with a lowering of temperature, and decreases with increasing temperature. The expression $T°K/\log \gamma$ is approximately constant, indicating that the heat of solution of ethyl acetate in water is negative. The relative volatility value shows a slight decrease with increasing temperature. The value of $K_{j,a\text{-}w}$, however, is markedly higher at higher temperatures, so that the higher the temperature of the wine being drunk, the greater will be the concentration of the ethyl acetate in the air head-space for a given concentration in the aqueous

Table I.5 Air–water partition coefficients for ethyl acetate (j) at different temperatures.

°C	Temperature $1/T°_k$	$K_{j,a-w}$ Mean	Precision	Vapour Pressure (j) mm	Pressure (w) mm H$_g$
80	2.83×10^{-3}	6.00×10^{-2}	±1.5	–	355
60	3.09	2.91	±0.26	436	149
40	3.19	1.11	±0.36	281	55
25	3.35	0.52	–	93	27.3
20	3.41	0.40	–	17.5	–
18	3.44	0.37	–	16.3	–

Table I.6 Effect of temperature on the physical properties of infinitely dilute solutions of ethyl acetate.

Temperature (°C)	Partition coefficient $K_{j,a-w}$	Activity coefficient γ_{j-w}^{∞}	Corresponding value of S mole/units	g/100	Relative volatility $(\alpha_{j,w}^{\infty})$
60	2.91×10^{-2}	68	0.0147	7.2	199
40	1.11	–	–	–	–
25	0.55	61	0.0164	8.0	208
20	0.40	58	0.0172	8.4	228
18	0.37	58	0.0172	8.4	232

Other independent data, such as Chandrasekaran & King (1972) by indirect determination of K_j for ethyl acetate gives $\gamma_{j-w}^{\infty} = 66$ at 20°C and $\alpha_{j,w}^{\infty} = 269$ but figures are based upon a higher vapour pressure figure of 74 mm.

liquid phase (Table I.6). This may not be true for the other volatile components, which would have to be examined on an individual basis for their solubility-temperature characteristics. Currently, there are few direct determinations of K_j for other alkanoates, except of the methyl alkanoates of limited wine interest, for comparison against estimates. These are presented in Table I.7.

The relevant physical properties for all other volatile compounds found in wines are presented in Tables II.1, II.2, II.3, II.4, II.5, II.6, II.7, II.8, II.9 and II.10, together with estimates where possible. Pierrotti correlations are available for alkyl ketones (n_i^I and $n_{I'}^{II}$, applicable) and aldehydes (saturated and unsaturated), but not for branched chain alcohols (saturated only, but including secondary and tertiary variants).

I.5 Grape varieties and cultivars

To the non-geneticist and non-horticulturalist, the terms 'variety' and 'cultivar' may well cause some confusion, as they are frequently used interchangeably in certain books on wine.

Table I.7 Comparative data partition coefficients $K_{j,a-w}$ for alkyl alkanoates determined by different methods.

Name of compound	Direct, $K \times 10^2$	Estimates					
	(1) Buttery (1969) (2) Pollien (2000) (3) Gretsch et al.	From solubility data at T (°C) γ^∞_{j-w}	$K \times 10^2$	Vapour pressure (mm H$_g$)	From Pierotti correlations at 20°C γ^∞_{j-w}	$K \times 10^2$
All at 25°C except where stated						
ETHYL						
Formate	–	37[18]	0.54	–	24	0.35
Acetate	(3) 0.40[20], 0.37[18]	58[18]	0.37	70	70	0.47
	(extrapolated)	57[25]	0.50	93	–	–
Propanoate	–	242[20]	0.69	28	275	0.69
Butanoate	(2) 1.87	1307[20]	1.56	15	1140	1.40
Pentanoate	–	3006[25]	2.10	5.5	4570	2.40
Hexanoate	–	–	–	3.2	20,900	6.70
Heptanoate	–	30,257[20]	4.4	1.5	89,700	13.00
Methyl						
Formate	(1)	–	–	–	–	–
Acetate	0.47	–	–	170	24	0.40
Propanoate	0.71	–	–	70	95	0.89
Butanoate	0.84	–	–	25	390	0.95
Pentanoate	1.3	–	–	7.3	1,300	0.92
Hexanoate	1.5	–	–	2.6	7,100	2.00
Octanoate	3.2	–	–	0.37	132,586	4.4

NB The data gives reasonable agreement between directly determined values for the methyl alkanoates, with a tendency to over-estimate for the higher homologues, especially the temperatures are slightly different (25°C v 20°C). There is little data for the ethyl alkanoates. The comparisons between estimates by the two methods are good for the lower homologues but not the higher homologues. There are practical difficulties in the determination of vapour pressure and solubilities for these compounds.

It is generally understood that the wine grape is the fruit of the vine of the *Vitis* genus and the *vinifera* species by the Linnean classification, with additional word 'sativa' to indicate a cultivated species, as distinct from the original wild state. The terms 'variety' and 'cultivar', or 'c.v.' are lower levels of botanical rank in determining characterizing differences between plants. To the wine chemist, characterization of a grape is most conveniently expressed by the name of the variety of the vine plant from which it was harvested. Plant variety is essentially determined by botanical features (taxonomy) and genetic make-up (ampelography, Gk. *ampelos* – vine), to which DNA profiling can now be included.

The domestic cultivation of vines has been practised now for many centuries, by initial selection from the wild, with differently appearing vines being given different varietal names, such as the familiar Chardonnay and Cabernet Sauvignon but many other locally given names. Pinot Noir is considered to be perhaps the variety to be first directly named, some 2000 years ago, and known to be planted in Burgundy in the fourth century AD (& Rand, 2001). Variety stability and difference has been largely sustained by the self-pollination characteristics of *Vitis vinifera* and by the modern use of clones, to be described. Mutations of genes, however, do occur over the centuries to give botanical sub-variants of the basic type; in fact, the variety Pinot Noir is particularly prone to this phenomenon.

To the horticulturalist in the vineyards, the term 'cultivar' gives an additional dimension of meaning over 'variety' to a particular vinestock or plant. A cultivar is variously defined as 'plant material of particular genetic characteristics produced by horticultural techniques'. One such technique is the production of clones by vegetative reproduction, which is taking cuttings from the plant and rooting them, to produce a next generation of vinestock that will be genetically identical. Clones, for example, can be prepared from different plants of the Chardonnay variety that show slightly different agronomic differences (fruiting yield, etc.) more favourable than others. These clones will be described as still of the Chardonnay variety, but will be additionally named, thus the Mendoza Chardonnay clone (Clarke & Rand, 2001). The use of clones is the way in which replanting of vineyards is carried out, though only necessary after about thirty years of fruit bearing.

Another horticultural technique is that of crossing varieties by manual pollination methods, which was especially practised in Germany in the last century and earlier. Müller-Thurgau and Scheurebe (both intra-specific, Riesling × Silvaner) are described in the main text of Chapter 5 and are still referred to as varieties. Clarke & Rand (2001) are somewhat disparaging about the success in terms of quality of such crossing, compared with 'natural' varieties – though they are more enthusiastic about the South African Pinotage grapes (Pinot Noir × Cinsault). Some crossings have also been made to produce inter-specific hybrids, though of limited interest for wine grapes.

The production of disease resistant plants is also important in enology, by cloning techniques from favourable vinestock. Laboratory growing techniques of tissue culture can be used. The most effective horticultural technique to be used was that following the *Phylloxera* epidemic in the 1860s in France, when grafting of branches of existing disease-free vines in France on root-stock from America (albeit of a different *Vitis* species) was successful. Varietal differences were not substantially altered. Since *Phylloxera* is ubiquitous in most vine growing regions, most vines are still grafted on a so-called American rootstock or a hybrid, which confers resistance to *Phylloxera* attack of the entire vine.

Recent studies in DNA profiling of vines have revealed the true origins for some of the traditionally named varieties. There appears to have been post-crossings (presumably some by chance inter-pollination); for example, in Chardonnay, which is really a crossing of Pinot Noir with an old variety, 'Gouais blanc' (at some genetic distance away). An investigation of this history, together with that of other grape varieties, has been given by Bowers *et al.* (2000).

References

Belitz, H.-D. & Grosch, W. (1987) *Food Chemistry*. Chapter 3, pp. 128–200. Springer-Verlag, Berlin.

Bowers, J., Boursiquot, J.M., This, P., Chu, K., Johansson, H. & Meredith, C. (2000) Historical Genetics: The Parentage of Chardonnay, Gamay, and Other Wine Grapes of Northeastern France. *Science*, **285**, 1562–1565.

Buttery, R.G., Guadagni, D.E. & Ling, L.C. (1969) Volatilities of aldehydes, ketones and esters in dilute solution. *Journal of Agricultural and Food Chemistry*, **17**(2), 385–389.

Chandrasekaran, S.K. & King, C.J. (1972) Multicomponent diffusion and vapor-liquid equilibria of dilute organic components in aqueous sugar solutions. *American Institute of Chemical Engineers Journal*, **18**, 513–520.

Clarke, O. & Rand, M. (2001) *Grapes and Wines*, Webster, London.

Clayden, D., Greeves, N., Warren, S. & Wothers, P. (2001) *Organic Chemistry*. Oxford University Press, Oxford.

Drawert, F. (1974) Wine making as a biotechnological sequence. In: *Chemistry of Wine Making*. Advances in Chemistry Series, 157, pp. 1–10. American Chemical Society, Washington.

Flament, I. (2001) *Coffee Flavour Chemistry*. John Wiley & Sons, Ltd, Chichester.

Gretsch, C., Grandjean, G., Maering, M., Liardon, K. & Westfall, S. (1995) Determination of the partition coefficients of coffee volatiles using static head-space. In: *Proceedings of the 16th ASIC Colloquium (Kyoto)*, pp. 326–31, ASIC, Paris, France.

The International Union of Pure and Applied Chemistry (1993) *A Guide to IUPAC Nomenclature of Organic Compounds, Recommendations 1993*. R. Panico, W.H. Powell & J.C. Richter (Eds.), Blackwell Publishing Ltd., Oxford.

Jackson, R.S. (2008) *Wine Science, Principles, Practice, Perception*, 3nd edn. Academic Press, San Diego.

Lide, D. (ed.) (2001) *Handbook of Chemistry and Physic*, 81st edn., CRC Press, Boca Raton, FL.

Macheix, J.J., Sapis, J.C. & Fleuriet, A. (1991) Phenolic compounds and polyphenol-oxidase in relation to browning in grapes and wines. *Critical Reviews in Food Science and Nutrition*, **30**, 441–486.

Pierotti, G.J., Deal, C.H. & Derr, E.L. (1959) Activity Coefficients and Molecular Structure. *Industrial & Engineering Chemistry,* **51**, 95-102.

Ribéreau-Gayon, P., Glories, Y., Maujean, A. & Dubourdieu, D. (2000) *Handbook of Enology, Vol I, Wine and Wine Making,* John Wiley & Sons, Ltd, Chichester.

Ribéreau-Gayon, P., Glories, Y., Maujean, A. & Dubourdieu, D. (2000) *Handbook of Enology, Vol II, The Chemistry of Wine, Stabilization and Treatments,* John Wiley & Sons, Ltd, Chichester.

Semmelroch, P. & Grosch, W. (1996) Studies in character impact compounds in coffee beans. *Journal of Agricultural and Food Chemistry,* **44**, 537-543.

Shriver, D.F., Atkins, P.W. & Langford, C.H. (1994) *Inorganic Chemistry, 2ⁿᵈ Edn.* Oxford University Press, Oxford.

Appendix II

Molecular Formula/Weight, Physical Properties (Boiling Point, Saturation, Water Solubility) and Partition Coefficients (Air–Water, Measured Direct or Estimated)

II.1   Units

Boiling point in °C:

(1)   at normal atmospheric pressure (760 mmHg or 101.3 kPa);
(2)   at sub-atmospheric pressure (in mmHg).

Saturation water solubility figure (S) expressed in either, %w/w (w/w_s-solutions), or parts of volatile compound ('j') per 100 parts of water ('w'), at T°C given in the superscript figure, though the difference in value is small for percentages below five parts. Qualitative data is given as soluble (s), very slightly soluble (v.sl.s) or insoluble (i).

Partition coefficient (air–water), $K_{j,a-w}$ either directly determined (value preferably with precision data) and given in the literature, or estimated from vapour pressure data/activity coefficient as described in Appendix I, at temperature T°C stated.

II.2   Data sources

Boiling point (vapour pressure) and solubility data are taken from Perry & Green (1997), Kirk-Othmer (1994), Lide (2001) and Burdock (2003).

Directly determined partition coefficients from Buttery et al. (1969, 1971), Gretsch et al. (1995) and Pollien & Yeretzian (2001), loc. cit.

Vapour pressures expressed in mmHg (1 mmHg = 0.133 kPa).

Tables II.1, II.2, II.3, II.4, II.5, II.6, II.7, II.8, II.9 and II.10 are compilations of the available data for each of the groups of the volatile compounds, already described as being present in wines in Chapter 4. The dashes in column positions mean that data is inadequate or not available for calculation. For certain groups of compounds, Pierotti correlations are not given, and calculations are not possible to provide estimates of partition coefficients ($K_{j,a-w}$).

Wine Flavour Chemistry, Second Edition. Jokie Bakker and Ronald J. Clarke.
© 2012 Blackwell Publishing Ltd. Published 2012 by Blackwell Publishing Ltd.

Table II.1 Volatile esters in wine (multi-referenced). Molecular formula and weight. Physical properties (boiling point; saturation water solubility (S); partition coefficients, $K_{j,a-w}$, measured direct or estimated).

Name of compound alcohol/acid moieties	Molecular Formula	Weight	Boiling point, °C at atmos. pressure, or other pressures (mm Hg)	S^t parts per 100 parts water or % w/w at T (°C)	$K_{j,a-w} \times 10^2$ Direct at 25°C		Estimated at 20°C	
					a	b	c	d
(1) *Ethyl alkanoates* $CH_3(CH_2)_nCOOC_2H_5$ where n = 0–14 for unbranched alkanoate radicals.								
Methanoate (formate)	$C_3H_6O_2$	74	54 20/200 5.4/100	11.8%[25]	–	–	0.64	–
Ethanoate (acetate)	$C_4H_8O_2$	88	77 27/100 25/93 20/70	8.1%[25]	–	0.50	0.50	0.48
Propanoate (propionate)	$C_5H_{10}O_2$	102	99 45/100 say 20/28	2.4[20]	–	–	0.67	0.69
Butanoate (n-butyrate)	$C_6H_{12}O_2$	116	121 62/100 say 25/17 20/13	0.68[25] 0.49[20]	– –	1.87 –	1.56 1.60	– 1.40
2-Methyl-propanoate (butyrate)	$C_6H_{12}O_2$	116	110 53/100 21/20	sl.sol	–	–	–	–
Pentanoate (n-valerate)	$C_7H_{14}O_2$	130	145–/100 say 25/74	0.24[25]	–	–	2.15 –	– 2.40
2-Methyl-butanoate	$C_7H_{14}O_2$	130	138–/100	–	–	5.21	–	–
3-Methyl-butanoate (iso valerate)	$C_7H_{14}O_2$	130	134 75/100	0.20[20]	–	–	2.47	–
Hexanoate (caproate)	$C_8H_{16}O_2$	144	166.5–/100 say 25/45 20/32	–	–	–	–	6.4
Heptanoate (oenanthate)	$C_9H_{18}O_2$	158	189–/100 say 20/1.15	0.029[20]	–	–	4.4	(13)
Octanoate (caprylate)	$C_{10}H_{20}O_2$	172	207–/100 say 20/1	–	–	–	–	(42)
Nonanoate (pelargonate)	$C_{11}H_{22}O_2$	186	228–/100	–	–	–	–	–
Decanoate (caprate)	$C_{12}H_{24}O_2$	200	245 (243)	0.002[20]	–	–	–	–
Dodecanoate (laurate)	$C_{14}H_{28}O_2$	228	269	–	–	–	–	–
(2) *Methyl alkanoates*, $CH_3(CH_2)_n COOCH_3$								
Ethanoate (acetate)	$C_3H_6O_2$	74	57.5–/100	–	0.47	–	–	–

Table II.1 (*Continued*)

Name of compound alcohol/acid moieties	Molecular Formula	Weight	Boiling point, °C at atmos. pressure, or other pressures (mm Hg)	S^t parts per 100 parts water or % w/w at T (°C)	$K_{j,a-w} \times 10^2$ Direct at 25°C		Estimated at 20°C	
					a	b	c	d
Hexanoate (caproate)	$C_7H_{14}O_2$	130	149 91/100 25/3.6 20/2.6	–	1.5	–	–	1.80
Heptanoate (oenanthate)	$C_8H_{16}O_2$	144	174–/100	–	–	–	–	–
Octanoate (caprylate)	$C_9H_{18}O_2$	158	192 128/100 20/0.4	–	3.20	–	–	4.40
(3) *Propyl alkanoates*								
Ethanoate (acetate)	$C_5H_{10}O_2$	102	102 48/100 say 20/22 18/20	1.6[18]	–	–	–	0.65
(4) *Butyl and isobutyl alkanoates*								
n-Butyl ethanoate	$C_6H_{12}O_2$	116	125–/100 20/12	0.7[20]	–	–	1.07	1.5
2-Methyl-propyl (isobutyl) ethanoate	$C_6H_{12}O_2$	116	118 60/100	–	–	–	–	–
2-Methyl-propyl octanoate	$C_{12}H_{24}O_2$	188	–	–	–	–	–	–
2-Methyl-propyl decanoate	$C_{14}H_{28}O_2$	216	204	–	–	–	–	–
(5) *Pentyl (amyl and isoamyl) alkanoates*								
3-Methyl-butyl (isoamyl) ethanoate (acetate)	$C_7H_{14}O_2$	130	142 83/100	0.2[25]	–	–	–	–
Pentyl ethanoate	$C_7H_{14}O_2$	130	148	0.07[20]	–	–	–	–
3-Methyl-butyl propanoate	$C_8H_{16}O_2$	144	160	0.1[18]	–	–	–	–
3-Methyl-butyl butanoate (n-butyrate)	$C_9H_{18}O_2$	158	178.6 113/100	–	–	–	–	–
3-Methyl-butylhexanoate	$C_{11}H_{22}O_2$	186	–	–	–	–	–	–

(*Continued*)

Table II.1 (*Continued*)

Name of compound alcohol/acid moieties	Molecular Formula	Weight	Boiling point, °C at atmos. pressure, or other pressures (mm Hg)	S^t parts per 100 parts water or % w/w at T (°C)	$K_{j,a\text{-}w} \times 10^2$ Direct at 25°C a	b	Estimated at 20°C c	d
(6) Hexyl alkanoates								
Ethanoate	$C_8H_{16}O_2$	144	169	0.02[20]	–	–	–	–
Hexanoate	$C_{12}H_{24}O_2$	200	–	–	–	–	–	–
Octanoate	$C_{14}H_{28}O_2$	228	–	–	–	–	–	–
(7) Ethyl hydroxyl and other alkanoates								
Hex-2-enoate (*trans*)(E)	$C_8H_{14}O_2$	142	63–64	–	–	–	–	–
Acetyl methanoate (pyruvate)	$C_5H_8O_3$	116	155	–	–	–	–	–
3-Hydroxy-butanoate	$C_3H_{13}O_3$	132	171	–	–	–	–	–
4-Hydroxy-butanoate	$C_6H_{13}O_3$	132	–	–	–	–	–	–
Benzoate	$C_9H_{10}O_2$	150	213.5 94/14 143/100	0.08[20] 0.08%[25]	–	–	–	–
3-Methyl-butyl succinate	$C_{11}H_{20}O_4$	216	–	–	–	–	–	–
Lactate (2-hydroxy-propanoate)	$C_5H_{10}O_3$	118	154	–	–	–	–	–
(8) Diethyl esters								
Malonate	$C_7H_{12}O_4$	160	199	2	–	–	–	–
Malate	$C_8H_{14}O_5$	190	253° 184/100	–	–	–	–	–
Tartrate	$C_8H_{14}O_6$	206	280	–	–	–	–	–
Succinate (butan-1,4-dioate)	$C_8H_{14}O_4$	174	218	2%[20]	–	–	–	–
(9) Other esters								
Ethyl phenyl acetate	$C_{10}H_{12}O_2$	164	229 100/10	–	–	–	–	–
2-Methyl propyl lactate	$C_7H_{14}O_3$	146	75/8	–	–	–	–	–

[a]Data from Buttery *et al.* (1969, 1999) for methyl alkanoates only. [b]Data from Pollien & Yeretzian (2001). [c]Calculated from solubility data. [d]Calculated from Pierotti (1959).

Table II.2 Volatile aldehydes in wines.

Name	Molecular Formula	Weight	Boiling point (°C)	St parts per 100 parts T (°C)	$K_{j,a\text{-}w} \times 10^3$ Direct at 25°C [a]	[b]	Estimated [c]	[d]
Ethanal (acetaldehyde)	C_2H_4O	44	20.8	vs	2.7	–	–	3.5
Propanal (propaldehyde)	C_3H_6O	58	49.5	20^{18} $30.6\%^{25}$	3.0	–	2.9	4.5
2-Methylpropanal (iso butanal)	C_4H_8O	72	64	11 $9.1\%^{25}$	–	11.4	8.1	–
Butanal (butyraldehyde)	C_4H_8O	72	76	$7.1\%^{25}$	4.7	–	8.3	9.6
2-Methyl butanal	$C_5H_{10}O$	86	–	–	–	16.7	–	–
3-Methyl butanal (iso pentanal) (iso valeraldehyde)	$C_5H_{10}O$	86	93	–	–	15.6	–	–
Hexanal (caproicaldehyde)	$C_6H_{12}O$	100	131	sl.sol	8.7	–	–	11.00
(E)-Hex-2-enal (trans-hexenal)	$C_6H_{10}O$	98	–	–	2.0	–	–	3.9
Octanal (caprylaldehyde)	$C_8H_{16}O$	128	168 23/1	sl.sol	21	–	–	18.0
Benzaldehyde (phenylmethanal)	C_7H_6O	106	179 50/5	$3\%^{20}$	–	–	–	–
Phenyl ethanal (phenylacetaldehyde)	C_8H_8O	120	193	vs.sol	–	–	–	–
4-Hydroxy-3-methoxy-1-benzaldehyde (vanillin)	$C_8H_8O_3$	152	285	1^{14}	–	–	–	–
Phenyl propenal	C_9H_8O	132	252	vs/s	–	–	–	–

[a]Buttery et al. (1969). [b]Pollien & Yeretzian (2001). [c]Calculated from S. [d]Calculated from Pierotti (1959).

Table II.3 Volatile ketones in wines.

Name of compound	Molecular Formula	Weight	Boiling point (°C)	S^t parts per 100 parts water at 25° or T°C	$K_{j,a\text{-}w} \times 10^3$ Direct At 25°C	Estimated At T°C	Estimated 20°C
Propanone (acetone)	C_3H_6O	58	56.5 22.7/100 25/230	∞	1.6[a]	–	1.6[d]
Butane-2,3-dione (diacetyl)	$C_4H_6O_2$	86	88	1:4 (25)	0.72[b]	–	–
3-Hydroxy-butan-2-one (acetoin)	$C_4H_8O_2$	88	148	∞	–	–	–
Pentane-2,3-dione (acetyl acetone)	$C_5H_8O_2$	100	140.5	1:8 (12.50)	1.37[b]	–	–
Hexan-2-one (methyl, butyl ketone)	$C_6H_{12}O$	100	127 39/10	1.75%[20]	–	0.9[c]	1.7[d]
Heptan-2-one	$C_7H_{14}O$	114	151.5 55/10	0.43%[25]	5.9	3.1[c]	2.4[d]
Octan-2-one	$C_8H_{16}O$	128	172.9 61/10	0.113%[25]	7.7[a]	4.9[c]	5.1[d]
Nonan-2-one	$C_9H_{18}O$	142	195 72/10	–	15[a]	–	15
4-Hydroxy-3-methoxy-aceto-phenone (1-phenyl-1-ethanone)	$C_9H_{10}O_3$	166	–	–	–	–	–
β-Ionone	$C_{13}H_{20}O$	192	271 140/18	sl.s	–	–	–
β-Damascenone	$C_{13}H_{18}O$	190	–	–	–	–	–

[a]Buttery *et al.* (1969). [b]Pollien & Yeretzian (2001). [c]Calculated from S. [d]Calculated from Pierotti (1959).

Table II.4 Volatile alcohols in wine (alkanols and alkenols).

Name	Molecular Formula	Weight	Boiling point (°C)	S^t parts per 100 parts	$K_{j,a\text{-}w} \times 10^4$ Direct at 25°C	Estimated At T°C 25°C a	b	c
Propan-l-ol	C$_3$H$_8$O	60	98	∞	2.8	–	–	–
Propan-2-ol	C$_3$H$_8$O	60	82.5 2.4/10	∞	–	–	–	–
Butan-l-ol	C$_4$H$_{10}$O	74	117	9 (7.4%)	3.5	3.6	4.3	4.5
2-Methyl-propan-l-ol (isobutyl alcohol) (fermentation butyl alcohol)	C$_4$H$_{10}$O	74	108 22/10	10^{15} $8.1\%^{25}$	–	–	–	–
Pentan-l-ol (pentyl alcohol)	C$_5$H$_{12}$O	88	138 45/10	2.7 $2.20\%^{25}$	5.3	–	5.3	6.5
2-Methyl butan-1-ol (active amyl alcohol)	C$_5$H$_{12}$O	88	128	3.6^{20} $3.0\%^{25}$	–	–	–	–
3-Methyl butan-1-ol (isoamyl alcohol)	C$_5$H$_{12}$O	88	132 41/10	2^{14} 2.7%	–	–	–	–
Hexan-1-ol (hexyl alcohol)	C$_6$H$_{14}$O	102	157 58/10	0.6^{20} $0.6\%^{20}$	6.3	7.0	9.1	11.4
3-Methyl pentan-1-ol	C$_6$H$_{14}$O	102	157 58/10	0.59	–	–	–	–
4-Methyl pentan-1-ol	C$_6$H$_{14}$O	102	157 58/10	0.59 $0.76\%^{25}$	–	–	–	–
cis-Hex-3-en-1-ol	C$_6$H$_{12}$O	100	156	–	–	–	–	–
Heptan-1-ol	C$_7$H$_{16}$O	116	176 86/20	0.10^{18} 0.17^{25}	7.7	–	13.2	9.3
2-Methyl-hexan-1-ol	C$_7$H$_{16}$O	116	176 86/20	0.10^{18} 0.18^{25}	–	–	–	–
Octan-1-ol	C$_8$H$_{18}$O	130	194.5°	$0.054\%^{25}$	9.9	10.0	9.3	11
Octan-2-ol (secondary alcohol)	C$_8$H$_{18}$O	130	179–180	0.096^{26} $0.4\%^{25}$	–	–	–	–
Nonan-1-ol	C$_9$H$_{20}$O	144	213° 114/20	0.014^{25}	–	–	–	–
Decan-1-ol	C$_{10}$H$_{22}$O	158	230° 126/20	0.0037^{25}	–	–	–	–
Benzyl alcohol (phenyl methyl)	C$_7$H$_8$O	108	206 106/20	4^{17} $0.123\%^{25}$	–	–	–	–
2-Phenyl ethanol	C$_8$H$_{10}$O	122	219	1.6	–	–	–	–
[2-(4-Hydroxyphenyl) ethanol]	C$_8$H$_{10}$O$_2$	138						

[a] Buttery (1969). [b] From solubility data. [c] From Pierotti (1959).

Table II.5 Volatile furanones/lactones in wines.

Name	Molecular Formula	Weight	Boiling point (°C)	Sat. sol in water S^t	$K_{j,a\text{-}w}$ Direct	Estimated
Secondary aromas						
-Furan-2-one						
Dihydro-3(H)- (γ-butyrolactone)	$C_4H_6O_2$	86	204 / 89/12	∞	–	–	–
Dihydro-3-methyl-3(H)- (3-methyl-γ butyro lactone)	$C_5H_8O_2$	100	–	–	–	–	–
Dihydro-5-methyl-3(H)- (γ-valerolactone)	$C_5H_8O_2$	100	–	–	–	–	–
5-Acetyl-dihydro-3(H)- (solarone)	$C_6H_8O_3$	112	102–104/1.2	–	–	–	–
2-Hydroxy-3,3-dimethyl-3(H)-	$C_6H_8O_3$	128	102–104/1.2	–	–	–	–
3-Hydroxy-4,5-dimethyl-5(H) (sotolon)	$C_6H_8O_3$	128		–	–	–	–
-Furan-3-one						
4-Hydroxy-2,5-dimethyl-2(H)- (furaneol; HDMF)	$C_6H_8O_3$	128	85(subl.) / mp 77–78	–	–	–	–
Tertiary aromas						
-Furan-2-one						
trans/cis 5-Butyl-4-methyl-dihydro-3(H)- (3-methyl-4-hydroxy-octanoic acid γ-lactone: (oak-lactones)	$C_9H_{16}O_2$	156	93–94	–	–	–	–
Dihydro-5-pentyl-3(H)- (γ-nonalone)	$C_9H_{16}O_2$	156	93–94	–	–	–	–
Dihydro-5-hexyl-3(H)- (γ-decalone)	$C_{10}H_{18}O_2$	170	–	–	–	–	–
Hydroxycoumarin (scopoletin)	$C_9H_7O_3$	163	–	–	–	–	–
Primary aromas						
-Furan-2-one						
5-Vinyl-5-methyl-dihydro- (in Riesling/Muscat)	$C_7H_{10}O_2$	126	–	–	–	–	–
5-Methyl-3-ethenyl- (2-pentenoic acid γ-lactone, the raisin lactone)	$C_7H_7O_2$	123	–	–	–	–	–

NB: Inadequate or non-available data currently for solubility and vapour pressure. Pierotti correlations not available.

Table II.6 Volatile acids in wines (n-alkanoic acids $CH_3 (CH_2)_n$ COOM, also some iso branched chain variants).

Name	Molecular Formula	Weight	Boiling point (°C)	S^t parts per 100 parts at temp T°C	$K_{j,a-w} \times 10^4$ Direct at 25°C a	b	Estimated c	d
Methanoic acid (formic acid, HCOOM)	CH_2O_2	46	100.6	∞	–	–	–	–
Ethanoic acid (acetic)	$C_2H_4O_2$	60	118	∞	–	–	–	0.36
Propionic acid (propanoic)	$C_3H_6O_2$	74	141	∞	–	–	–	0.4
Butanoic (butyric)	$C_4H_8O_2$	88	163	∞	–	–	–	–
2-Methyl propanoic (iso-butanoic)	$C_4H_8O_2$	88	154	20[20] 22.8%[20]	–	–	–	–
Pentanoic acid (valeric)	$C_5H_{10}O_2$	102	184 42/1	3.3[18] 2.4%[20]	–	–	–	–
2-Methyl butanoic	$C_5H_{10}O_2$	102	187	–	–	–	–	–
3-Methyl butanoic (iso-valeric)	$C_5H_{10}O_2$	102	176	4.2[20]	–	–	–	–
Hexanoic acid (caproic)	$C_6H_{12}O_2$	116	202	1.1[20] 0.96%[20]	–	–	–	–
Heptanoic acid (oenanthoic)	$C_7H_{14}O_2$	130	223° 187/256	0.25[18]	–	–	–	–
Octanoic acid (caprylic)	$C_8H_{16}O_2$	144	240	0.07[20] 0.08%[25]	–	–	–	–
Nonanoic acid (pelargonic)	$C_9H_{18}O_2$	158	253	0.03%[25]	–	–	–	–
Decanoic acid (capric)	$C_{10}H_{20}O_2$	172	268	0.003[25]	–	–	–	–
Dodecanoic acid (lauric)	$C_{12}H_{24}O_2$	200	225/100	insol.	–	–	–	–
Tetradecanoic acid (myristic)	$C_{14}H_{28}O_2$	228	250/100	0.0012[20]	–	–	–	–
Hexadecanoic acid (palmitic)	$C_{16}H_{32}O_2$	256	271/100	insol.	–	–	–	–

[a,b]Data not available. [c]Data inadequate. [d]Pierotti correlations.

Table II.7 Volatile phenols in wines.

Name of compound (phenol)	Molecular Formula	Molecular Weight	Bp (°C)	S^t parts or % at 25°C or *T*°C	$K_{j,a\text{-}w}$ Direct [a]	$K_{j,a\text{-}w}$ Direct [b]	$K_{j,a\text{-}w}$ Calculated [c]	$K_{j,a\text{-}w}$ Calculated [d]
Hydroxybenzene (phenol)	C_6H_6O	94	181 86/20	8.2^{20} $8.6\%^{25}$	–	–	–	–
2-Methyl- (*o*-cresol)	C_7H_8O	108	191	$3.08\%^{40}$	–	–	–	–
3-Methyl- (*m*-cresol)	C_7H_8O	108	203 88/10	0.5 $2.51\%^{40}$	–	–	–	–
4-Methyl- (*p*-cresol)	C_7H_8O	108	202	$2.26\%^{40}$	–	–	–	–
4-Vinyl-(4-ethenyl)	C_8H_8O	120	–	–	–	–	–	–
4-Ethyl	$C_8H_{10}O$	122	218	–	–	–	–	–
2-Ethyl	$C_8H_{10}O$	122	–	–	–	–	–	–
2-Methoxy- (guaiacol)	$C_7H_8O_2$	124	205 53/4	1.7^{15}	–	–	–	–
2-Methoxy-4-ethyl-(ethyl guaiacol)	$C_9H_{12}O_2$	152	234	–	–	–	–	–
2,6,-Dimethoxy-phenol (syringol)	$C_8H_{10}O_3$	154	–	–	–	–	–	–
Eugenol	$C_{10}H_{12}O_2$	164	–	–	–	–	–	–

[a,b]Data not available. [c]Data inadequate. [d]Pierotti correlations not available.

Table II.8 Volatile terpenes in wines.

Name	Molecular Formula	Molecular Weight	Bp (°C)	Sol. in water at *T*°C	$K_{j,a\text{-}w}$ Direct	$K_{j,a\text{-}w}$ Calculated
Limonene [1-methyl-4-(2-but-1-en)-cyclo-hex-1-en]	$C_{10}H_{16}$	136	175 84/40	0.00138^{25}	–	–
Linalool (3,7-dimethyl-octa-1,6-dien-3-ol)	$C_{10}H_{18}O$	154	198 98/25	v.sl.s	–	–
Geraniol (3,7-dimethyl-octa-2,6-dien-1-ol)	$C_{10}H_{18}O$	154	230 142/40 121/17	insol.	–	–
-Nerol (isomer of above)	$C_{10}H_{18}O$	154				
α-Terpineol (*p*-menth-1-en-8-ol)	$C_{10}H_{18}O$	154	220/27 126/40	insol.	–	–
Ho-trienol (3,7-dimethyl-octa-1,5,7-trien-3-ol)	$C_{10}H_{16}O$	152			–	–
Citronellol (3,7 dimethyl-octa-6-en-1-ol)	$C_{10}H_{20}O$	156	224 108/10	insol.	–	–

Table II.9 Volatile methoxy pyrazines in wines.

Name	Molecular Formula	Weight	Bp (°C)	Saturation sol. in water S^t	$K_{j,a-w}$ Direct at 25°C[a]	Estimated
2-Methoxy-3-isobutyl (2-methyl-propyl)-	$C_9H_{14}ON_2$	166	120 116/61 83/12	–	2×10^{-3}	–
2-Methoxy-3-isopropyl-(3-methyl-ethyl-)	$C_8H_{12}ON_2$	152	94–100/61 80/25	–	–	–
2-Methoxy-3-sec-butyl-(1-methyl-propyl)	$C_9H_{14}ON_2$	166	105–115/55 73/10	–	–	–
2-Methoxy-3-ethyl-	$C_7H_{10}ON_2$	138	–	–	6×10^{-4}	–

[a]Data from Buttery et al. (1971).

Table II.10 Volatile sulfur compounds in wines.

Name	Molecular Formula	Weight	Bp (°C)	Sat. sol. S^t at T°C	$K_{j,a-w} \times 10^3$ Direct at 25°C[a] (a)	(b) Estimated (c)	(d)
Hydrogen sulfide	H_2S	34	−61	0.125 molar[28]	–	–	–
Methane thiol (methyl mercaptan)	CH_4S	48	7 −39/100	2.2%	4.44	–	–
Ethane thiol (ethyl mercaptan)	C_2H_6S	62	35 −13/100	1.8% 1.5[25]	–	–	–
Dimethyl sulfide (methyl thio methane)	C_2H_6S	62	36 25/500	2% w/w	8.46	–	–
2-methyl-thiolane-3-ol	C_5H_6OS	118	–	–	–	–	–
Ethyl propan-3-thiol-oate (ethyl 3-mercapto-propanoate)	$C_5H_{10}O_2S$	134	162	–	–	–	–
4-mercapto 4-methyl-pentan-2-one (4MMP)	$C_6H_{12}OS$		132	–	–	–	–
Ethan-2-thio-1-ol (2-mercapto-ethanol)	C_2H_6OS	78	84	–	–	–	–
Methionol [3-(methyl-thio)-propan-1-ol)]	$C_4H_{10}OS$	106	90	–	–	–	–

[a]Pollien & Yeretzian (2001). [b]No other data. [c]Calculated from S^t. [d]Calculated from Pierotti (1959).
[c,d]Data not available or inadequate.

References

Buckingham, J. & McDonald, F. (eds.) (1982, 1996) *Dictionary of Organic Compounds* 6th edn. Chapman and Hall/CRC Press, London.

Kirk-Othmer (1994) *Kirk-Othmer Encyclopaedia of Chemical Technology*, 4th edn. John Wiley & Sons, Inc., New York.

Lide, D. (ed.) (2001) *Handbook of Chemistry and Physic*, 81st edn., CRC Press, Boca Raton, FL.

Burdock, G.H. (2003) *Fenaroli's Handbook of Food Flavour Ingredients*, 4th edn. CRC Press, Boca Raton, FL.

Buttery, R.G., Guadagni, D.E. & Ling, L.C. (1969) Volatilities of aldehydes, ketones and esters in dilute solution. *Journal of Agricultural and Food Chemistry*, **17**(2), 385–389.

Buttery, R.G., Bomben, J.L., Guadagni, D.G. & Ling, L.C. (1971) Volatilities of organic compounds in foods. *Journal of Agricultural and Food Chemistry*, **19**, 1045–1048.

Buttery, R.G. (1999) Flavour chemistry and odour thresholds. In: *Flavour Chemistry: Thirty years of progress* (eds. R. Teranishi, E.L. Wick & J. Hornstein), pp. 353–365. Kluwer Academic/Plenum, New York.

Gretsch, C., Grandjean, G., Maering, M., Liardon, K. & Westfall, S. (1995) Determination of the partition coefficients of coffee volatiles using static head-space. In: *Proceedings of the 16th ASIC Colloquium (Kyoto)*, pp. 326–31, ASIC, Paris, France.

Perry, R.H. & Green, D.W. (eds.) (1997) *Perry's Chemical Engineer's Handbook*, 7th edn. McGraw-Hill.

Pierotti, G.J., Deal, C.H. & Derr, E.L. (1959) Activity Coefficients and Molecular Structure. *Industrial & Engineering Chemistry*, **51**, 95–102.

Pollien, P. & Yeretzian, C. (2001) Measurement of partition coefficients. *Proceedings of the 19th ASIC Colloquium on Coffee*, CD-ROM, ASIC, Paris.

Index

Wine Flavour Chemistry, Second Edition. Jokie Bakker and Ronald J. Clarke.
© 2012 Blackwell Publishing Ltd. Published 2012 by Blackwell Publishing Ltd.

Food Science and Technology

GENERAL FOOD SCIENCE & TECHNOLOGY, ENGINEERING AND PROCESSING

Organic Production and Food Quality: A Down to Earth Analysis	Blair	9780813812175
Handbook of Vegetables and Vegetable Processing	Sinha	9780813815411
Nonthermal Processing Technologies for Food	Zhang	9780813816685
Thermal Procesing of Foods: Control and Automation	Sandeep	9780813810072
Innovative Food Processing Technologies	Knoerzer	9780813817545
Handbook of Lean Manufacturing in the Food Industry	Dudbridge	9781405183673
Intelligent Agrifood Networks and Chains	Bourlakis	9781405182997
Practical Food Rheology	Norton	9781405199780
Food Flavour Technology, 2nd edition	Taylor	9781405185431
Food Mixing: Principles and Applications	Cullen	9781405177542
Confectionery and Chocolate Engineering	Mohos	9781405194709
Industrial Chocolate Manufacture and Use, 4th edition	Beckett	9781405139496
Chocolate Science and Technology	Afoakwa	9781405199063
Essentials of Thermal Processing	Tucker	9781405190589
Calorimetry in Food Processing: Analysis and Design of Food Systems	Kaletunç	9780813814834
Fruit and Vegetable Phytochemicals	de la Rosa	9780813803203
Water Properties in Food, Health, Pharma and Biological Systems	Reid	9780813812731
Food Science and Technology (textbook)	Campbell-Platt	9780632064212
IFIS Dictionary of Food Science and Technology, 2nd edition	IFIS	9781405187404
Drying Technologies in Food Processing	Chen	9781405157636
Biotechnology in Flavor Production	Havkin-Frenkel	9781405156493
Frozen Food Science and Technology	Evans	9781405154789
Sustainability in the Food Industry	Baldwin	9780813808468
Kosher Food Production, 2nd edition	Blech	9780813820934

FUNCTIONAL FOODS, NUTRACEUTICALS & HEALTH

Functional Foods, Nutraceuticals and Degenerative Disease Prevention	Paliyath	9780813824536
Nondigestible Carbohydrates and Digestive Health	Paeschke	9780813817620
Bioactive Proteins and Peptides as Functional Foods and Nutraceuticals	Mine	9780813813110
Probiotics and Health Claims	Kneifel	9781405194914
Functional Food Product Development	Smith	9781405178761
Nutraceuticals, Glycemic Health and Type 2 Diabetes	Pasupuleti	9780813829333
Nutrigenomics and Proteomics in Health and Disease	Mine	9780813800332
Prebiotics and Probiotics Handbook, 2nd edition	Jardine	9781905224524
Whey Processing, Functionality and Health Benefits	Onwulata	9780813809038
Weight Control and Slimming Ingredients in Food Technology	Cho	9780813813233

INGREDIENTS

Hydrocolloids in Food Processing	Laaman	9780813820767
Natural Food Flavors and Colorants	Attokaran	9780813821108
Handbook of Vanilla Science and Technology	Havkin-Frenkel	9781405193252
Enzymes in Food Technology, 2nd edition	Whitehurst	9781405183666
Food Stabilisers, Thickeners and Gelling Agents	Imeson	9781405132671
Glucose Syrups – Technology and Applications	Hull	9781405175562
Dictionary of Flavors, 2nd edition	De Rovira	9780813821351
Vegetable Oils in Food Technology, 2nd edition	Gunstone	9781444332681
Oils and Fats in the Food Industry	Gunstone	9781405171212
Fish Oils	Rossell	9781905224630
Food Colours Handbook	Emerton	9781905224449
Sweeteners Handbook	Wilson	9781905224425
Sweeteners and Sugar Alternatives in Food Technology	Mitchell	9781405134347

FOOD SAFETY, QUALITY AND MICROBIOLOGY

Food Safety for the 21st Century	Wallace	9781405189118
The Microbiology of Safe Food, 2nd edition	Forsythe	9781405140058
Analysis of Endocrine Disrupting Compounds in Food	Nollet	9780813818160
Microbial Safety of Fresh Produce	Fan	9780813804163
Biotechnology of Lactic Acid Bacteria: Novel Applications	Mozzi	9780813815831
HACCP and ISO 22000 – Application to Foods of Animal Origin	Arvanitoyannis	9781405153669
Food Microbiology: An Introduction, 2nd edition	Montville	9781405189132
Management of Food Allergens	Coutts	9781405167581
Campylobacter	Bell	9781405156288
Bioactive Compounds in Foods	Gilbert	9781405158756
Color Atlas of Postharvest Quality of Fruits and Vegetables	Nunes	9780813817521
Microbiological Safety of Food in Health Care Settings	Lund	9781405122207
Food Biodeterioration and Preservation	Tucker	9781405154178
Phycotoxins	Botana	9780813827001
Advances in Food Diagnostics	Nollet	9780813822211
Advances in Thermal and Non-Thermal Food Preservation	Tewari	9780813829685

For further details and ordering information, please visit www.wiley.com/go/food

Food Science and Technology from Wiley-Blackwell

SENSORY SCIENCE, CONSUMER RESEARCH & NEW PRODUCT DEVELOPMENT

Title	Author	ISBN
Sensory Evaluation: A Practical Handbook	Kemp	9781405162104
Statistical Methods for Food Science	Bower	9781405167642
Concept Research in Food Product Design and Development	Moskowitz	9780813824246
Sensory and Consumer Research in Food Product Design and Development	Moskowitz	9780813816326
Sensory Discrimination Tests and Measurements	Bi	9780813811116
Accelerating New Food Product Design and Development	Beckley	9780813808093
Handbook of Organic and Fair Trade Food Marketing	Wright	9781405150583
Multivariate and Probabilistic Analyses of Sensory Science Problems	Meullenet	9780813801780

FOOD LAWS & REGULATIONS

Title	Author	ISBN
The BRC Global Standard for Food Safety: A Guide to a Successful Audit	Kill	9781405157964
Food Labeling Compliance Review, 4th edition	Summers	9780813821818
Guide to Food Laws and Regulations	Curtis	9780813819464
Regulation of Functional Foods and Nutraceuticals	Hasler	9780813811772

DAIRY FOODS

Title	Author	ISBN
Dairy Ingredients for Food Processing	Chandan	9780813817460
Processed Cheeses and Analogues	Tamime	9781405186421
Technology of Cheesemaking, 2nd edition	Law	9781405182980
Dairy Fats and Related Products	Tamime	9781405150903
Bioactive Components in Milk and Dairy Products	Park	9780813819822
Milk Processing and Quality Management	Tamime	9781405145305
Dairy Powders and Concentrated Products	Tamime	9781405157643
Cleaning-in-Place: Dairy, Food and Beverage Operations	Tamime	9781405155038
Advanced Dairy Science and Technology	Britz	9781405136181
Dairy Processing and Quality Assurance	Chandan	9780813827568
Structure of Dairy Products	Tamime	9781405129756
Brined Cheeses	Tamime	9781405124607
Fermented Milks	Tamime	9780632064588
Manufacturing Yogurt and Fermented Milks	Chandan	9780813823041
Handbook of Milk of Non-Bovine Mammals	Park	9780813820514
Probiotic Dairy Products	Tamime	9781405121248

SEAFOOD, MEAT AND POULTRY

Title	Author	ISBN
Handbook of Seafood Quality, Safety and Health Applications	Alasalvar	9781405180702
Fish Canning Handbook	Bratt	9781405180993
Fish Processing – Sustainability and New Opportunities	Hall	9781405190473
Fishery Products: Quality, safety and authenticity	Rehbein	9781405141628
Thermal Processing for Ready-to-Eat Meat Products	Knipe	9780813801483
Handbook of Meat Processing	Toldra	9780813821825
Handbook of Meat, Poultry and Seafood Quality	Nollet	9780813824468

BAKERY & CEREALS

Title	Author	ISBN
Whole Grains and Health	Marquart	9780813807775
Gluten-Free Food Science and Technology	Gallagher	9781405159159
Baked Products -- Science, Technology and Practice	Cauvain	9781405127028
Bakery Products: Science and Technology	Hui	9780813801872
Bakery Food Manufacture and Quality, 2nd edition	Cauvain	9781405176132

BEVERAGES & FERMENTED FOODS/BEVERAGES

Title	Author	ISBN
Technology of Bottled Water, 3rd edition	Dege	9781405199322
Wine Flavour Chemistry, 2nd edition	Bakker	9781444330427
Wine Quality: Tasting and Selection	Grainger	9781405113663
Beverage Industry Microfiltration	Starbard	9780813812717
Handbook of Fermented Meat and Poultry	Toldra	9780813814773
Microbiology and Technology of Fermented Foods	Hutkins	9780813800189
Carbonated Soft Drinks	Steen	9781405134354
Brewing Yeast and Fermentation	Boulton	9781405152686
Food, Fermentation and Micro-organisms	Bamforth	9780632059874
Wine Production	Grainger	9781405113656
Chemistry and Technology of Soft Drinks and Fruit Juices, 2nd edition	Ashurst	9781405122863

PACKAGING

Title	Author	ISBN
Food and Beverage Packaging Technology, 2nd edition	Coles	9781405189101
Food Packaging Engineering	Morris	9780813814797
Modified Atmosphere Packaging for Fresh-Cut Fruits and Vegetables	Brody	9780813812748
Packaging Research in Food Product Design and Development	Moskowitz	9780813812229
Packaging for Nonthermal Processing of Food	Han	9780813819440
Packaging Closures and Sealing Systems	Theobald	9781841273372
Modified Atmospheric Processing and Packaging of Fish	Otwell	9780813807683
Paper and Paperboard Packaging Technology	Kirwan	9781405125031

For further details and ordering information, please visit www.wiley.com/go/food